PHOSPHOLIPID RESEARCH
AND THE NERVOUS SYSTEM
BIOCHEMICAL AND MOLECULAR PHARMACOLOGY

PHOSPHOLIPID RESEARCH AND THE NERVOUS SYSTEM
BIOCHEMICAL AND MOLECULAR PHARMACOLOGY

Edited by

Lloyd A. Horrocks

> Department of Physiological Chemistry
> College of Medicine
> The Ohio State University
> Columbus, Ohio

Louis Freysz

> Centre de Neurochimie du CNRS
> Strasbourg, France

Gino Toffano

> Fidia Research Laboratories
> Abano Terme, Italy

FIDIA
RESEARCH
SERIES
Volume 4

Springer-Verlag Berlin Heidelberg GmbH

FIDIA RESEARCH SERIES

An open-end series of publications on international biomedical research, with special emphasis on the neurosciences.

The series will be devoted to advances in basic and clinical research in the neurosciences and other fields.

The aim of the series is the rapid and worldwide dissemination of up-to-date, interdisciplinary data as presented at selected international scientific meetings and study groups.

Each volume is published under the editorial responsibility of scientists chosen by organizing committees of the meetings on the basis of their active involvement in the research of the field concerned.

© 1986 Springer-Verlag Berlin Heidelberg
Originally published by Springer-Verlag Berlin Heidelberg New York in 1986
Softcover reprint of the hardcover 1st edition 1986

ISBN 978-1-4899-0492-8   ISBN 978-1-4899-0490-4 (eBook)
DOI 10.1007/978-1-4899-0490-4

PREFACE

The purpose of this book is to describe the latest findings relating to biochemical and molecular pharmacology of the nervous system and phospholipids and to report the proceedings of the fourth symposium on phospholipids. These Symposia have been satellite meetings of the International Society for Neurochemistry. This meeting was held on May 26-29, 1985 in the Teatro Bibiena in Mantova, Italy. Preceding meetings were held in Cortona, Italy in 1975, in Birmingham, England in 1981, and in British Columbia, Canada in 1983. As was the case for the proceedings of those meetings, this volume presents information that is new and important from the researchers most involved in advancing our knowledge of the function of membranes and of lipid metabolism in the nervous system.

The presence of phosphorus in the brain was reported in 1719 by Hensing at the University of Giessen in Germany. Tower[1] has translated this pioneering work. The rather long, philosophical preface contains the following paragraph.

Regardless of what may be thought of this matter, the brain is certainly that part of the animate body in which that subtlest and most penetrating [substance] alone is received, so often circulated from the heart and cleansed by the remaining viscera, and firmly held where thereafter life flourishes and the motions of the lower parts endure. Indeed the brain is truly the Throne of the soul and the abode of wisdom, from whose nature the former is the recipient of the virtues of health, and the latter of brilliance. Whence Hippocrates asserts the brain: "to be the mediator and interpreter of understanding and knowledge." And it is no less true, whence our Excellent Dieterich analogizes, as he says: "among the virtues of the soul, the comparison, where that is reasonable, is like a priest, living in the head as if it were a chapel, whose altar is the brain, on which functions like cognition and reasoning are accomplished. The brain is the capital of thought, the workshop of judgment, the repository of memory, the source of all senses, the office and universe of the animal spirits, etc."

This passage recognizes the importance of the brain and anticipates the effects of hormones (endogenous drugs) on the brain.

Research has advanced far since then, as can be derived from this volume.

The Editors

---

[1] Tower, D.B. 1983 Hensing, 1719. An Account of the First Chemical Examination of the Brain and the Discovery of Phosphorus Therein. Raven Press, New York.

SCIENTIFIC COMMITTEE

N.G. Bazan (*New Orleans, LA, U.S.A.*)
A. Bruni (*Padova, Italy*)
L. Freysz (*Strasbourg, France*)
L.A. Horrocks (*Columbus, OH, U.S.A.*)
G. Toffano (*Abano Terme, Italy*)

LOCAL ORGANIZING COMMITTEE

L. Binaglia
G. Calderini
L. Freysz
A. Gaiti
G. Goracci

ACKNOWLEDGMENT

The organizers of the satellite meeting on Phospholipids Research in the Nervous System of the 10th congress of the International Society for Neurochemistry would like to thank FIDIA research Laboratories and their staff for their support to the meeting and this publication. The Editors wish to express their gratitude to Mrs Lisa Dainese for her editorial assistance.

CONTENTS

ABBREVIATIONS

Unless otherwise indicated, the following abbreviations are used throughout the book

AA, arachidonic acid
ACh, acetylcholine
DAG,DG, diacylglycerol
GroPCho, GPC, *sn*-glycero-3-phosphocholine
InsP,IP, inositol monophosphate
InsP$_2$, IP$_2$, inositol bisphosphate
InsP$_3$, IP$_3$, inositol trisphosphate
LysoPtdCho, LPC, LysoPC, lisophosphatidylcholine
LysoPtdEtn, LPE, LysoPE, lysophosphatidylethanolamine
PtdCho, PC, phosphatidylcholine
PtdEtn, PE, phosphatidylethanolamine
PtdGro, PG, phosphatidylglycerol
PtdH, PA, phosphatidic acid
PtdIns, PI, phosphatidylinositol
Ptd Ins4P, PIP, DPI, phosphatidylinositol
PtdIns(4,5)P$_2$,PIP$_2$, TPI, phosphatidylinositol 4,5-bisphosphate
PtdMeEtn, phosphatidylmonomethylethanolamine
PtdMe$_2$Etn, phosphatidyldimethylethanolamine
PtdSer, PS, phosphatidylserine

IN MEMORY OF GIUSEPPE PORCELLATI

Neapolitans say: *O parlà chiaro è fatto p'amici*, which means one can only speak clearly with friends and this is what I am trying to do today, because we are here to remember the life and work of Giuseppe Porcellati, our Chairman, who sadly died in July 1984 after a long and painful illness, bravely borne. We should perhaps recall the words of Machiavelli: *Tutte le cose del mondo hanno il termine della vita loro* (there is a term fixed for the life of all things in this world).

We remember him as a scientist and teacher, and as an organiser, national and international, for this meeting is a living if ephemeral memorial to him, as is the satellite meeting on his beloved phospholipids in Mantova. Mostly, however, we remember him as a person, a friendly humour-loving Neapolitan devoted to his family.

He was born in Naples in 1929 and in 1952 took his degree in Medicine and Surgery *summa cum laude* in Naples. Not content with this he took further degrees in Biology in 1956 and Natural Sciences in 1958. Still not content with these qualifications he spent two years in the grimy fastness of South East London at Guy's Hospital Medical School and gained a Ph.D. in Biochemistry under Professor Robert Thompson. Robert Thompson recalls him with pleasure; he was known affectionately in his department as "Porchy". Thus armed, he returned to this sunny clime to take teaching posts first in Pavia and then in Perugia and became full Professor of Biochemistry in Perugia in 1970, a post he held until his death. There he built up a vigorous institute with many members of staff.

I will now detail his research achievements. He started publishing in the nineteen fifties and published for 34 years, more than 500 papers in all. His early work was on nicotinic acid and NAD which was published in Italian journals. He then worked on amino acids in the CSF. With Robert Thompson he worked on Wallerian degeneration in hen nerves caused by organophosphorus compounds, with particular respect to free amino acids, but also to the phosphate esters of ethanolamine and choline, which is how I first encountered his work, for I too had studied these compounds in the brain. In the early nineteen sixties he became interested in the curious phosphodiesters of serine and ethanolamine which are found in large amounts in earthworms.

It was then that he really commenced what was to become his life's work, not only in Italy, but with collaborators in the United States, France, Poland, Germany and Sweden, namely the phospholipid metabolism of the brain. In the late nineteen sixties and early seventies he produced important data on the cytidine pathway in brain and our paths crossed again. He had a sideline, the inhibition of cholesterol biosynthesis by derivatives of butyric acid. Then in 1971 he and his collaborators published an extensive study in the *Journal of Neurochemistry* on the base exchange reaction by which ethanolamine, choline and serine are incorporated into phospholipids by a calcium-dependent pathway, a process still not understood and difficult to demonstrate in vivo but he did so in 1976. He extended this work to neurones and glia and the plasmalogens. By this time he was collaborating with Woelk in Köln, and Freysz in Strasbourg and was examining endogenous phospholipases in subcellular fractions of brain, a rather neglected area. With Hamberger in Sweden he looked at possible roles of base exchange as modulators of GABA transport.

All this time he was becoming a major force in his native land, promoting phospholipid biochemistry and neurochemistry. He contributed not only to the Italian Encyclopaedia of Science but to the Italian Encyclopaedias of Medicine and of Chemistry.

In 1977 with Lloyd Horrocks and others he showed the reversibility of the ethanolamine and choline phosphotransferase reaction which led to interesting discoveries on the mechanism of ischaemia and the production of diacylglycerols. They showed that CDPcholine can "reverse" some processes induced by ischaemia. With Droz he did fascinating studies on the axonal flow of phospholipids and an important paper on this subject was published in *Brain Research* in 1981. There were further papers on the effect of aging on phospholipid metabolism on brain function. Very recently he was concerned with the base exchange and cytidine pathways in various parts of cellular membranes and became interested in the topology of brain phospholipid metabolism and the biosynthesis of membranes. Though single-minded in his research he always took account of, and incorporated, new developments elsewhere because, for him, there was a grand design, the explanation of the role in the CNS of the major phospholipids.

In all his work he relied on his long-standing Italian colleagues Arienti, Binaglia, Gaiti, Goracci, and others too numerous to mention but he always encouraged young research workers and was very proud of his international collaborations. He was also a teacher and was awarded the Gold Medal of the Italian Ministry of Education. He was an administrator of some note, becoming Dean of Pharmacy and held other positions in Perugia. But he was also a tremendous organiser of many meetings in Italy and elsewhere including several phospholipid satellite meetings for ISN in which many of us participated. I do not have to mention to this audience that he was the Secretary of the ISN for 4 years before becoming Chairman.

Now I would like to talk about Giuseppe as a person who followed the precept of Fra Bartolomeo da San Concordio: *L'anima dell'uomo apprendendo si nutrisce, siccome il corpo per lo cibo* (the soul of man is nourished by learning as the body is by food).

He was very much a Neapolitan. Naples, the Queen of the Mediterranean, is Greek in origin; it has had a chequered history, and has often been called "a republic of philosophers". Its population is easy-going but full of life and capable of tremendous heroism as in the *Quattro Giornate di Napoli* in World War 2. Giuseppe carried with him his Neapolitan *brio* and his unmistakable *blasone*, his metaphorical coat-of-arms, as it were, and he remained attached to his native city throughout his life. The song *Santa Lucia Luntana* reminds us of the indissoluble link between Naples and its children:

> Se giro o munno sano
> Se va a cercà fortuna
> Ma quanno sponte a luna
> Luntana a Napoli
> Nun se po sta

Roughly this means that, when you see the moon shining on the water, wherever you are, it reminds you of the moon shining on the Bay of Naples. Giuseppe loved Neapolitan music. Robert Thompson tells me he sang favourite songs for his family when in London at Christmas. He loved other music too, particularly that of Monteverdi, and it is ironical that Monteverdi lived in Mantova and composed for the Duke of Mantova. Perhaps Giuseppe had that in mind when he planned the location of the forthcoming satellite meeting. His other diversion was *scopone scientifico*, a card game for four players which relies on bluff to win. Giuseppe was very good at it but he played it mainly for the contact it gave him with friends and it is as a friendly person that we remember him (he often signed his letters "very friendly"). He was devoted to his family who are here today and to them we express our condolences and thank them for sharing Giuseppe with us. He will long be remembered by us all.

G.B. Ansell

This presentation was made by G.B. Ansell at the meeting of the International Society for Neurochemistry in Riva del Garda, Italy in May of 1985. That meeting immediately preceded the symposium reported in this book. Giuseppe Porcellati was very active in the organization of this symposium and its predecessors. We miss him very much. This volume is dedicated to his memory.

Lloyd A. Horrocks
Louis Freysz
Gino Toffano

Phospholipid research and the nervous system
Biochemical and molecular pharmacology
L.A. Horrocks, L. Freysz, G. Toffano (eds)
Fidia Research Series, vol. 4.
Liviana Press, Padova. © 1986

POLYPHOSPHOINOSITIDE BREAKDOWN WITHOUT CALCIUM MOBILIZATION: STUDIES WITH ADRENAL CHROMAFFIN CELLS AND RETINA

J.N. Hawthorne, F.A. Millar, A.M.F. Swilem and H. Yagisawa

Department of Biochemistry, University of Nottingham Medical School,
Queen's Medical Centre, Nottingham NG7 2UH, UK

The polyphosphoinositides (PtdIns4P and PtdIns(4,5)P_2) have a considerable affinity for divalent cations and many years ago a physiological role in connection with calcium ions was suggested (Hawthorne and Kemp, 1964). In nervous tissue at least, the monoesterase and diesterase enzymes hydrolysing these lipids are much more active than the kinases responsible for their synthesis. The original description of PtdIns4P kinase by Kai et al. (1968) contains the following comment: "It (PtdIns(4,5)P_2) resembles a transmitter substance in that the capacity for its destruction is very great".

Though the rapid labelling of the polyphosphoinositides by ^{32}P was described quite early (Dawson, 1954), most workers concentrated on receptor-linked metabolism of PtdIns since agonists appeared to have little effect on the polyphosphoinositides (review of Hawthorne and Pickard, 1979). The present widespread interest in phosphoinositides probably dates from Michell's concept that their metabolism is linked to receptors which use calcium as a second messenger (Michell, 1975), but his theory that PtdIns breakdown led to calcium gating continued the emphasis on PtdIns.

The theory was unsatisfactory for several reasons (Cockroft, 1981; Hawthorne, 1982) and the current theory that the polyphosphoinositides are important in receptor activation began with the work of Abdel-Latif et al. (1977) and Kirk et al. (1981). The Michell group and that of Putney (Weiss et al., 1982) provided evidence that PtdIns(4,5)P_2 hydrolysis led to calcium mobilization and the most popular view now is that the second messenger responsible for this is inositol trisphosphate (InsP_3) released from PtdIns(4,5)P_2 (Berridge, 1983). The diacylglycerol released by phospholipase C action on the phosphoinositides is postulated to activate protein kinase C (Nishizuka, 1984) though the evidence is indirect, so that it also plays a part in signal transduction (Berridge and Irvine, 1984).

PROBLEMS FOR THE INSP_3-DIACYLGLYCEROL DUAL MESSENGER THEORY

Though InsP_3 has remarkable physiological effects in mimicking the action of light on *Limulus* photoreceptors (Brown et al., 1984) and the early stages of sea urchin egg fertilization (Whitaker and Irvine, 1984) for instance, there are difficulties about the dual messenger theory as presented by Berridge and Irvine (1984). These can be summarised under three headings. (1) DAG can be released by phospholipase C action on any of the three phosphoinositides and it would be less expensive in metabolic terms to obtain it from PtdIns rather than from PtdIns(4,5)P_2. It is in any case a rather odd second messenger molecule, since it is an intermediate in phospholipid and triacylglycerol synthesis. (2) There is little evidence of selective hydrolysis of PtdIns(4,5)P_2 in response to activation of receptors. Equally rapid loss of PtdIns4P is usually seen and the purified phospholipase C is not specific for PtdIns(4,5)P_2 (Wilson et al., 1984). It follows that both the supposed calcium-mobilizing messenger, InsP_3, and the product of its inactivation, InsP_2, are released simultaneously. This requires explanation. Litosch et al. (1984) have shown that 5-hydroxytryptamine stimulates loss of all three phosphoinositides from blowfly salivary glands. (3) In a number of cells polyphosphoinositide hydrolysis follows receptor activation but there is no mobilization of calcium. Examples are the inhibitory muscarinic receptors of bovine adrenal medulla as reported below and similar receptors in the heart (Brown and Brown, 1983).

POLYPHOSPHOINOSITIDES OF BOVINE CHROMAFFIN CELLS

Acetylcholine acts through nicotinic receptors to cause release of catecholamines from bovine adrenal medulla. The process involves entry of Ca^{2+} through voltage-sensitive channels but is not accompanied by changes in phosphoinositide metabolism. There is a cholinergic phosphoinositide effect however, first shown in guinea pig adrenal medulla by Hokin et al. (1958). This effect is muscarinic rather than nicotinic (Adnan and Hawthorne, 1981; Fisher et al., 1981) and in bovine chromaffin cells (Derome et al., 1981) as well as in the perfused gland (Swilem and Hawthorne, 1983), muscarinic receptors inhibit the release of catecholamines due to nicotine. We have therefore studied the muscarinic phosphoinositide changes in more detail and also monitored changes in intracellular Ca^{2+} due to cholinergic agonists.

Chromaffin cells from bovine adrenal medulla were isolated by the method of Knight and Baker (1983b) and cultured according to Fisher et al. (1981). The cultured cells were prelabelled (1.5×10^6 cells per 60 mm diameter plate) for 24 h in 3 ml culture medium containing 3 μCi/ml [^{3}H]inositol (Table 1) or for 75 min with 50 μCi/ml [^{32}P]orthophosphate (Fig. 1). The cells were then washed twice with cold Locke's solution to remove excess radioactivity. Where Ca^{2+}-free conditions were to be used, the Locke's solution contained no $CaCl_2$. The cells were then incubated in the presence of 0.3 mM carbachol (methacholine gave similar results) for various periods of time. At the end of the incubation 3 ml 20% trichloroacetic acid and 20 mg wet weight unlabelled adrenal medulla homogenate (as carrier) were added. The precipitate was centrifuged, washed once with 1 ml 5% trichloroacetic acid containing 1 mM EDTA and then with 2 ml distilled water. Lipid was extracted from the pellet with 1.5 ml chloroform/

Figure 1. Changes in ^{32}P-labelled phospholipids of chromaffin cells in response to 0.3 mM carbachol added at zero time. Details in the text. Each point is a mean with S.D. of results from three separate 4-plate experiments. Abbreviations are PA, PtdH; PI, PtdIns; PIP, PtdIns4P; and PIP$_2$, PtdIns(4,5)P_2.

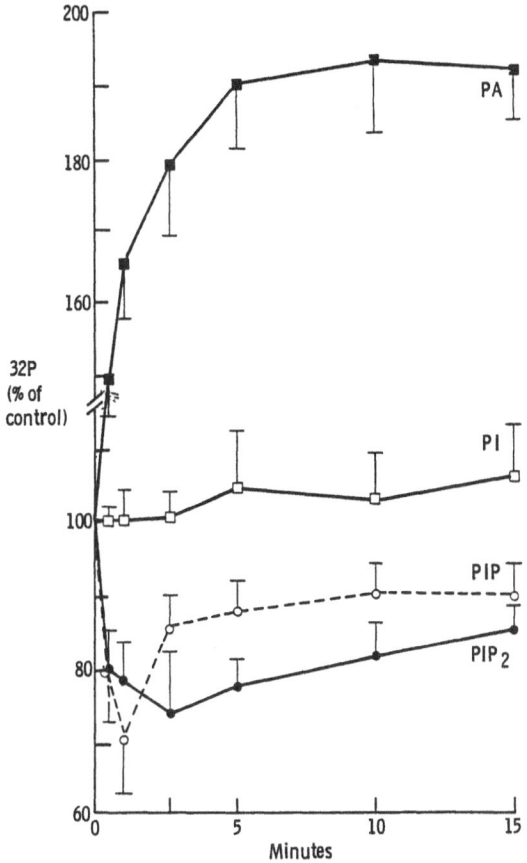

Table 1. *Changes in [³H]inositol labelled phosphoinositides of bovine adrenal chromaffin cells in response to carbachol*

	CaCl$_2$ (mM)	PtdIns(4,5)P_2	PtdIns4P	PtdIns
Control	2.2	802 ± 247	2074 ± 435	8058 ± 2028
Carbachol (0.3 mM)	2.2	379 ± 38*	1348 ± 155*	6108 ± 2270
	0; 0.5 mM EGTA	728 ± 459	1709 ± 454	8567 ± 1330

Each figure (d.p.m.) is the mean with S.D. of results from four plates. Using Student's t-test *P < 0.02. Experimental details are given in the text.

methanol/conc. HCl (100:100:1 by vol.) followed by 1.5 ml chloroform/methanol/conc. HCl (200:100:1). After adding 1.5 ml 0.1 M HCl and 1.5 ml chloroform to the combined extracts and shaking, the lower phase was removed for concentration under nitrogen and application to thin-layer plates of silica gel 60 H spread in 3% magnesium acetate. Two-dimensional chromatography used chloroform/methanol/ conc. ammonia (65:25:5) followed by chloroform/acetone/methanol/glacial acetic acid/water (30:40:10:10:5). After iodine staining, spots were scraped off for scintillation counting (Yagihara et al., 1973).

Table 1 shows that there was a loss of all three [³H]inositol-labelled phosphoinositides in 30 sec, though the PtdIns loss was not significant statistically. Using ^{32}P, the loss of PtdIns is not apparent, but there is rapid initial loss of PtdIns4P and PtdIns(4,5)P_2, accompanied by increased labelling of PtdH which was already apparent at 30 sec. The PtdH may be formed by phosphorylation of DAG released from the phosphoinositides.

Fig. 2 shows the separation of the inositol phosphates released from [³H]inositol-labelled chromaffin cells by carbachol. The drug did not affect glycerophosphoinositol but increased the radioactivity due to Ins$P$, InsP_2 and InsP_3. The major increase was in the InsP_2. Stimulation in the presence of 10 mM LiCl to inhibit inositol phosphatase produced a larger increase in Ins$P$ but no greater increases in the InsP_2 or InsP_3.

Fisher et al. (1981) showed that nicotinic but not muscarinic drugs stimulated the influx of Ca^{2+} into bovine chromaffin cells. Since Ca^{2+} might be released from an intracellular store in response to the muscarinic stimulus, the fluorescent indicator Quin 2 (Tsien, 1981) has been used to monitor cytoplasmic free Ca^{2+}. Fig. 3 shows a sharp rise in this intracellular Ca^{2+} in response to nicotine, but no change with the muscarinic drug methacholine. Since methacholine may be hydrolysed by acetylcholinesterase, pilocarpine was also used, but without any effect on Ca^{2+} concentration. These results differ from recent reports by Kao and Schneider (1985) and Cheek and Burgoyne (1985) of a small (50-100 nM) increase in cytosolic Ca^{2+} with 0.3 mM methacholine. In neither case however, had the cell preparations been purified by density gradient centrifugation. This and other differences in technique may explain the discrepancy. Since concentrations as low as 10 μM nicotine increase cytoplasmic Ca^{2+} from 100 nM to 1 μM or higher, it is difficult to see how a further small increase could explain the muscarinic inhibition of secretion.

The inhibitory muscarinic effect could be related to a rise in cyclic GMP but we are left with the role of the polyphosphoinositide loss and production of InsP_3 in chromaffin cells. If the InsP_3 has no calcium-mobilizing function it is still possible that the diacylglycerol could activate protein kinase C and inhibit the secretion due to nicotinic Ca^{2+} influx. Phorbol esters, however, which are considered to have similar effects to this diacylglycerol, appear to make the secretory response more rather than less sensitive to Ca^{2+} (Knight and Baker, 1983a). Nevertheless, proteins of the chromaffin granule can be phosphorylated (Burgoyne, 1984) and it is possible that some balance between PtdIns and protein phosphorylation could regulate exocytosis. Our present results though, suggest that in the bovine adrenal medulla InsP_3 is a messenger without a message - or perhaps without a listener.

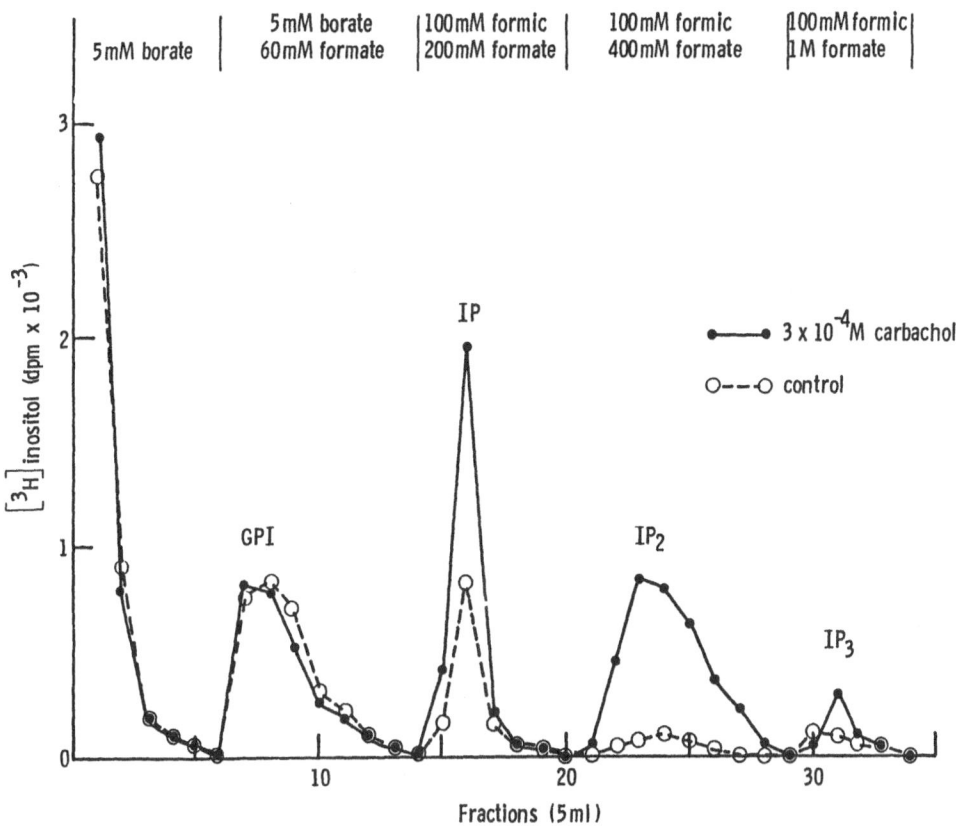

Figure 2. Water-soluble [³H]inositol phosphates released from chromaffin cells in response to carbachol. Details are given in the text.

Incubation for 90 sec in medium containing 2.2 mM CaCl₂. The trichloroacetic acid supernatant was extracted five times with diethyl ether to remove most of the acid, then neutralized with NaOH. For loading on to a column of Dowex 1×8 (200 mesh, formate form) 5 mm diameter, volume 1 ml, Na₂B₄O₇ was added to give a final concentration of 5 mM. Elution followed the method of Ellis et al. (1963). From left to right the peaks represent glycerophosphoinositol, GPI; inositol phosphate, IP; inositol bisphosphate, IP₂; and inositol trisphosphate, IP₃.

POLYPHOSPHOINOSITIDES IN THE RETINA

It has been known for some years that light affects the labelling of phosphoinositides in the retina and a recent study shows a specific loss of ^{32}P-labelled PtdIns(4,5)P_2 during a 5s light flash (Ghalayini and Anderson, 1984). As in the experiment of Table 2, there was no effect on PtdIns4P. Ghalayini and Anderson used frog retina and prepared rod outer segments after quenching with trichloroacetic acid, the whole retina having been exposed to light. Preparation of rod outer segments before light exposure

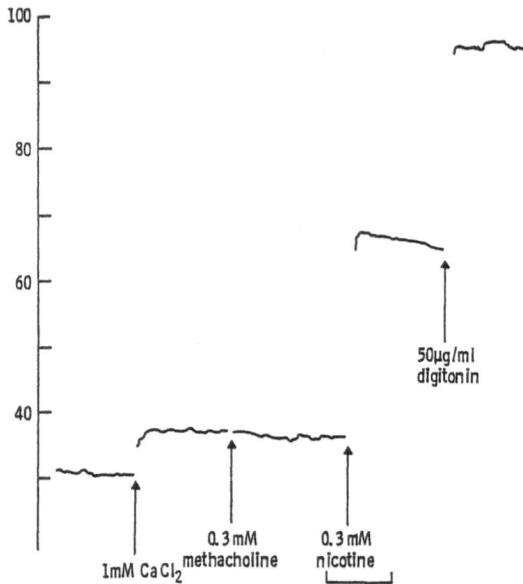

Figure 3. Quin-2 fluorescence response to cholinergic agonists in chromaffin cells. The method of Tsien (1981) was used and fluorescence is expressed in arbitrary units. The horizontal bar represents 2 min.

Table 2.*Polyphosphoinositide responses to light in rod outer segments from bovine retina*

	PtdIns(4,5)P_2	PtdIns4P
	(d.p.m. ^{32}P)	
Dark	163 ± 54	127 ± 39
Light	81 ± 44*	145 ± 30

After labelling with [^{32}P]orthophosphate for 2 h the rod outer segments were prepared and exposed to bright light for 30 sec. After quenching in 10% trichloroacetic acid phosphoinositides were extracted in acidified chloroform-methanol and separated by thin-layer-chromatography. Each figure is a mean with S.D. from five experiments. *By Student's t-test, this result differs from the dark result (P < 0.05).

caused loss of polyphosphoinositides during fractionation. This is clear from the low radioactivity in our studies of bovine outer segments, which were prepared and then exposed to light. In view of what has been said above, it is interesting that there seems to be a selective loss of PtdIns(4,5)P_2 and no significant change in PtdIns4P radioactivity. Further experiments with rat retina are in progress and the freshly excised tissue appears able to incorporate much more radioactivity into its phosphoinositides than bovine retina.

Fein et al. (1984) and Brown et al. (1984) have shown that InsP_3 injected into the ventral photoreceptor cells of the crab *Limulus* caused a depolarization similar to the effect of light. They suggest that InsP_3 is a messenger in the light transduction process. Exposure of the ventral eyes to light for 30 sec caused a loss of [^{3}H]inositol-labelled PtdIns(4,5)P_2 (Brown et al., 1984). No effect of light was seen however, when the phosphoinositides were labelled with ^{32}P.

In all the photoreceptor cells studied therefore, light causes a rapid breakdown of PtdIns(4,5)P_2 and presumably release of InsP_3. The physiological function of this release is not at all clear at present. An obvious suggestion is that InsP_3 is mobilizing calcium but there are several difficulties about this. In *Limulus* (Brown et al., 1984) both light and InsP_3 depolarize the photoreceptor cells, but injection of Ca^{2+} does not. Furthermore, while PtdIns(4,5)P_2 breakdown is a general response to light, vertebrate photoreceptor cells are hyperpolarized by light while those of *Limulus* are depolarized. A simple theory that Ca^{2+} release is essential for transduction is not valid and many workers consider that the key event is hydrolysis of cyclic GMP (Fesenko et al., 1985). Contrary to previous reports, light reduces the cytoplasmic free Ca^{2+} in retinal rod outer segments of the toad (Yau and Nakatani, 1985). These workers suggest that Ca^{2+} may be involved in adaptation to light.

In summary, the polyphosphoinositides respond rapidly to light in all photoreceptor cells studied, but there is no evidence that they act by mobilizing cytoplasmic Ca^{2+}.

ACKNOWLEDGMENTS

We thank the Medical Research Council, The Wellcome Trust and the Japan Society for the Promotion of Science (H.Y.) for financial support.

REFERENCES

Abdel-Latif AA, Akhtar RA, Hawthorne JN (1977) Biochem J 162: 61-73.
Adnan NAM, Hawthorne JN (1981) J Neurochem 36: 1858-1860.
Berridge MJ (1983) Biochem J 212: 849-858.
Berridge MJ, Irvine RF (1984) Nature 312: 315-321.
Brown SL, Brown JH (1983) Molec Pharmacol 24: 351-356.
Brown JE, Rubin LJ, Ghalayini GJ, Tarver AP, Irvine RF, Berridge MJ, Anderson RE (1984) Nature 311: 160-163.
Burgoyne RD (1984) Biochim Biophys Acta 779: 201-216.
Cheek TR, Burgoyne RD (1985) Biochim Biophys Acta 846: 167-173.
Cockroft S (1981) Trends Pharmac Sci 2: 340-342.
Dawson RMC (1954) Biochem J 57: 237-245.
Derome G, Tseng R, Mercier P, Lemaire I, Lemaire S (1981) Biochem Pharmacol 30: 855-860.
Ellis RB, Galliard T, Hawthorne JN (1963) Biochem J 88: 125-131.
Fein A, Payne R, Corson DW, Berridge MJ, Irvine RF (1984) Nature 311: 157-160.
Fesenko EE, Kolesnikov SS, Lyubarsky AL (1985) Nature 313: 310-313.
Fisher SK, Holz RW, Agranoff BW (1981) J Neurochem 37: 491-497.
Ghalayini A, Anderson RE (1984) Biochem Biophys Res Commun 124: 503-506.
Hawthorne JN (1982) Nature 295: 281-282.
Hawthorne JN, Kemp P (1964) Adv Lipid Res 2: 127-166.
Hawthorne JN, Pickard MR (1979) J Neurochem 32: 5-14.
Hokin MR, Benfey BG, Hokin LE (1958) J Biol Chem 233: 814-817.
Kai M, Salway JG, Hawthorne JN (1968) Biochem J 106: 791-801.
Kao LS, Schneider AS (1985) J Biol Chem 260: 2019-2022.
Kirk CJ, Creba JA, Downes CP, Michell RH (1981) Biochem Soc Trans 9: 377-379.

8

Knight DE, Baker PF (1983a) FEBS Lett 160: 98-100.
Knight DE, Baker PF (1983b) Quart J Exp Physiol 68: 123-143.
Litosch I, Lee HS, Fain JN (1984) Am J Physiol 246: C141-C147.
Michell RH (1975) Biochim Biophys Acta 415: 81-147.
Nishizuka Y (1984) Nature 308: 693-698.
Swilem AMF, Hawthorne JN (1983) Biochem Pharmacol 32: 3873-3874.
Tsien RY (1981) Nature 290: 527-528.
Weiss SJ, McKinney JS, Putney Jr JW (1982) Biochem J 206: 555-560.
Whitaker MJ, Irvine RF (1984) Nature 312: 636-639.
Wilson DB, Bross TE, Hofmann SL, Majerus PW (1984) J Biol Chem 259: 11718-11724.
Yagihara Y, Bleasdale JE, Hawthorne JN (1973) J Neurochem 21: 173-190.
Yau KW, Nakatani K (1985) Nature 313: 579-582.

Phospholipid research and the nervous system
Biochemical and molecular pharmacology
L.A. Horrocks, L. Freysz, G. Toffano (eds)
Fidia Research Series, vol. 4.
Liviana Press, Padova. © 1986

INOSITOL LIPID METABOLISM IN RECEPTOR-STIMULATED AND DEPOLARIZED SYMPATHETIC GANGLIA AND ADRENAL GLANDS

Robert H. Michell and Elisabeth A. Bone*

Department of Biochemistry, University of Birmingham, P.O. Box 363,
Birmingham B15 2TT, UK

INTRODUCTION

Recent investigations of the classical phosphatidylinositol response, first described in exocrine pancreas and brain by Hokin and Hokin in 1953 and later extensively characterised by the same workers, have identified a widespread receptor-stimulated signalling process that is important to the regulation of processes as diverse as fibroblast growth, hormone secretion, activation of the immune system, vision in horseshoe crabs, and fertilization in toads. The central reaction of this signalling system is hydrolysis of $PtdIns(4,5)P_2$ to 1, 2-DAG and $Ins(1,4,5)P_3$, with both products acting as cellular second messengers, 1, 2-DAG activates a protein kinase (protein kinase C, Nishizuka, 1984) and $Ins(1,4,5)P_3$ mobilizes Ca^{2+} from an intracellular membrane-sequestered pool (Berridge and Irvine, 1984). The development of these ideas can be traced in numerous reviews since 1975 (Michell, 1975, 1979, 1982a; Michell et al., 1977, 1981, 1984; Berridge, 1980, 1981, 1984; Downes and Michell, 1982, 1985; Nishizuka, 1984; Berridge and Irvine, 1984).

Although acetylcholine-stimulated inositol lipid metabolism in brain was reported in the Hokins' first paper in 1953, the precise function of the inositol lipid signalling system in neural tissues remains uncertain (for reviews, see Hokin, 1969; Hawthorne and Pickard, 1979; Michell, 1981, 1982b; Downes, 1982, 1984). Many of the experiments on inositol lipid metabolism in stimulated nervous tissues have employed tissue slices prepared from particular regions of the CNS, most often cerebral cortex. The major outcome of these studies has been the demonstration that many of the receptors whose activation causes $PtdIns(4,5)P_2$ hydrolysis in peripheral tissues are coupled to the same biochemical signalling system in the nervous system. These include receptors for the

---

* Present address: Lab. of Biochem. Pharmacology, NIADDK, National Institutes of Health, Bethesda, Md 20205, USA.

following stimuli: muscarinic cholinergic (some, but maybe not all), α_1-adrenergic, S-2 serotonin, H_1-histamine, V_1-vasopressin, neurotensin, substance P and bradykinin (for references, see Hawthorne and Pickard, 1979; Michell, 1981; Downes, 1984).

Early in these studies of stimulated inositol lipid metabolism in the nervous system, it was realised that clear indications as to the neural function(s) of this response would probably come only from studies of tissues preparations which were less complex than brain slices and which retained physiological functions that could be monitored in parallel with the studies of lipid metabolism. As a result, the Hokins and their colleagues initiated work on cholinergic stimulation of sympathetic ganglia (Hokin, et al., 1960) and of the adrenal medulla, a closely related endocrine secretory organ (Hokin, et al., 1958). Subsequent work on sympathetic ganglia has shown that: a) ganglionic inositol lipid turnover can be stimulated by electrical stimulation of the cholinergic innervation of the ganglia (Larrabee et al., 1963; Larrabee and Leicht, 1965) or by the normal signal throughput of the ganglion in vivo (Larrabee, 1968); b) that much of the response to cholinergic stimulation is located in the major noradrenergic cell bodies of the ganglia (Hokin, 1965); and c) that inositol lipid metabolism can also be stimulated by depolarization of ganglia in a high K^+ medium (Nagata et al., 1973). Initial studies suggested that the response to nerve stimulation was mediated through nicotinic cholinergic receptors, but later investigations demonstrated that the responses both to nerve stimulation and to added carbamylcholine can be abolished by muscarinic cholinergic antagonists (Lapetina et al., 1976; Pickard et al., 1977). Similarly, the cholinergic response of the adrenal medulla is atropine-sensitive and is therefore mediated by muscarinic cholinergic receptors (Hokin et al., 1958), whereas catecholamine secretion from the adrenal medullae of many species is mainly controlled by nicotinic cholinergic receptors (for review, see Hawthorne and Swilem, 1982 and this volume).

When these studies were initiated about 25 years ago, sympathetic ganglia were regarded as rapid unidirectional neural relays in which cholinergic input from the CNS caused rapid activation of neurones that were mostly noradrenergic and that innervated many peripheral tissues. Similarly, adrenal medullary chromaffin cells were considered to be relatively simple adrenaline-secreting cells whose secretion was regulated through a similar cholinergic nerve input. Since then, it has become clear that both of these systems are much more complex: ganglia include several types of neurones, with a variety of transmitters, and adrenal chromaffin secretion is modulated through receptors for a variety of stimuli.

Here we briefly describe studies which have revealed that sympathetic ganglia employ inositol lipid hydrolysis as a signalling mechanism in their responses to a variety of peptides that are putative neurotransmitters. At least one of these peptides, which has not yet been identified, is released and acts within depolarized ganglia.

VASOPRESSIN STIMULATES INOSITOL LIPID HYDROLYSIS IN ISOLATED MAMMALIAN SYMPATHETIC GANGLIA

In 1982, we were screening a variety of putative neurotransmitters for possible effects on lipid metabolism in isolated superior cervical sympathetic ganglia from rats. Amongst the agents tested was arginine-vasopressin, which stimulates inositol lipid hydrolysis in a wide variety of tissues, always through receptors of the V_1-type that regulate Ca^{2+}-mediated cellular responses such as contraction of vascular smooth muscle and activation of hepatic glycogen phosphorylase (see Table 1). At that time, vasopressin was not known to have any effect upon the activity of the sympathetic nervous system, so we were surprised to find that it caused a much more rapid accumulation of inositol phosphates in ganglia stimulated in the presence of Li^+ (Berridge et al., 1982) than did muscarinic cholinergic stimulation (Bone et al. 1984; see Table 2).

Table 1. *Cells and tissues in which vasopressin stimulates inositol lipid metabolism*

Hepatocytes	Kirk et al., 1979, 1981; Michell et al. 1979, 1981, 1984; Creba et al., 1983; Thomas et al., 1983.
Aorta smooth muscle	Takhar and Kirk, 1981.
WRK-1 mammary tumour	Monaco, 1982; Monaco and Woods, 1983; Kirk et al., 1985.
Renal mesangial cells	Troyer et al., 1984.
Platelets	MacIntyre, 1985.
Adipocytes	R Rubio, P Newsholme and RH Michell, unpublished.
Fibroblasts	Brown et al., 1984.
Sympathetic ganglia	Hanley et al., 1984; Bone et al., 1984; Bone and Michell, 1985.
Hippocampus	Stephens and Logan, 1985.

Table 2. *Accumulation of inositol phosphates in ganglia exposed to muscarinic cholinergic stimulation or to vasopressin*

	^3H dpm/10^5 dpm in:			
Conditions	GroPIns	InsP	InsP$_2$	InsP$_3$
Control (n = 7)	3150 ± 250	21500 ± 2800	13800 ± 2900	990 ± 100
100 μM Bethanechol (n = 9)	3610 ± 290	40600 ± 3700**	18800 ± 2300	1350 ± 110*
10 μM Atropine (n = 5)	3530 ± 570	25800 ± 3400	12300 ± 3200	1220 ± 240
100 μM Bethanechol and 10 μM Atropine (n = 7)	3020 ± 210	11500 ± 1700‡‡	7400 ± 1500	740 ± 110‡
0.32 μM [Arg^8] Vasopressin (n = 8)	2470 ± 230	130300 ± 12800**	47200 ± 5200**	6830 ± 780**

Ganglia were labelled with [2-^{3}H]inositol. After preincubation for 5 min with 10 mM LiCl in the presence or absence of 10 μM-atropine, ganglia were stimulated with 100 μM bethanechol or 0.32 μM [Arg^8]vasopressin for 60 min. The values given are means ± SEM corrected to a standard incorporation of 10^5 dpm into the lipids of each ganglion. Significantly different from control incubations: * $P<0.05$, ** $P<0.001$; significantly different from incubations with atropine: + $P\sim0.05$, ++ $P<0.001$. (Reproduced, with the permission of the Biochemical Journal, from Bone et al., 1984).

Detailed investigation of the properties of this response showed it to be essentially identical to that observed in hepatocytes and other cells when stimulated by this peptide (Bone et al., 1984). It was initiated within seconds and sustained for at least 60 min, was a response to activation of V_1-receptors, and still occurred when ganglia were stimulated during incubation in a low-Ca^{2+} medium. InsP$_3$ and InsP$_2$ accumulated

12

Table 3. *Time-course of the accumulation of inositol phosphates in [Arg^8] vasopressin-stimulated sympathetic ganglia*

	^{3}H (dpm) in:					
	Ins$P$		InsP_2		InsP_3	
Time	Control	+ [Arg^8] vasopressin	Control	+ [Arg^8] vasopressin	Control	+ [Arg^8] vasopressin
15 s	5850 ± 660 (15)	6900 ± 880 (15)	2010 ± 200 (15)	4350 ± 380** (15)	470 ± 45 (15)	1520 ± 155** (15)
60 s	15200 ± 1710 (23)	13500 ± 1170 (28)	7750 ± 880 (23)	15500 ± 1060** (28)	1460 ± 190 (28)	3390 ± 310* (28)
10 min	13900 ± 1060 (9)	54300 ± 2490** (9)	8470 ± 1110 (9)	47200 ± 1990** (11)	860 ± 90 (9)	7250 ± 420** (11)

Isolated ganglia prelabelled with [2-^{3}H] inositol were incubated in the presence and absence of 0.32 μM-[Arg^8] vasopressin for the appropriate time. Incubations were terminated and the inositol phosphates were separated. The values are means ± SEM (numbers of ganglia analysed in parentheses) and are not corrected for incorporation of [^{3}H] inositol into ganglionic lipids. Significantly different from control incubations: * $P<0.01$, ** $P<0.001$. (Reproduced, with the permission of the Biochemical Journal, from Bone et al., 1984).

before Ins$P$ (Bone et al., 1984; Table 3). This result is compatible with the dominant view that receptor activation stimulates the hydrolysis of PtdIns(4,5)P_2 (Michell et al., 1981, 1984; Berridge, 1984), yielding Ins(1,4,5)P_3 that mobilizes intracellular Ca^{2+} (Berridge and Irvine, 1984) and is inactivated and returned to the cellular free inositol pool by dephosphorylation (Storey et al., 1984; Seyfred et al., 1984; Joseph and Williams, 1985). However, it should be noted that we have not yet determined whether the accumulated InsP_3 is the expected 1,4,5-isomer or the novel and unexplained 1,3,4-isomer that has recently been detected in stimulated salivary glands (Irvine et al., 1984, 1985).

Two of the most striking aspects of the ganglionic response to vasopressin were its magnitude and the tissue's resistance to inositol depletion. When ganglia were prelabelled with [^{3}H]inositol and then stimulated during incubation in an unlabelled medium, labelled inositol phosphates corresponding to the entire labelled inositol lipid pool accumulated about every 30 min, without any accompanying decrease in the labelling of the ganglionic lipid pool (Bone et al., 1984). Presumably, therefore, this tissue is capable of retaining labelled free inositol for a considerable period, maybe through the activity of an active uptake system. Moreover, the pool of inositol lipid susceptible to receptor-stimulated hydrolysis must be resynthesised with extreme rapidity: either the metabolically active inositol pool of the entire ganglion is renewed every 30 min or a smaller pool has an even shorter turnover time.

A VASOPRESSIN-LIKE PEPTIDE IS PRESENT IN THE MAJOR NEURONES OF MAMMALIAN SYMPATHETIC GANGLIA

Discovery of the dramatic response of ganglia to vasopressin provoked, in the London laboratories of Michael Hanley and of Stafford Lightman, a successful search in these ganglia for vasopressin-like material (Hanley et al., 1984). In summary, the results of this study, which immunologically identified a material that was designated vasopressin-like peptide (VLP), were as follows. Sympathetic ganglia of several mammalian species, from mouse to monkey, contain immunoreactive VLP that can be detected immunocytochemically or by solution immunoassays. This material reacts with several antisera to vasopressin and oxytocin, and the oxytocin-like and vasopressin-like immunoreactive materials coelute from an HPLC column with a much larger apparent molecular weight than either of the authentic peptides. VLP is present both in superior cervical and coeliac ganglia, with more in the former. Immunocytochemical staining, inhibitable by an excess of authentic vasopressin or oxytocin, demonstrates that VLP is present in most, and probably all, of the large dopamine beta-hydroxylase-positive neurones of superior cervical ganglia, and also within a diffuse plexus of nerve fibres within the ganglia. Examination of peripheral tissues reveals VLP-positive nerve fibres in sympathetically innervated structures such as the renal artery, with the distribution of VLP-positive nerve fibres similar to that of established sympathetic neurotransmitters (Hanley et al., 1984). At present, it is still not known whether the VLP detected in peripheral nerve fibres is of the same large molecular weight as the ganglionic material: it is possible that the ganglionic material might be processed to a smaller material en route to the periphery.

Surprisingly, the quantity of immunoreactive VLP in ganglia was only modestly reduced in the Brattleboro rat, a mutant that suffers from hereditary diabetes insipidus because authentic vasopressin is absent from the neurohypophysis (Hanley et al., 1984). However, it was subsequently recognised that immunoreactive vasopressin also exists in ovarian luteal cells and some adrenal medullary chromaffin cells, and that here again it persists in Brattleboro rats (Ang and Jenkins, 1984; Lim et al., 1984; Nussey et al., 1984). At present, it appears that the vasopressin precursor is encoded by only a single gene and that in the Brattleboro rat this shows a point mutation in a sequence remote from that encoding vasopressin itself (Schmale and Richter, 1984). It is not yet known whether the vasopressin and VLP present in the tissues of Brattleboro animals are encoded by the mutant vasopressin gene, with peptide production sustained as a result of tissue-specific processing of the mutant precursor glycopeptide, or by a second, and as yet undetected, gene (Bonner and Brownstein, 1984).

Whatever the answers to the questions raised above, it is clear that many peripheral tissues may normally be exposed, either continuously or intermittently, to considerably greater stimulation of their vasopressin receptors than that achieved by circulating vasopressin of neurohypophyseal origin. The vasopressin receptors of these peripheral tissues may therefore be of greater physiological importance, particularly in responses to locally released peptides, that has been recognised hitherto.

STIMULATION OF INOSITOL LIPID BREAKDOWN IN DEPOLARIZED SYMPATHETIC GANGLIA IS PROBABLY MEDIATED BY A NOVEL NEUROPEPTIDE RATHER THAN VLP

Some years ago, Nagata et al. (1973) showed that depolarization stimulates ganglionic inositol lipid metabolism, and that this is probably not due to cholinergic activation of the major noradrenergic neurones. Once we knew that VLP is present in an intraganglionic plexus of nerve fibres (which also contains immunoreactive dopamine beta-hydroxylase) and that vasopressin stimulates ganglionic inositol lipid hydrolysis, we anticipated that the response of ganglia to depolarization would be mediated through intraganglionic release of VLP. However, our experiments suggest that this is not the case, and the evidence summarized below suggests that another neurotransmitter, which is capable of stimulating inositol lipid hydrolysis, is released within depolarized ganglia (Bone and Michell, 1985).

We first confirmed that the stimulation of inositol lipid metabolism in depolarized ganglia does involve the liberation of inositol phosphates (Fig. 1), and that it is therefore a result of activation of a phosphoinositidase C. Secondly, we demonstrated that this

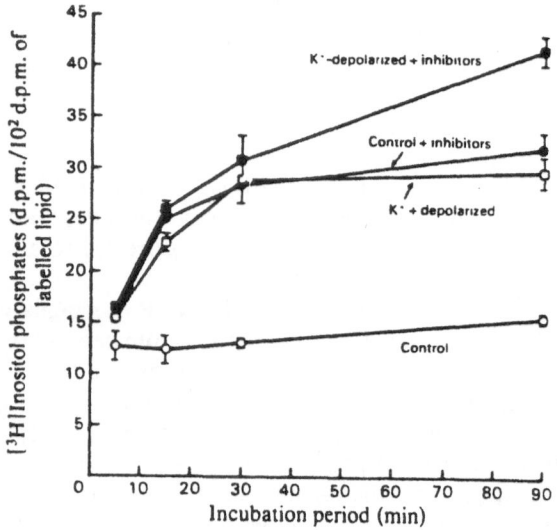

Figure 1. Effect of proteinase inhibitors on inositol phosphate accumulation in control and depolarized ganglia.

After labelling with [³H]inositol, ganglia were preincubated for 5 min with 10 mM LiCl and antipain, amastatin and bestatin (10 μg/ml of each). Ganglia were then incubated for a further 90 min, in the presence or absence of 80 mM KCl, together with the inhibitors. (Reproduced, with the permission of the Biochemical Journal, from Bone and Michell, 1985).

response, unlike the response to added vasopressin, is abolished by removal of Ca^{2+} from the incubation medium (Bone and Michell, 1985). The most likely interpretation of this result is that depolarization causes Ca^{2+}-dependent release within the ganglia of a neurotransmitter that is a potent stimulator of inositol lipid hydrolysis. The first three obvious candidates for this role, all of which stimulate inositol lipid hydrolysis in appropriate tissues and all of which are present within the ganglion, were acetylcholine, noradrenaline and VLP. Studies with antagonists indicated that none of these agents is the major mediator of the response to depolarization: the response was reduced by only about one-quarter in the presence of atropine (to block muscarinic receptors) and was unchanged by an α_1-adrenoceptor antagonist (phenoxybenzamine) or a V_1-vasopressin receptor antagonist (Bone and Michell, 1985). Studies with other agonists and antagonists ruled out a major role for histamine, 5-hydroxytryptamine, GABA, or prostaglandins.

Since neither VLP nor any of the classical small amine neurotransmitters appeared to be responsible, we wondered whether depolarization might be causing the release within ganglia of an unidentified peptide neurotransmitter. This was investigated in two ways: by testing candidate peptides and by investigating whether the response could be prolonged or potentiated by protease inhibitors which might prolong the lifetime of a released peptide within the ganglia. The peptides to be tested (at a concentration of either 1 μM or 10 μM) were chosen either because of their ability to stimulate inositol lipid breakdown in other cells or because they had been detected in sympathetic ganglia (for references, see Bone and Michell, 1985). Some of these peptides were inactive (somatostatin, LHRH, angiotensin), whilst others produced responses much smaller than that evoked by depolarization (bombesin, pancreozymin, bradykinin, and a metabolically stabilized substance P analogue) (Bone and Michell, 1985). No single peptide gave a stimulation nearly as large as that produced by depolarization, though the possibility remains that the response to depolarization might be the sum of the effects of several endogenously released peptide stimuli.

Despite the inconclusive results obtained with these known peptides, the studies with protease inhibitors gave some support for the view that a peptide (or mixture of peptides) is involved: a mixture of three protease inhibitors (antipain, bestatin and amastatin) both increased the rate of accumulation of inositol phosphates during depolarization and partially reproduced the response even in undepolarized ganglia (Bone and Michell, 1985; Fig. 1). Peptides that are present in ganglia but have not yet been tested include neuropeptide Y, enkephalins, and neurotensin (Bone and Michell, 1985).

INOSITOL LIPID HYDROLYSIS IN RAT ADRENAL GLANDS IS STIMULATED BY MUSCARINIC STIMULI BUT NOT BY VASOPRESSIN

The adrenal medulla and sympathetic ganglia share several important developmental and functional features: a) both are derived from the neural crest during development; b) both synthesize catecholamine whose secretion is largely controlled by cholinergic nerve input; and c) both synthesize, and presumably secrete, vasopressin or a vasopressin-like peptide (Hanley et al., 1984; Ang and Jenkins, 1984; Nussey et al., 1984). It has been known for many years that inositol lipid hydrolysis in the adrenal medulla is stimulated by muscarinic cholinergic stimulation, so we wondered whether the similarities between these cells would extend to possession of V_1-vasopressin receptors coupled to inositol lipid hydrolysis. This was tested by a simple experimental protocol which measured the total labelled inositol phosphates that accumulated in halved adrenal glands stimulated in the presence of Li^+.

The results, shown in Table 4, demonstrate that the established response of adrenals to muscarinic cholinergic stimulation does indeed involve inositol lipid hydrolysis. However, these glands show no response to stimulation with high concentrations of either vasopressin or adrenaline, to both of which they are presumably exposed whenever they secrete in response to a physiological stress.

THE POSSIBLE ROLES OF VASOPRESSIN, VLP, AND OTHER PEPTIDES IN THE REGULATION OF SYMPATHETIC FUNCTION AND IN THE RESPONSES OF END-ORGANS TO SYMPATHETIC INNERVATION

For many years, the role of vasopressin receptors in tissues other than renal tubules has been enigmatic, since the Ca^{2+}-mobilizing V_1-receptors present in other tissues appeared only to provoke appreciable physiological responses at vasopressin concentra-

Table 4. *Accumulation of labelled inositol phosphates in rat hemi-adrenals stimulated for 60 min*

	dpm/10^5 dpm in lipids
Control	3750 ± 390 (24)
10 μM muscarine	8250 ± 600 (21)
10 μM muscarine and 10 μM atropine	3900 ± 1100 (6)
0.17 μM arg-vasopressin	3500 ± 670 (10)
10 μM adrenaline	4300 ± 330 (14)

Adrenal glands were dissected from female rats (220-250g) anaesthetized with ether. The connective tissue capsules were removed and the glands bisected longitudinally to expose the medullae. Each hemi-adrenal was incubated for 1 h in 0.3 ml of gassed Krebs-Ringer bicarbonate medium containing 0.9 mM Ca^{2+}, 10 mM glucose and 7.5 μCi of [2-^{3}H]inositol, and incubations were gassed at 30 min intervals. 10 mM LiCl was then added to inhibit inositol 1-phosphatase (Berridge et al., 1982), and 10 μM atropine was added to some incubations. The hemi-adrenals were transferred 5 min later to similar medium containing, when appropriate, a neurotransmitter or vasopressin. Incubations were terminated 20 min later with $HClO_4$, and the extracts were analysed as described by Bone et al. (1984). The total radioactivity of the accumulated inositol phosphates was expressed in terms of a standard incorporation of 10^5 dpm into the lipids of a single hemi-adrenal. The values are means ± SEM (number of adrenals analysed).

tions higher than those that are usually achieved in the circulation. The discovery of vasopressin and VLP in non-neurohypophyseal tissues outside the CNS, and of V_1-vasopressin receptors in an ever-increasing number of peripheral tissues, raises the intriguing possibility that many of the physiological responses to stimulation of these receptors are responses to peptides released locally from sympathetic nerve endings or other structures. The next few years should see a detailed exploration of this idea.

Additional questions for which we lack a clear answer are: what is the function of V_1-receptor-stimulated inositol lipid hydrolysis in sympathetic ganglia; and what are the nature and the function of the intrinsic, probably peptide, mediator of the ganglionic response to depolarization? Both of these responses involve rapid hydrolysis, and compensating resynthesis, of substantial fractions of the labelled inositol lipids of prelabelled ganglia. Presumably the Ins(1,4,5)P_3 and 1, 2-DAG produced in the as yet unidentified responsive structures (neurones?) in the ganglia serve as signals that modify the functions of these cells in some manner.

Since the evidence summarized above clearly indicates that endogenously released VLP does not appreciably stimulate the V_1-receptors of isolated ganglia, we must consider whether ganglia might be sensitive to regulation by circulating vasopressin. Several pieces of evidence, none of them convincing in isolation, now make a modest case for such control. First, there is some evidence, particularly from spinally transected animals, for some influence of near-normal concentrations of circulating vasopressin on the sympathetic nervous system (Cowley and Barber, 1983). Secondly, it has recently been shown that vasopressin modifies transmission through intact sympathetic ganglia: this involves both a presynaptic inhibitory component and excitation of a small proportion of the major postsynaptic neurones (Peters and Kreulen, 1984; Wali, 1984; and personal communications from J-J Dreifuss and S D Logan). Thirdly, the inositol lipid hydrolysis evoked in stimulated ganglia by vasopressin is so large that an appreciable signal is likely to be delivered to the stimulated cells within the ganglia even at the low concentrations of vasopressin that circulate in unstressed conditions.

The nature and function of the endogenously released activator of inositol lipid hydrolysis is a problem for the future. Although the available evidence suggests that it is probably a peptide, our experiments appear to have ruled out the most obvious candidates (VLP, substance P, LHRH, angiotensin). One possibility is that this material acts as an intraganglionic facilitation of transmission in rather the same way that sustained exposure to bradykinin stimulates the firing of certain neuroblastoma cells in culture (Yano et al., 1984).

ACKNOWLEDGMENTS

These studies were supported by a Project Grant from the Medical Research Council.

REFERENCES

Ang VTY and Jenkins JS (1984) J Clin Endocrinol Metab 58: 688-691.
Berridge MJ (1980) Trends Pharmacol Sci 1: 419-420.
Berridge MJ (1981) Mol Cell Endocrinol 24: 115-140.
Berridge MJ (1984) Biochem J 220: 345-360.
Berridge MJ and Irvine RF (1984) Nature 312: 315-321.
Berridge MJ, Downes CP and Hanley MR (1982) Biochem J 206: 587-595.
Bone EA and Michell RH (1985) Biochem J 227: 263-269.
Bone EA, Fretten P, Palmer S, Kirk CJ and Michell RH (1984) Biochem J 221: 803-811.
Bonner TI and Brownstein MJ (1984) Nature 310: 17.
Brown KD, Blay J, Irvine RF, Heslop JP and Berridge MJ (1984) Biochem Biophys Res Commun 123: 377-384.
Cowley AW and Barber BJ (1983) In: Cross BA and Leng C (eds): The Neurohypophysis: Structure, Function and Control. Elsevier, Amsterdam, pp. 415-424.
Creba JA, Downes CP, Hawkins PT, Brewster G, Michell RH and Kirk CJ (1983) Biochem J 212: 733-747.
Downes CP (1982) Cell Calcium 3: 413-428.
Downes CP (1984) Trends Pharmacol Sci 6: 313-316.
Downes CP and Michell RH (1982) Cell Calcium 3: 467-502.
Downes CP and Michell RH (1985) In: Cohen P and Houslay MD (eds): Molecular Aspects of Cellular Regulation, Vol 4, Molecular Mechanisms of Transmembrane Signalling. Elsevier/North Holland, Amsterdam, pp. 3-56.
Hawthorne JN and Pickard MR (1979) J Neurochem 32: 5-14.
Hawthorne JN and Swilem AF (1982) Cell Calcium 3: 351-358.
Hokin LE (1965) Proc Natl Acad Sci USA 53: 1369-1376.
Hokin LE (1969) In: Bourne G (ed): Structure and Function of Nervous Tissue. Vol 3. Academic Press, New York, pp. 161-184.
Hokin MR, Benfey BG and Hokin LE (1958) J Biol Chem 233: 814-817.
Hokin MR, Hokin LE and Shelp WD (1960) J Gen Physiol 44: 217-226.
Irvine RF, Letcher AJ, Lander DJ and Downes CP (1984) Biochem J 223: 237-243.
Irvine RF, Anggard EE, Letcher AJ and Downes CP (1985) Biochem J, 229: 505-512.
Joseph SK and Williams RJ (1985) FEBS Lett 180: 150-154.
Kirk CJ, Rodrigues LM and Hems DA (1979) Biochem J 178: 493-496.

Kirk CJ, Michell RH and Hems DA (1981) Biochem J 194: 155-165.

Kirk CJ, Guillon G, Balestre M-N, Creba JA, Michell RH and Jard S (1985) Biochimic 67: 1161-1167.

Lapetina EG, Brown WE and Michell RH (1976) J Neurochem 26: 649-651.

Larrabee MG (1968) J Neurochem 26: 649-651.

Larrabee MG and Leicht WS (1965) J Neurochem 12: 1-13.

Larrabee MG, Klingman JD and Leicht WS (1963) J Neurochem 10: 549-560.

Lim ATW, Lolait SJ, Barlow JW, Autelitano DJ, Toh BH, Boublik J, Abraham J, Johnston CI and Funder JW (1984) Nature 310: 61-64.

MacIntyre E (1985) Mechanisms of Stimulus-Response Coupling in Platelets. Plenum, in press.

Michell RH (1975) Biochim Biophys Acta 415: 81-147.

Michell RH (1979) Trends Biochem Sci 4: 128-131.

Michell RH (1981) Neurosci Res Prog Bull 20: 338-350.

Michell RH (ed) (1982a) Inositol Phospholipids and Cell Calcium, special issue of Cell Calcium 3: 285-502.

Michell RH (1982b) In: Horrocks LA, Ansell GB and Porcellati G (eds): Phospholipids in the Nervous System, Vol 1. Raven Press, New York, pp. 315-325.

Michell RH (1986) In: Mechanisms of Receptor Regulation. Plenum Press, pp. 75-94.

Michell RH, Jafferji SS and Jones LM (1977) In: Bazan NG, Brenner RR and Giusto NM (eds): Function and Biosynthesis of Lipids. Plenum Press, New York, pp. 447-464.

Michell RH, Kirk CJ and Billah MM (1979) Biochem Soc Trans 7: 861-865.

Michell RH, Kirk CJ, Jones LM, Downes CP and Creba JA (1981) Phil Trans Roy Soc Ser B 296: 123-137.

Michell RH, Hawkins PT, Palmer S and Kirk CJ (1984) In: Endo M (ed): Calcium Regulation in Biological Systems. Takeda Science Foundation, Tokyo, pp. 85-103.

Monaco ME (1982) J Biol Chem 157: 2173-2179.

Monaco ME and Woods D (1983) J Biol Chem 258: 15125-15129.

Nagata Y, Mikoshiba K and Tsukada Y (1973) Brain Res 56: 259-269.

Nishizuka Y (1984) Nature 308: 693-698.

Nussey SS, Ang VTY, Jenkins JS, Chowdrey HS and Bisset GW (1984) Nature 310: 64-66.

Peters S and Kreulen L (1984) Fed Proc 43: 96.

Pickard MR, Hawthorne JN, Hayashi E and Yamada S (1977) Biochem Pharmacol 26: 448-450.

Schmale H and Richter D (1984) Nature 308: 705-709.

Seyfred MA, Farell LE and Wells WW (1984) J Biol Chem 259: 13204-13208.

Stephens LR and Logan SD (1985) J Neurochem, submitted.

Storey DJ, Shears SB, Kirk CJ and Michell RH (1984) Nature 312: 374-376.

Takhar APS and Kirk CJ (1981) Biochem J 194: 164-172.

Thomas AP, Marks JS, Coll KE and Williamson JR (1983) J Biol Chem 258: 5716-5725.

Troyer DA, Kreisberg JI, Schwertz DW and Venkatachalem MA (1983) Fed Proc 42: 1259.

Wali FA (1984) Pharmacol Res Commun 16: 55-62.

Yano K, Higashida H, Inoue R and Nozawa Y (1984) J Biol Chem 259: 10201-10207.

Phospholipid research and the nervous system
Biochemical and molecular pharmacology
L.A. Horrocks, L. Freysz, G. Toffano (eds)
Fidia Research Series, vol. 4.
Liviana Press, Padova. © 1986

SECOND MESSENGER GENERATION AND SECRETION IN PC12 CELLS. ROLE OF Ca++ AND PHOSPHOINOSITIDE HYDROLYSIS

L.M. Vicentini[1], A. Ambrosini[1,2], F. Di Virgilio[3,4], T. Pozzan[3,4] and J. Meldolesi[1,2]

[1]Department of Pharmacology, [2]National Research Council
Center of Cytopharmacology, University of Milano, Italy,
[3]Institute of General Pathology, [4] National Research Council Center
for the Physiology of Mitochondria, University of Padova, Italy

PC12 CELLS, A MODEL OF NEUROSECRETORY CELLS

PC12 is a line of neurosecretory cells originally developed by Greene and Tischler (1976) from a rat pheochromocytoma. This line has been extensively used in a number of laboratories around the world. Morphologically, growing PC12 cells resemble chromaffin cells of the adrenal medulla, although their secretion granules are smaller and less numerous (Greene and Tischler, 1976, 1982; Tischler and Greene, 1978; Watanabe et al., 1983). Because of this and other differences with respect to chromaffin cells, PC12 are now considered to be undifferentiated sympatoblasts. Their major catecholamine is dopamine, accumulated within secretory granules together with noradrenaline (Greene and Tischler, 1976, 1982; Rebois et al., 1980). Other neurotransmitters are synthesized in PC12 cells. Among these, acetylcholine is stored in organelles (not yet identified unambiguously), which are heavier than regular synaptic vesicles and lighter than the bulk of dopamine containing granules (Schubert et al., 1977). Peptides (enkephalins; neurotensin; Tischler et al., 1983; Panerai and Meldolesi, in preparation), as well as proteins typical of chromaffin and other secretory organelles

---

Abbreviations: NGF, nerve growth factor; PMA, Phorbol-12 myristate - 13 - acetate; DAG, diacylglycerol.

(chromogranin A; secretogranin; Lee and Huttner, 1983) are present in PC12 cells, but their possible colocalization with dopamine has not been established with certainty yet.

When PC12 cells are treated with nerve growth factor (NGF), they stop growing within 1-2 days, and then undergo profound changes in both their morphology and physiology. Cells treated for 10-15 days (PC12$^+$) are about 3 times larger than the untreated (PC12$^-$) cells and possess numerous neurites by which they interact with neighbour cells (Greene, 1984). Part of the neurosecretory granules are redistributed to these neurites, where they accumulate in varicosities and terminals (Greene and Tischler, 1982; Saito et al., 1985). Quantized release of acetylcholine has been observed from these terminals in cocultures of PC12 and rat myoblasts (Schubert et al., 1977). The granules remaining in the cell body are primarily accumulated at the neurite hillock and in the cytoplasmic rim immediately adjacent to the plasma membrane (Greene and Tischler, 1982; Saito et al., 1985).

NGF-induced differentiation implies also extensive physiological modifications of PC12 cells. The density (number/unit area) of the voltage-dependent Na$^+$ channels is greatly increased in PC12$^+$ cells, i.e., the cells are more excitable (Rudy et al., 1982); receptors for some peptides and neurotoxins are also increased in density. Of particular interest for our work is the behaviour of cholinergic receptors. Many, but not all, PC12 subclones possess nicotinic receptors (Greene and Tischler, 1982). Muscarinic receptors increase considerably in number (about 15 fold when calculated per single cell) after NGF-induced differentiation (Jumblatt and Tischler, 1982). As a consequence the size of the postreceptor events triggered by the activation of the muscarinic receptor (phosphoinositide hydrolysis; increase of cytosolic Ca^{2+} concentration, [Ca^{2+}]$_i$) is markedly greater in PC12$^+$ than in PC12$^-$ cells (Vicentini et al., 1985a; our unpublished observations).

In conclusion, we believe that PC12 cells constitute an interesting and versatile cell model that can be used for the study of the process of neurotransmitter release. As already mentioned above, these cells are certainly more neuronal than chromaffin cells of the adrenal medulla. Thus, although the results obtained with PC12 must be regarded at the present time as pertaining uniquely to the cell line (or even to the subclone used, because subclones in the various laboratories are now recognized to be very heterogeneous), they nevertheless promise to ultimately lead to interesting developments for our understanding of nerve cells.

REGULATION OF TRANSMITTER RELEASE IN PC12 CELLS

Extensive evidence indicates that evoked neurotransmitter release occurs by exocytosis, which can be triggered by an increase of [Ca^{2+}]$_i$ in axon terminals and neurosecretory cells (for a recent review, see Reichardt and Kelly, 1983). Release induced by depolarization occurs via the activation of the voltage-dependent Ca^{2+} channels of the plasma membrane, and requires therefore Ca^{2+} to be present in the extracellular fluid. However, many treatments (listed in Meldolesi et al., 1984; Pozzan et al., 1984) evoke neurotransmitter release even in Ca^{2+}-free, EGTA-containing media, in which the actual concentration of Ca^{2+}, [Ca^{2+}]$_o$, can be estimated to be of the order of 10^{-8}M or less. At these low concentrations there is no inward electrochemical Ca^{2+} gradient, and therefore Ca^{2+} influx is impossible. Release responses in Ca^{2+}-free media have therefore been explained not in terms of Ca^{2+} influx, but of Ca^{2+} redistribution from intracellular, membrane bounded stores. Cells and terminals contain in fact large amounts of Ca^{2+} segregated within organelles, and leakage from these stores could certainly increase $[Ca^{2+}]_i$ from the resting level, $\sim$ 100 nM, to the activated level, and thus account for the responses in Ca^{2+}- free media. Until recently no procedures were available for measuring $[Ca^{2+}]_i$ of small neurosecretory cells and nerve terminals, and therefore the question as to whether the increase of $[Ca^{2+}]_i$ is the only mechanism necessary for triggering the evoked release of neurotransmitters had remained unanswered.

During the past 3 years, we have carried out detailed studies on the effects of a variety of agents which stimulate transmitter release from PC12 cells even when applied in Ca^{2+}-free, EGTA-containing media (Meldolesi et al., 1984; Pozzan et al., 1984). Clear results were obtained with two such agents: the neurotoxic component of the black widow spider venom, α-latrotoxin, and the diterpene tumor promoter phorbol 12-myristate, 13-acetate (PMA). The fluorescent indicator quin-2 (Tsien et al., 1982) was used to monitor $[Ca^{2+}]_i$ in treated cells. The results with α-latrotoxin demonstrated that $[Ca^{2+}]_i$ is increased substantially when the toxin is applied in Ca^{2+}-containing medium, but remains unchanged in the Ca^{2+}-free medium, although, as already mentioned, transmitter release is clearly stimulated even in the latter experimental condition (one-third of the massive stimulation elicited by the toxin in the Ca^{2+}-containing medium, Meldolesi et al., 1984).

These results were interpreted to indicate that in the neurosecretory cells (as well as in brain synaptosomes, investigated in parallel) the process of transmitter release is under a multiple (at least dual) control. A Ca^{2+}-dependent mechanism seems to coexist with one or more Ca^{2+}-independent mechanisms*, and activation of either one of these mechanisms seems sufficient to trigger the release of neurotransmitters. The physiological role of the Ca^{2+}-dependent and Ca^{2+}-independent mechanisms seems however different. The $[Ca^{2+}]_i$ increase probably accounts for the fast, synchronous release triggered by individual action potentials, which typically occurs at the neuromuscular junction. The Ca^{2+}-independent mechanism(s), on the other hand, could play a major role in long-term modulation processes, such as synaptic potentiation (Meldolesi et al., 1984; Pozzan et al., 1984).

The results with PMA provided further support to these concepts. PMA is now believed to work as a specific allosteric activator of protein kinase C, (otherwise referred to as the phospholipid-dependent, Ca^{2+}-activated protein kinase) (Castagna et al., 1982; Mitchell, 1982). When bound to the enzyme, PMA causes an increase in its affinity for Ca^{2+}, so that protein kinase C can now be fully active even at a Ca^{2+} concentration in the resting $[Ca^{2+}]_i$ range (around 100 nM) (Nishizuka, 1981). The physiological counterpart of PMA is diacylglycerol (DAG) that is generated in intact

---

* The possibility of multiple Ca^{2+}-independent release mechanisms emerges when our results obtained with α-latrotoxin and PMA (Meldolesi at al., 1984; Pozzan et al., 1984) are considered together. PMA is believed to act through the activation of protein kinase C.; the endogenous activator of the enzyme is DAG, generated from the hydrolysis of phosphoinositides (see below). Initially we thought that the stimulation of transmitter release induced by α-latrotoxin in Ca^{2+} media could correlate with phosphoinositide hydrolysis. Indeed, the toxin causes phosphoinositide hydrolysis in PC12 cells (Vicentini and Meldolesi, 1984), but this effect is dependent on $[Ca^{2+}]_o$ and disappears when Ca^{2+} is withdrawn from the medium.

cells by receptor-triggered phosphoinositide hydrolysis (Nishizuka, 1984; Michell, 1982; Berridge and Irvine, 1984). Application of either PMA or DAG to PC12 cells (no matter whether suspended in Ca^{2+}-containing or Ca^{2+}-free media) was found to trigger slow, but sustained release of dopamine (Fig. 1), without concomitant changes of $[Ca^{2+}]_i$ (Fig. 2, Pozzan et al., 1984). In our work, another type of drug was used, namely Ca^{2+} ionophores, such as ionomycin, which induce transport of Ca^{2+} down its electrochemical gradient, i.e. from both the medium and the stores to the cytosol (when cells are incubated in Ca^{2+}-containing media). Thus, Ca^{2+} ionophores cause $[Ca^{2+}]_i$ to rise. The release responses induced by these drugs had a time course distinctly different from those induced by PMA (Fig. 1). They started without delay, proceeded at a high rate for 1-2 min, and then inactivated. When activators of protein kinase C (PMA or DAG) were applied together with ionomycin, the release responses were synergistic (i.e., greater than the sum of the responses caused by the two types of drugs applied separately) and sustained (Fig. 1, Pozzan et al., 1984).

These results, which are analogous to those reported for other, non-nervous cells (originally platelets, Nishizuka, 1984; Rink et al., 1983, then a variety of other cell types reviewed by Nishizuka, 1984; Berridge and Irvine, 1984), confirm the complexity of the

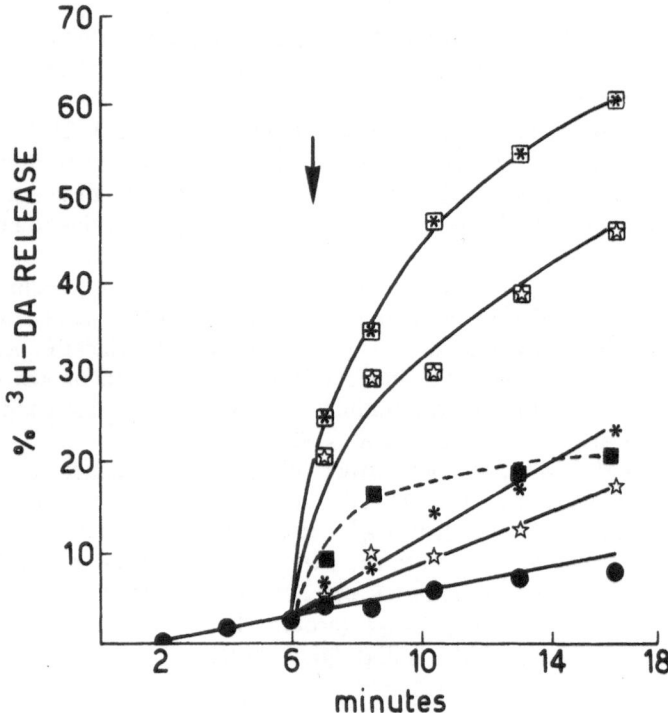

Figure 1. Time course of $[^3H]$ dopamine release induced in PC12 cells by treatment with PMA and DAG or the combination of these drugs with ionomycin. Cells suspended in Krebs-Ringer medium buffered with HEPES were incubated at 37°.
After 6 min (arrow) PMA (10^{-8}M, -- ✳ --), DAG (1.5×10^{-4} M, -- ☆ --) or either one of the two together with ionomycin (10^{-7}M, -- ⊞ -- and -- ⊠ --, respectively) were added. --●-- controls, --■-- ionomycin (10^{-7}M). Values shown are averages from 3 experiments.

regulation of the neurotransmitter process. Hypotheses as to how the two (or more) mechanisms are interconnected have been recently put forth (Baker, 1984) but the experimental evidence available at the present time is not yet sufficient to draw final conclusions.

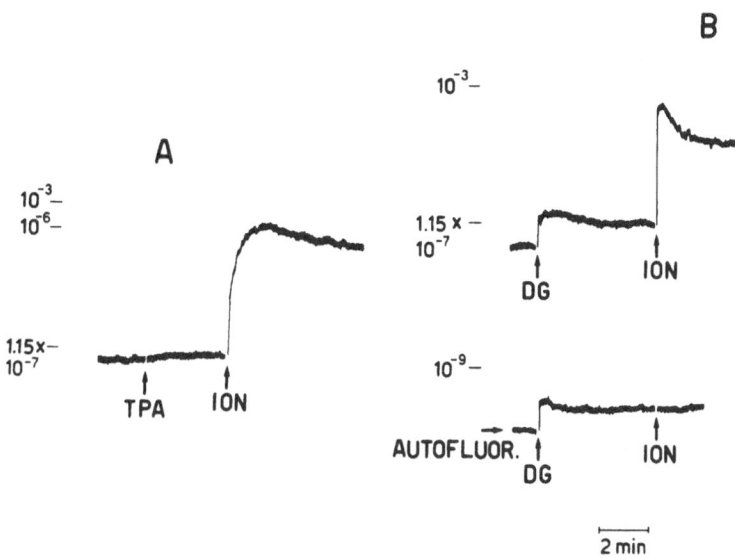

Figure 2. Effects of PMA, DAG (DG) and ionomycin on $[Ca^{2+}]_i$ in PC12⁻ cells. (A): Suspensions of cells loaded with quin-2 (0.25 nmol/10^6 cells) were treated with PMA (TPA, 10^{-7} M) followed by ionomycin (ION, 3×10^{-7} M). (B) : DAG (DG, 1.5×10^{-4} M) followed by ionomycin was applied to parallel aliquots of quin-2 loaded cells (upper trace) and unloaded cells (lower trace). Note that the apparent rise of $[Ca^{2+}]_i$ induced by DAG is accounted for by an increase of autofluorescence.

ACTIVATION OF THE MUSCARINIC RECEPTOR IN PC12 CELLS

Recent studies of our group have been aimed at clarifying the physiological importance of the regulatory mechanisms of neurotransmitter release ($[Ca^{2+}]_i$-and protein kinase C-dependent) in PC12 cells. Treatment with muscarinic agonists is expected to result in the activation of both these mechanisms, inasmuch as it induces in the cells both phosphoinositide hydrolysis (with generation of DAG) and $[Ca^{2+}]_i$ increase. These postreceptor events were conveniently investigated in PC12⁺ cells because, as already mentioned, these cells are endowed with a large number of muscarinic receptors (Jumblatt and Tischler, 1982). In addition, in preliminary experiments we discovered that exposure of our PC12 cells to nicotinic agonists (and/or antagonists) remained without detectable consequences, indicating that the cell subclone that we have available is devoid of a functioning nicotinic receptor. Thus, the effects of muscarinic receptor activation could be investigated by employing a mixed muscarinic-nicotinic agonist, such as carbachol.

Application of carbachol (in the 10^{-5} -10^{-3}M range) to suspensions of PC12 cells induced atropine inhibitable, hexamethonium and verapamil unaffected increases of $[Ca^{2+}]_i$. As will be discussed in detail elsewhere (Pozzan, Di Virgilio, Vicentini and Meldolesi, in preparation), such increases have a dual origin: redistribution from Ca^{2+} stores and increased influx from the plasmalemma. Although a direct proof in our cellular system is not yet available, we believe that redistribution is mediated by inositol 1,4,5-trisphosphate (InsP_3), which has been shown to cause release of Ca^{2+} from non-mitochondrial intracellular store(s) (see below and Berridge and Irvine, 1984 for a review). Influx, on the other hand, was initially suspected to contribute to the carbachol-induced $[Ca^{2+}]_i$ increases we observed because the $[Ca^{2+}]_i$ increases induced in the Ca^{2+}-containing medium were not only more sustained, but also greater than those in the Ca^{2+}-free medium. However, for quite sometime we were unable to demonstrate a significant carbachol-induced atropine-inhibitable increase of ^{45}Ca influx. Therefore the doubt remained that the partial dependence on $[Ca^{2+}]_o$ of $[Ca^{2+}]_i$ increases was due to an effect (direct or indirect) of carbachol on the Ca^{2+} efflux mechanisms (Ca^{2+}-ATPase and/or Na^+/Ca^{2+} exchange). Increased influx was finally demonstrated when a protocol was employed in which cell suspensions in Ca^{2+}-free, EGTA-containing medium were first treated with carbachol for 2 min, and then exposed to ^{45}Ca (final $[Ca^{2+}]_o$: 1.2 mM). Under these experimental conditions the initial ^{45}Ca influx (measured 10, 25 and 45 sec after the addition of ^{45}Ca) was over 60% greater in carbachol-treated than in control cells. Also at steady state significantly more radioactivity was accumulated in treated cells. $[Ca^{2+}]_i$ measurements, carried out in cell suspensions treated according to the above protocol, documented the dissociation of the two processes, redistribution and influx. A first $[Ca^{2+}]_i$ increase after addition of carbachol in the Ca^{2+}-free medium (redistribution) was followed by a second increase occurring immediately after introduction of Ca^{2+} in the medium (influx) (Vicentini et al., 1985a). Both the early and the late increases depend on the treatment with carbachol, and both were wiped out by atropine and unaffected by verapamil. Taken together these results document the existence in the PC12 cell plasma membrane of a Ca^{2+} pathway different from the voltage-dependent Ca^{2+} channel, and opened as a consequence of the activation of the muscarinic receptor, i.e., of a receptor operated channel (Pozzan et al., in preparation). It should be noted that the existence of a muscarinic receptor-coupled channel had never been demonstrated in neurosecretory cells (McKinney and Richelson, 1984).

Phosphoinositide hydrolysis was stimulated by carbachol applied to PC12 cells (Vicentini et al., 1985a) at concentrations in the 10^{-5} -10^{-3} M range.

The effect was unmodified by nicotinic blockers, and readily inhibited by atropine (IC_{50}: 6 nM). The muscarinic blocker pirenzepine, which has been proposed as a tool to differentiate muscarinic receptor subclasses, was active in the 10^{-8} -10^{-7} M range ($IC_{50} \sim 100$ nM). Among the hydrophilic metabolites generated from phosphoinositide hydrolysis, InsP_3 was greatly accumulated at short times after the application of carbachol, while the accumulation of InsP_2 and Ins$P$ was more delayed. These results suggest that PtdIns(4,5)P_2 is a major substrate of the muscarinic receptor coupled hydrolytic reaction, and that at least part of the InsP_2 and Ins$P$ formed could originate not from the hydrolysis of the corresponding phosphoinositides, but from the sequential dephosphorylation of InsP_3 (Vicentini et al., 1985a).

One problem which was still open in the Ca^{2+}-phosphoinositides hydrolysis field was the relationship between the two processes. In the experimental conditions used up to recently, no temporal dissociation between $[Ca^{2+}]_i$ and $InsP_3$ generation could be obtained. Evidence (primarily from experiments on subcellular fractions and permeabilized cells) had suggested the possibility of $InsP_3$ playing the role of a second messenger, responsible for the redistribution of Ca^{2+} from non-mitochondrial store(s) (see Berridge and Irvine, 1984 for a review). On the other hand, evidence from intact cells had cast doubts on this conclusion (see Vicentini et al., 1985a for discussion) and suggested that phosphoinositide hydrolysis had to be preceded, and thus probably caused, by a $[Ca^{2+}]_i$ increase. Were this interpretation correct, the role of $InsP_3$ generation would have been much more limited (i.e., modulatory) than that of a second messenger. In order to solve this problem we employed a direct approach. PC12 were suspended in a Ca^{2+}-free, EGTA-containing medium and then exposed to ionomycin. Because of the very low $[Ca^{2+}]_o$, and because of the Ca^{2+} ionophore-induced discharge of Ca^{2+} stored within cytoplasmic organelles, neither influx nor redistribution are possible in those experimental conditions. Thus, application of carbachol failed to raise $[Ca^{2+}]_i$ (as shown by quin-2 measurements), yet phosphoinositide hydrolysis, in particular $InsP_3$ generation, was still stimulated.

These results demonstrate for the first time unambiguously the $[Ca^{2+}]_i$ independence of phosphoinositide hydrolysis, and give therefore strong support to the second messenger role of $InsP_3$ (Vicentini et al., 1985a).

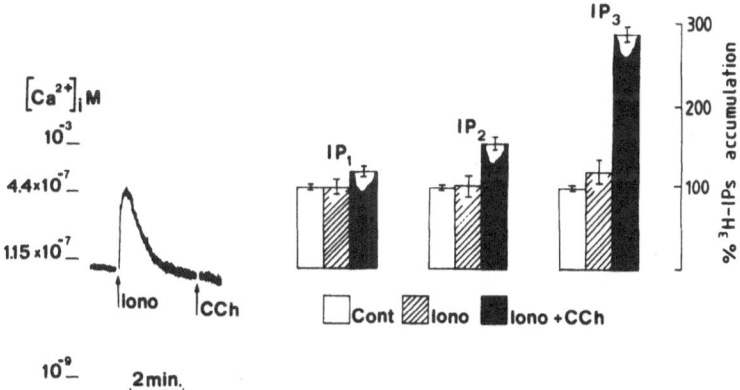

Figure 3. Hydrolysis of phosphoinositides induced by muscarinic receptor activation at resting $[Ca^{2+}]_i$. Suspensions of PC12$^+$ cells, labelled for 24 hrs with [^{3}H] inositol, were loaded with quin-2 (0.8 nmol/10^6 cells) and incubated in Ca^{2+}-free, EGTA-containing medium. Parallel aliquots of these suspensions were used for measuring $[Ca^{2+}]_i$ and accumulation of inositol phosphates.

Left: quin-2 fluorometric trace: additions of ionomycin (Iono, 0.2 μM) and carbachol (CCh, 0.5 mM) were made as indicated. Right: recoveries (averages of 3 determinations $\pm$ SE) of InsP (IP$_1$), InsP$_2$ (IP$_2$) and InsP$_3$ (IP$_3$) measured as described by Vicentini and Meldolesi, 1984. Cells (600 μg protein/sample) were incubated in Ca^{2+}-free, EGTA-containing medium and treated with ionomycin for 3 min followed by $\pm$ carbachol for 1 min.

□ cont = controls ▨ iono = ionomycin; ■ iono + CCh = ionomycin and then carbachol.

ACTIVATION OF THE MUSCARINIC RECEPTOR
AND TRANSMITTER RELEASE

Based on the considerations summarized in the preceding sections, the activation of the muscarinic receptor in PC12 cells was expected to turn on both the $[Ca^{2+}]_i$ - and protein kinase C-dependent mechanisms of control of exocytosis, and thus to induce appreciable stimulation of transmitter release. We were therefore surprised to find that even high concentrations of carbachol (up to 0.5 mM) were without a detectable release effect. The only explanation that came to our mind in order to account for this result was that, although the two postreceptor events (phosphoinositide hydrolysis and $[Ca^{2+}]_i$ increase) are both potentially competent to cause stimulation, they remain subthreshold when induced by muscarinic receptor activation because of their small size. This explanation appeared reasonable in the case of PC12⁻ cells, that are endowed with a small number of muscarinic receptors (Jumblatt and Tischler, 1982). In these cells, treatment with optimal concentrations of carbachol induced $[Ca^{2+}]_i$ to increase only by about 60%, and phosphoinositide hydrolysis (5 min of stimulation) by 50%. In PC12⁺ cells, however, the $[Ca^{2+}]_i$ increases were much greater (approximately 3 fold) and therefore, in order for our explanation to be valid, we had to postulate a decreased sensitivity of exocytosis to its regulatory processes occurring during differentiation by NGF. That this might indeed be the case is suggested by the results of K⁺ and PMA-induced release experiments. We observed that the release responses induced by various concentrations of KCl were approximately the same in PC12⁺ and PC12⁻ cells, although the causative corresponding rises of $[Ca^{2+}]_i$ were more pronounced in the former than in the latter preparations. Even greater was the difference observed in experiments with PMA. The concentrations needed to elicit maximal responses were 3×10^{-8} M in PC12⁻ and 3×10^{-7} M for PC12⁺ cells.

If the intracellular signals elicited by the activation of the muscarinic receptor are indeed subthreshold with respect to transmitter release stimulation, attempts could be made to reveal them by taking advantage of the synergism between $[Ca^{2+}]_i$ and protein kinase C-dependent mechanisms previously demonstrated in the experiments with ionomycin and PMA (Pozzan et al., 1984). A new series of experiments was therefore carried out in which carbachol was applied to PC12 cells together with various concentrations of either ionomycin or PMA. These combined treatments were expected to yield synergistic responses, i.e. responses greater than those elicited by either ionomycin or PMA alone. Our expectations were borne out by the results obtained with carbachol + ionomycin. Indeed the effects of the combined treatments were moderately (20-40%), but consistently greater than those with the Ca^{2+} ionophore alone. Quin-2 measurements carried out in parallel excluded the possibility that the greater effect of the two drugs together was due to greater $[Ca^{2+}]_i$ increases, i.e. the $[Ca^{2+}]_i$ increases caused by carbachol + ionomycin were not appreciably different from those caused by ionomycin alone. Thus, the most likely explanation of these data is synergism between the small degree of protein kinase C activation (by carbachol, through phosphoinositide hydrolysis) with the large $[Ca^{2+}]_i$ increase (by ionomycin). In contrast, with the combined carbachol-PMA treatment no synergism was observed, i.e. the responses were as large as with PMA alone. An explanation for this unexpected result is given in the following section.

PROTEIN KINASE C REGULATION OF MUSCARINIC RECEPTOR ACTIVITY

Recent data in a variety of cell systems clearly indicate that the function of various receptors can be metabolically regulated. It appears now that classical pharmacological phenomena, such as receptor desensitization and down regulation, can be the result of covalent modifications of either the receptor molecules, or the molecules operating in receptor coupling or in postreceptor transduction. Protein kinase C seems to play a major role in these processes. In particular, it has been demonstrated that phosphorylation by protein kinase C induces decreased affinity of insulin and growth factor receptors for their ligands (Jacobs et al., 1983; Cochet et al., 1984) and uncoupling of receptors from adenylate cyclase (Sibley et al., 1984; Kelletier et al., 1984). Of special interest is the possibility that protein kinase C affects receptors coupled to phosphoinositide hydrolysis, i.e. the process responsible for the physiological activation of the enzyme in intact cells. Results of experiments in PC12 cells treated with PMA, and then exposed to carbachol, demonstrate for the first time that this is indeed the case (Vicentini et al., 1985b). We found that PMA is without effect on basal $[Ca^{2+}]_i$ and phosphoinositide hydrolysis. However, the rises induced by carbachol via the activation of the muscarinic receptor were substantially inhibited (by 75 and 40%, respectively). Inhibition occurred at the low concentrations of PMA which are considered specific for protein kinase C activation, and was detected as soon as 1 min after the application of the phorbol diester (Fig. 4; Vicentini et al., 1985b). These results readily explain the lack of synergism in the stimulation of transmitter release observed when carbachol and PMA were applied together as described in the preceding section. We believe that in those experiments the phorbol ester had two opposing effects. On the one hand, it stimulated release (via the activation of the protein kinase C-dependent mechanism); on the other hand, it inhibited the $[Ca^{2+}]_i$ increase induced by carbachol, and its potential stimulatory effect with the activation of the $[Ca^{2+}]_i$ -dependent

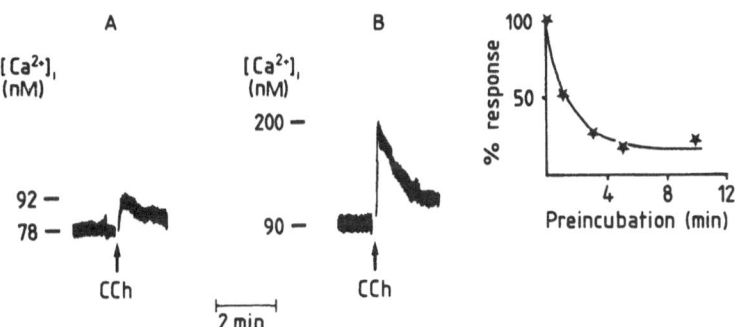

Figure 4. Effect of preincubation with PMA on $[Ca^{2+}]_i$ rise induced by carbachol in PC12$^+$ cells. Suspensions of PC12$^+$ cells were loaded with quin-2 (0.34 nmol/10^6 cells), washed, resuspended in KRH (1.2 × 10^6 cells/ml) and preincubated for 5 min with either PMA (100 nM) in dimethyl sulfoxide (0.3%) (A), or with the solvent only (B). Additions of CCh (0.5 mM) are marked by arrows. Calibration of $[Ca^{2+}]_i$ is indicated to the left of each trace. The inset shows the effect of the length of preincubation with PMA (100 nM) on the maximal rise of $[Ca^{2+}]_i$ induced by CCh (0.5 mM).

mechanism. The stringent feedback regulation of muscarinic receptor activity by protein kinase C might be arranged so as to preclude the persistent activation of the enzyme. Protein kinase C is known to be involved in a variety of processes, among which is cell growth. Long term activators of the enzyme, such as PMA, are known as tumor promoters (Nishizuka, 1984). Thus, persistent activation of protein kinase C appears to be a potentially dangerous event for the cell.

CONCLUSION

The regulation of neurotransmitter release is a complex process which implies the finely tuned interplay of at least two and possibly more mechanisms, one dependent on, another independent of $[Ca^{2+}]_i$, and probably operating through the activation of protein kinase C. Feedback and feedforward functions appear to tie together receptor, postreceptor, and effector events. Although numerous and important links are still missing (for example: how is the process of exocytosis regulated by $[Ca^{2+}]_i$ and protein kinase C activity?) a general framework has begun to emerge. Studies by us and others on PC12 cells have contributed substantially to our present understanding of the problem, and promise to yield further developments in the near future.

REFERENCES

Baker PF (1984) Nature 310:629-630.

Berridge MJ and Irvine RF (1984) Nature 312:315-321.

Castagna M, Takai Y, Kaibuchi K, Sano K, Kikkawa U and Nishizuka Y (1982) J Biol Chem 257:7847-7851.

Cochet G, Gill GN, Meisenhelder J, Cooper JA, Hunter T, (1984) J Biol Chem 259: 2553-2558.

Greene LA and Tischler AS (1976) Proc Natl Acad Sci USA 73:2424-2428.

Greene LA and Tischler AS (1982) Adv Cell Neurobiol 3:373-414.

Greene LA (1984) Trends Neurosci 9:91-94.

Jacobs S, Sahyoun NE, Saltiel AR and Cuatrecasas P (1983) Proc Natl Acad Sci USA 80:6211-6213.

Jumblatt JE and Tischler AS (1982) Nature 297:152-154.

Kelleher DJ, Pessin JE, Ruho AE and Johnson GL (1984) Proc Natl Acad Sci USA 81:4316-4320.

Lee RWH and Huttner WB (1983) J Biol Chem 258:11326-11334.

McKinney M and Richelson E (1984) Ann Rev Pharmacol Toxicol 24:121-146.

Meldolesi J, Hattner WB, Tsien RY and Pozzan T (1984) Proc Natl Acad Sci USA 81:620-624.

Michell RH (1982) Trends Biochem Sci 8:263-265.

Nishizuka Y (1984) Nature 308:693-698

Pozzan T, Gatti G, Dozio N, Vicentini ML and Meldolesi J (1984) J Cell Biol 99:628-638.

Rebois RV, Reynolds EE, Toll L and Howard BD (1980) Biochemistry 19:1240-1248.

Reichardt LF and Kelly RB (1983) Annu Rev Biochem 52:871-926.

Rink TJ, Sanchez A and Hallam TJ (1983) Nature 305:317-319.

Rudy B, Kinsehenbaum B and Greene LA (1982) Neuroscience 2:1405-1411.

Saito I, Dozio N and Meldolesi J (1985) Neuroscience 14:1163-1174.

Schubert D, Heinemann S and Kidokoro Y (1977) Proc Natl Acad Sci USA 74:2579-2583.

Sibley DR, Nambi P, Peters JR and Lefkowitz RJ (1984) Biochem Biophys Res Comm 121:973-979.

Tischler AS and Greene LA (1978) Lab Invest 39:77-89.

Tischler AS, Lee YC, Slayton VW, Bloom SR (1983) Life Sci 33:347-351.

Tsien RY, Pozzan T and Rink TJ (1982) J Cell Biol 94:325-334.

Vicentini LM and Meldolesi J (1984) Biochim Biophys Res Comm 121:538-544.

Vicentini LM, Ambrosini A, Di Virgilio F, Pozzan T and Meldolesi J (1985a) J Cell Biol 100:1330-1335.

Vicentini LM, Di Virgilio F, Ambrosini A, Pozzan T and Meldolesi J (1985b) Biochim Biophys Res Comm 127:310-317.

Watanabe O, Torda M and Meldolesi J (1983) Neuroscience 10:1011-1024.

Phospholipid research and the nervous system
Biochemical and molecular pharmacology
L.A. Horrocks, L. Freysz, G. Toffano (eds)
Fidia Research Series, vol. 4.
Liviana Press, Padova. © 1986

PHOSPHOPROTEIN B-50
AND POLYPHOSPHOINOSITIDE-DEPENDENT
SIGNAL TRANSDUCTION IN BRAIN

W.H. Gispen, P.N.E. de Graan, L.H. Schrama and J. Eichberg*

Division of Molecular Neurobiology, Rudolf Magnus Institute for Pharmacology,
Institute of Molecular Biology, State University of Utrecht,
3584 CH Utrecht, The Netherlands

INTRODUCTION

During the past several years, an impressive body of information has been accumulating that polyphosphoinositides (PPtdIns) metabolism plays a central role in cellular signal transduction (Berridge and Irvine, 1984). The prevailing view is that in response to receptor activation by a variety of hormones, neurotransmitters and other external stimuli, a plasma membrane pool of phosphatidylinositol 4,5-bisphosphate (PtdIns(4,5)P_2) undergoes rapid phosphodiesteratic cleavage to yield two biologically active products: 1,2 diacylglycerol (DAG) and inositol 1,4,5-trisphosphate (InsP_3). Diacylglycerol is considered to activate the widely distributed Ca^{2+} and phospholipid-dependent protein kinase C which is capable of phosphorylating many, largely uncharacterized, cellular proteins, whereas InsP_3 is believed to interact to trigger the release of sequestered Ca^{2+} from non-mitochondrial sites into the cytosol. The simultaneous processes of protein phosphorylation and Ca^{2+} mobilization are thought to constitute synergistic events which are integral to a large number of cellular responses (Nishizuka, 1984).

The removal of these putative second messengers is accomplished in the case of DAG either by phosphorylation to phosphatidic acid or by lipase-catalyzed degradation, and for InsP_3 by sequential removal of phosphate moieties by a series of specific phosphatases: InsP_3 → inositol 1,4-bisphosphate → inositol 1-phosphate → myo-inositol. Replenishment of the depleted PtdIns(4,5)P_2, supply occurs via the conversion of phosphatidic acid → CDP-diacylglycerol → phosphatidylinositol → phosphatidylinositol

---

* On sabbatical leave of absence from the Department of Biochemical and Biophysical Sciences, University of Houston, Houston, Texas, USA.

4-phosphate (PtdIns4P) → PtdIns(4,5)P_2, the last two steps being catalyzed by ATP-utilizing kinases (Fisher and Agranoff, 1985). While the existence and importance of such a phosphoinositide cycle is well-established, unresolved complications and questions remain. Among these are whether the active form of the PPtdIns phosphodiesterase (phospholipase C) is membrane-bound or soluble and the precise divalent ion requirements for this activity, as well as the significance of the recent finding that at least one other InsP_3 isomer, namely inositol 1,3,4-trisphosphate may be generated in cells (Irvine et al., 1984).

The nervous system is characterized not only by a rapid PPtdIns turnover, especially in non-myelin structures (Gonzales-Sastre et al., 1971), but also by the highest known level of protein phosphorylation an dephosphorylation of any tissue (Weller, 1979). These phenomena are particularly prominent in the synaptic regions where mechanisms for dynamic modulation of information transmitted between neurons must be in continual operation. At the molecular level, an interplay between the state of phosphorylation of inositol-containing phospholipids and specific synaptic membrane proteins could contribute to changes in ion permeability or catalytic activity of selected enzymes. Such rapid effects might thus underlie short-term adaptation of synaptic function to prevailing environmental conditions.

Our laboratory has been interested for some time in the potential interrelationship between lipid and protein phosphorylation in synaptic plasma membranes and has amassed evidence that the extent of phosphorylation of a membrane protein influences the extent of PtdIns(4,5)P_2 formation. Our current understanding of this system and the possible implications which our findings suggest for mechanisms of synaptic plasticity are the subjects of this paper.

STIMULATED PPtdIns METABOLISM IN THE NERVE ENDING

It has been known for some time that muscarinic agonists stimulate |^{32}P|-incorporation into phosphatidic acid and phosphatidylinositol in isolated nerve endings (Schacht and Agranoff, 1972; Fisher and Agranoff, 1981) and that the enhanced labeling is blocked by atropine and other classical muscarinic antagonists. This effect is now interpreted as likely due to synthesis of these lipids secondary to PPtdIns breakdown and apparently takes place at a postsynaptic site, possibly resealed dendritic fragments (Fisher et al., 1980, 1981). Muscarinic ligands were first shown to stimulate PPtdIns degradation in iris smooth muscle (Abdel-Latif et al., 1977) and subsequently this effect was demonstrated in nerve endings (Fisher and Agranoff, 1981) and brain slices (Berridge et al., 1982; Brown et al., 1984; Janowsky et al., 1984; Fisher et al., 1984), as shown by measurement of the disappearance of label from PtdIns4P and PtdIns(4,5)P_2 as well as of the accumulation of inositol phosphates in the presence of Li$^+$, an agent that inhibits inositol 1-phosphatase (Sherman et al., 1981). The presence in well-washed nerve ending membranes of a PPtdIns phosphodiesterase stimulated by high (1.5 mM) Ca^{2+} concentrations that cleaves endogenous substrate has been demonstrated (Van Rooijen et al., 1983). Indications are that these membranes also possess PPtdIns phosphomonoesterase activity since under certain conditions a rapid loss of these phospholipids occurs without accumulation of inositol phosphates (Van

Rooijen, 1984). Synaptic plasma membrane preparations also contain phosphatidylinositol kinase and PtdIns4P kinase activities inasmuch as they incorporate label from $[\gamma\text{-}^{32}P]ATP$ into PtdIns4P and PtdIns(4,5)P_2. We have recently purified and characterized the latter enzyme from brain (Van Dongen et al., 1984). Thus the nerve ending membranes contain all of the enzymatic machinery required for the synthesis and at least the initial steps in the degradation of PPtdIns.

THE B-50 PROTEIN: PHOSPHORYLATION, PROPERTIES AND LOCALIZATION

A protein which we have termed B-50 undergoes marked phosphorylation under appropriate conditions when synaptic plasma membranes are incubated with $|\gamma\text{-}^{32}P|$-ATP. B-50 is a 48,000 dalton, strongly acidic (pI 4.5) protein that is intimately associated with the membranes since it can only be solubilized in the presence of detergent (Zwiers et al., 1979). Solubilized B-50 has been purified together with endogenous B-50 kinase activity by DEAE cellulose chromatography followed by ammonium sulfate precipitation and isoelectric focusing (Zwiers et al., 1980). The purified protein displays microheterogeneity upon isoelectric focusing in the pH range 3.5-5.0 and upon two-dimensional polyacrylamide gel electrophoresis it can be resolved into four forms that are partially interconvertible either by exhaustive phosphorylation or dephosphorylation. These results indicate that B-50 contains at least two phosphorylatable sites (Zwiers et al., 1985).

Considerable evidence indicates that B-50 is phosphorylated by a kinase that is indistinguishable from protein kinase C and distinct from cyclic nucleotide-dependent or calmodulin-stimulated protein kinases (Aloyo et al., 1982a, 1983; Gispen and Zwiers, 1985). Indeed, of several kinases tested for their ability to phosphorylate purified B-50, only kinase C was able to do so; moreover, when added to synaptic plasma membranes, kinase C preferentially phosphorylated B-50.

Previous work using anti-B-50 immunoglobulins (IgG's) had led us to conclude that the protein was present in detectable quantities only in the nervous system and in the adult animal confined to areas rich in synaptic structures (Oestreicher et al., 1981, 1983). The ultrastructural localization of the protein has been investigated recently at the electron microscopic level and by means of immunostaining using purified anti-B-50 IgG's and protein A-coated gold particles, and was found to be predominantly associated with the inner surface of the presynaptic membrane (Gispen et al., 1985). Both our work (Kristjansson et al., 1982) and that of others (Sorensen et al., 1981) has provided evidence that among subcellular fractions B-50 phosphorylation is most active in synaptic membrane preparations and, though detected throughout rat brain, activity is most pronounced in septum > hippocampus > neocortex > thalamus > cerebellum > medulla oblongata > spinal cord (Kristjansson et al., 1982). We have recently developed a sensitive radioimmunoassay for B-50 using the ^{32}P-labeled protein in combination with affinity-purified anti-B-50 IgG's. This has enabled us to determine that there is approximately 10 μg B-50 per mg rat brain synaptic membrane protein and to ascertain that the relative abundance of B-50 in brain regions correlates well with the distribution of B-50 phosphorylating activity (Kristjansson et al., 1982; Oestreicher et al., 1983, 1985).

MODULATION OF B-50 PHOSPHORYLATION
AND PtdIns4P KINASE ACTIVITY BY ACTH

It has been known for some time that the addition of $ACTH_{1-24}$ to synaptic plasma membranes reduced the endogenous phosphorylation of several proteins, one of which is the B-50 protein (Zwiers et al., 1976, 1978; Mahler et al., 1982). Some years ago, we also observed that $ACTH_{1-24}$ in concentrations of 10^{-7} to 10^{-4}M inhibited the incorporation of radioactivity into B-50 when preparations of the protein, purified through the DEAE and ammonium sulfate precipitation steps, were incubated with $|\gamma\text{-}^{32}P|$-ATP (Jolles et al., 1980). Using this preparation, which is devoid of PPtdIns phosphodiesterase activity, we found that concentrations of the neuropeptide, which inhibited B-50 phosphorylation, simultaneously stimulated the conversion of exogenously added PtdIns4P to PtdIns(4,5)P_2. These results prompted us to undertake detailed studies which have convinced us that a reciprocal relationship exists between the extent of B-50 phosphorylation and the degree of PtdIns4P kinase activity. The additional supporting evidence for this conclusion may be summarized as follows.

a. If partially purified B-50 preparations were exposed to $[\gamma\text{-}^{32}P]ATP$ for increasing periods of time and PtdIns4P was then added, progressively more prephosphorylation of the protein decreased the labeling of PtdIns(4,5)P_2.

b. The addition of $ACTH_{1-24}$ to synaptic plasma membranes likewise stimulated the uptake of ^{32}P into endogenous PtdIns(4,5)P_2 and also reduced entry of isotope into phosphatidic acid (Jolles et al., 1981b). The rate of loss of prelabeled PtdIns(4,5)P_2 in the membranes was unaffected by the peptide, again suggesting that the effect is due to increased PtdIns(4,5)P_2 synthesis (Jolles et al., 1981a).

c. Affinity-purified anti-B-50 IgG's added to synaptic plasma membranes markedly and specifically inhibited B-50 phosphorylation and simultaneously enhanced PtdIns(4,5)P_2 labeling several-fold (Oestreicher et al., 1983).

d. Purified PtdIns4P kinase (M_r 45,000; pI 5.8) has been purified 67-fold and further identified by means of specific immunostaining and accompanying reduction of the enzyme activity by interaction with affinity-purified anti-45,000 dalton protein antibodies (Van Dongen et al., 1984, 1985). When the effect of B-50 preparations enriched in either the phosphorylated or dephosphorylated protein was tested on the activity of purified PtdIns4P kinase, the dephosphorylated form had no effect, whereas the identical amount of phosphorylated B-50 substantially diminished the formation of PtdIns(4,5)P_2. To minimize non-specific protein-protein interactions, the experiments were conducted in the presence of bovine serum albumin (Van Dongen et al., 1985).

e. A similar inverse relationship resulting in decreased B-50 phosphorylation and elevated PtdIns(4,5)P_2 labeling was obtained when rat hippocampal slices were incubated with 5×10^{-4}M dopamine, synaptic plasma membranes prepared, and post hoc phosphorylation then performed. These effects were antagonized by the presence of haloperidol in the incubation medium, consistent with a receptor-mediated process (Jork et al., 1984).

Similar reciprocal effects of $ACTH_{1-24}$ on the phosphorylation of certain proteins and of PtdIns4P in subcellular fractions from rabbit iris smooth muscle have been reported by Abdel-Latif et al. (1983). Of relevance also may be the observations by Deshmukh et al. (1984) that in a solubilized myelin preparation containing both myelin basic protein and PtdIns4P phosphorylating activities, when exogenously added basic protein and PtdIns4P (compounds that easily associate in vitro) are present together, the phosphorylation of each is greatly increased. Such mutual activation may represent another class of interactions involving rapidly reversible phosphorylation of certain proteins and phosphoinositides.

These findings have led us to propose that B-50, B-50 kinase, and PtdIns4P kinase exist together in a multimolecular complex in the presynaptic plasma membrane and that phosphorylation of B-50 exerts a regulatory effect on PtdIns4P kinase. The exact nature of the suggested interaction between these entities in the membrane remains to be elucidated. Possibly direct effects mediated by B-50 binding to PtdIns4P kinase are involved or, alternatively, the degree of B-50 phosphorylation could affect membrane topography such that PtdIns4P is rendered more or less accessible to the enzyme.

The action of ACTH$_{1-24}$ in this system may ultimately provide important clues as to the precise mechanism involved. So far evidence is lacking that this neuropeptide exerts its effects through binding to a specific membrane receptor (Witter, 1979). ACTH$_{1-24}$ increases the fluidity of synaptic plasma membranes but not that of liposomes prepared from synaptic plasma membranes as judged by fluorescence depolarization experiments using diphenylhexatriene as probe (Hershkowitz et al., 1982; Verhallen et al., 1984). Physico-chemical studies have shown that whereas ACTH$_{1-24}$ penetrates model membranes composed of a mixture of neutral and anionic phospholipids, neither ACTH$_{1-10}$, the hydrophobic portion, nor ACTH$_{11-24}$, the strongly cationic hydrophilic segment, does so, although the latter fragment presumably interacts electrostatically with negatively charged polar head groups on the liposome surface (Gremlich et al., 1983; Gysin and Schwyzer, 1984). Structure-activity studies in our laboratory have revealed that ACTH$_{1-10}$ and ACTH$_{11-24}$ are ineffective in altering either B-50 phosphorylation or polyphosphoinositide metabolism (Zwiers et al., 1978; Jolles et al., 1981b). Moreover, the structural requirements for the ACTH molecule to fluidize synaptic plasma membranes correlate with those necessary to alter PPtdIns and protein phosphorylation (Van Dongen et al., 1983). Taken together, these findings indicate that ACTH$_{1-24}$ most likely acts in an amphipathic manner by simultaneously forming complexes with anionic groups at the membrane surface and disrupting the hydrophobic acyl chains in the interior (Verhallen et al., 1984). It is quite reasonable that both phospholipids and proteins should take part in these interactions, but it is unknown whether the primary effect of the neuropeptide is on B-50, either of the kinases or a lipid substrate (Verhallen et al., 1984).

IS B-50 PHOSPHORYLATION PART
OF A FEEDBACK MECHANISM AFFECTING PtdIns(4,5)P_2 AVAILABILITY?

Evidence gathered to date indicated that the extent of B-50 phosphorylation influences PtdIns(4,5)P_2 synthesis in synaptic plasma membranes. We suggest that this action of B-50 forms part of a negative feedback loop that contributes to the regulation of the PtdIns(4,5)P_2 pool available for generation of DAG and InsP_3 as shown in Fig. 1 (Gispen, 1986). In this scheme, receptor-mediated activation of PtdIns(4,5)P_2 phosphodiesteratic hydrolysis gives rise to DAG and this substance in turn stimulates B-50 kinase, thereby tending to decrease PtdIns4P kinase activity. The outcome would be to decrease the production of PtdIns(4,5)P_2 hydrolysis

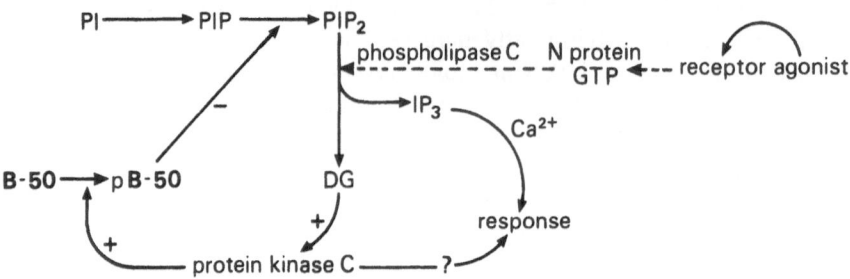

Figure 1. Working model for the role of phosphoprotein B-50 in the feedback control of phosphatidylinositol 4,5-bisphosphate turnover (for explanation see text).
Abbreviations: PI, PtdIns; PIP, PtdIns4P; PIP$_2$, PtdIns(4,5)P_2; IP$_3$, InsP$_3$; DG, DAG.

products and their ensuing biological effects when these are no longer needed by the cell. In support of the feedback concept are observations that phorbol esters, which are known to activate protein kinase C, decrease the carbamylcholine-induced accumulation of inositol phosphates in hippocampal slices and PC12 cells (Labarca et al., 1984; Vicentini et al., 1985). Our model predicts the complementary effect, namely that stimulated PtdIns(4,5)P_2 breakdown should bring about increased B-50 phosphorylation. In addition, assuming that B-50 phosphorylation plays a role in the "off" signal for stimulated PtdIns(4,5)P_2 hydrolysis, we speculate that factors which affect phosphorylation of the protein might do so by interfering with the coupling between receptor-ligand association and the stimulation of PtdIns(4,5)P_2 hydrolysis. Very recent findings support earlier hints (cf. Berridge and Irvine, 1984) that, analogous to the generally accepted mechanism for activation of adenylate cyclase, one or more GTP-binding proteins (N proteins) may be involved in the activation of PPtdIns phosphodiesterase. The presence of GTP analogs has been shown to enhance PPtdIns breakdown in the absence of added Ca^{2+} in plasma membranes from the neutrophils and blowfly salivary gland (Cockcroft and Gomperts, 1985; Litosch et al., 1985). It may be noteworthy that ACTH$_{1-24}$ (10^{-5}M) selectively extracts a 41 kilodalton protein from synaptic plasma membranes (Aloyo et al., 1982b). While the removal of this protein does not affect the ability of ACTH$_{1-24}$ to inhibit B-50 phosphorylation in the treated membranes, the possibility must be considered that the protein, which has an M$_r$ similar to those of previously characterized GTP-binding proteins, may be an obligatory component in the activation of PtdIns(4,5)P_2 hydrolysis. If so, this would point to an additional site of action of ACTH in the regulation of PtdIns(4,5)P_2 metabolism.

InsP_3 AND PROTEIN PHOSPHORYLATION

The wide interest which has been aroused in the proposed second messenger functions of InsP_3 in phenomena as diverse as secretion, phototransduction, and cellular proliferation and development has generally centered on the ability of this compound to mobilize intracellular Ca^{2+}. However, other roles of InsP_3 in cellular regulation may be postulated and one such is a direct effect on protein phosphorylation. Recently, Whitman et al. (1984) reported that 0.5-2.0 μM InsP_3 stimulated phosphorylation of a 62 kilodalton protein in bovine brain homogenates and monkey CV1 fibroblast lysates and that the enhancement was still seen when Ca^{2+} was depleted by EGTA. Lapetina et al. (1984) also observed that $InsP_3$ (5-20 μM) stimulated phosphorylation of the 20,000 and 40,000 dalton proteins in saponin-permeabilized human platelets, but could not exclude the possibility that the compound acts by releasing bound intracellular Ca^{2+}. A demonstration that $InsP_3$ affects protein phosphorylation in the synaptic plasma membrane might shed light on novel regulatory pathways in synaptic function,

Accordingly, we have investigated the effect of $InsP_3$ in putative concentrations as high as 25 μM on protein phosphorylation in synaptic plasma membranes incubated with $[\gamma\text{-}^{32}P]ATP$ and in the presence or absence of Ca^{2+} (Fig. 2). Two sources of $InsP_3$

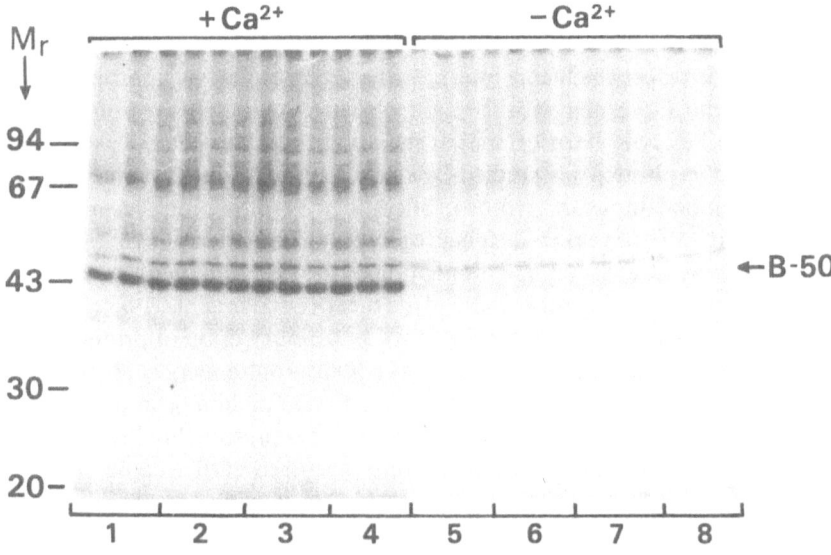

Figure 2. Effect of $InsP_3$ on the phosphorylation of proteins in purified synaptic plasma membranes. Light synaptosomal plasma membranes were isolated from total rat brain. The phosphorylation assay mixture contained 10mM sodium acetate, 10 mM magnesium acetate, either 0.1 mM calcium acetate or 1.0 mM EGTA, 10 μg protein, and 7.5 μM$[\gamma\text{-}^{32}P]ATP$ (1μCi, specific activity 3000 Ci/mmol) in a final volume of 25 μl. The mixture was preincubated at 30°C for 5 min in the presence or absence of $InsP_3$. The reaction was started by addition of $[\gamma\text{-}^{32}P]ATP$ and terminated 15 sec later by addition of a denaturing solution (Zwiers et al., 1976). The proteins were separated on a one-dimensional 11% polyacrylamide gel according to Zwiers et al. (1976). The resulting autoradiogram after incubation with various concentrations of $InsP_3$ in the presence (lanes 1-4) or absence (lanes 5-8) of calcium ion is shown. Lanes 1 and 5: buffer; lanes 2 and 6: 25 μM $InsP_3$; lanes 3 and 7: 5 μM $InsP_3$; lanes 4 and 8: 1 μM $InsP_3$.

were used in these experiments: a commercial product obtained from Sigma Chemical Co. and a preparation generously provided by Dr. Bernard Agranoff. Furthermore, we conducted similar incubation using rat brain homogenates prepared and incubated exactly as described by Whitman et al. (1984) (data not shown). In no case did we observe an effect of $InsP_3$ on the extent of phosphorylation of any protein resolved by one-dimensional polyacrylamide gel electrophoresis. We have also examined the ability of InsP_3 to affect phosphorylation of histones by purified rat brain protein kinase C, again with negative results. Thus we could obtain no evidence in support of the idea that InsP_3 at levels generally thought to prevail physiologically can alter protein phosphorylation in cell-free rat brain preparations.

CONCLUDING REMARKS

Synaptic plasticity is usually defined as a series of adaptive changes in neuronal connectivity in response to either internal or external environmental cues that can result in altered function of the neuronal network involved and ultimately in altered behavior of the organism. Various neurophysiological and neurochemical alterations in synaptic function have been described and interpreted as aspects of synaptic plasticity, but mechanisms at the molecular level remain obscure, although the role of synaptic phosphoproteins has long been of special interest in this respect.

Often synaptic plasticity is studied by tetanic stimulation of monosynaptic subsystems of the hippocampus in vivo or in slices. The resulting long-term potentiation of the synapses involved is taken as a neurophysiological index of synaptic plasticity. Recently, we have shown that tetanic stimulation of the perforant pathway in hippocampal slices results in marked changes in PPtdIns turnover (Bär et al., 1984). In addition, evidence is accumulating that also the degree of phosphorylation of protein B-50 (protein F1) is affected by such treatment (Bär et al., 1980; Routtenberg et al., 1985). Hence it would appear that differences in the functional state of hippocampal synapses are accompanied by changes in the presynaptic machinery, presumably involved in signal transduction. The negative B-50 feedback loop discussed earlier could contribute to maintenance of the appropriate levels of PtdIns(4,5)P_2 available to generate DAG and InsP_3 needed for receptor-activated Ca^{2+} mobilization and possibly membrane fusion and exocytosis (see Baker, 1984; Das and Rand, 1984; Gispen, 1985).

REFERENCES

Abdel-Latif AA (1983) Metabolism of phosphoinositides. In: Lajtha A (ed): Handbook of Neurochemistry. Plenum Press, New York; Vol. 3, pp. 91-131.

Abdel-Latif AA, Akhtar RA, Hawthorne JN (1977) Acetylcholine increases the breakdown of triphosphoinositide of rabbit iris muscle prelabelled with |^{32}P|-phosphate. Biochem J 162: 61-73.

Aloyo VJ, Zwiers H, Gispen WH (1982a) B-50 protein kinase and kinase C in rat brain. Progr Brain Res 56: 303-315.

Aloyo VJ, Zwiers H, Gispen WH (1982b) ACTH(1-24) releases a protein from synaptosomal plasma membranes. J Neurochem 38: 871-875.

Aloyo VJ, Zwiers H, Gispen WH (1983) Phosphorylation of B-50 protein by calcium-activated phospholipid-dependent protein kinase and B-50 protein kinase. J Neurochem 41: 649-653.

Bär PR, Schotman P, Gispen WH, Tielen AM, Lopes da Silva FH (1980) Changes in synaptic membrane phosphorylation after tetanic stimulation in the dentate area of the rat hippocampal slices. Brain Res 198: 478-484.

Bär PR, Wiegant F, Lopes da Silva FH, Gispen WH (1984) Tetanic stimulation affects the metabolism of phosphoinositides in hippocampal slices. Brain Res 321: 381-385.

Baker PF (1984) Multiple controls for secretion? Nature 310: 629-630.

Berridge MJ, Irvine RF (1984) Inositol trisphosphate, a novel second messenger in cellular signal transduction. Nature 312: 315-321.

Berridge MJ, Downes CP, Hanley MR (1982) Lithium amplifies agonists-dependent phosphatidylinositol responses in brain and salivary glands. Biochem J 206: 587-595.

Brown E, Kendall DA, Nahorski SR (1984) Inositol phospholipid hydrolysis in rat cerebral cortical slices. I. receptor characterisation. J Neurochem 42: 1379-1387.

Cockcroft S, Gomperts BD (1985) Role of guanine nucleotide binding protein in the activation of polyphosphoinositide phosphodiesterase. Nature 314: 534-536.

Das S, Rand RP (1984) Diacylglycerol causes major structural transitions in phospholipid bilayer membranes. Biochem Biophys Res Commun 124: 491-496.

Deshmukh DS, Kuizon S, Brockerhoff H (1984) Mutual stimulation by phosphatidylinositol 4-phosphate and myelin basic protein of their phosphorylation by the kinases solubilized from rat brain myelin. Life Sci 34: 259-264.

Fisher SK, Agranoff BW (1981) Enhancement of the muscarinic synaptosomal phospholipid effect by the ionophore A23187. J Neurochem 37: 968-977.

Fischer SK, Agranoff BW (1985) The biochemical basis and functional significance of enhanced phosphatidate and phosphoinositide turnover. In: Eichberg J (ed): Phospholipids in nervous tissues. John Wiley and Sons, New York, pp. 241-296.

Fisher SK, Boast CA, Agranoff BW (1980) The muscarinic stimulation of phospholipid labeling in hippocampus is independent of its cholinergic input. Brain Res 189: 284-288.

Fisher SK, Frey KA, Agranoff BW (1981) Loss of muscarinic receptors and of stimulated phospholipid labeling in ibotinate-treated hippocampus. J Neurosci 1: 1407-1413.

Fisher SK, Figueiredo JC, Bartus RT (1984) Differential stimulation of inositol phospholipid turnover in brain by analogs of oxotremorine. J Neurochem 42: 1171-1179.

Gispen WH (1986) Phosphoprotein B-50 and phosphoinositides in brain synaptic plasma membranes: a possible feedback relationship. Trans Biochem Soc UK, 14: 163-165.

Gispen WH, Zwiers H (1985) Behavioral and neurochemical effects of ACTH. In: Lajtha A (ed): Handbook of Neurochemistry. Plenum Press, New York; Vol 8, pp. 375-412.

Gispen WH, Leunissen JLM, Oestreicher AB, Verkleij AJ, Zwiers H (1985) Presynaptic localization of B-50 phosphoprotein: the ACTH sensitive protein kinase substrate involved in rat brain polyphosphoinositide metabolism. Brain Res 328: 381-385.

Gonzalez-Sastre F, Eichberg J, Hauser G (1971) Metabolic pools of polyphosphoinositides in rat brain. Biochim Biophys Acta 248: 96-104.

Gremlich H-U, Fringeli U-P, Schwyzer R (1983) Conformational changes of adrenocorticotropin peptides upon interaction with lipid membranes revealed by infra-red attenuated total reflection spectroscopy. Biochemistry 22: 4257-4263.

Gysin B, Schwyzer R (1984) Hydrophobic and electrostatic interactions between adrenocorticotropin-(1-24)-tetracosapeptide and lipid vesicles. Amphiphilic primary structures. Biochemistry 23: 1811-1818.

Hershkowitz M, Zwiers H, Gispen WH (1982) The effects of ACTH on rat brain synaptic plasma membrane lipid fluidity. Biochim Biophys Acta 692: 495-497.

Irvine RF, Letcher AJ, Laner DJ, Downes CP (1984) Inositol trisphosphates in carbachol-stimulated rat parotid glands. Biochem J 223: 237-243.

Janowsky A, Labarca R, Paul SM (1984) Characterization of neurotransmitter receptor-mediated phosphatidylinositol hydrolysis in the rat hippocampus. Life Sci 35: 1953-1961.

Jolles J, Schrama LH, Gispen WH (1981a) Calcium-dependent turnover of brain polyphosphoinostides after prelabelling *in vivo*. Biochim Biophys Acta 666: 90-98.

Jolles J, Zwiers H, Dekker A, Wirtz KWA, Gispen WH (1981b) Corticotropin-(1-24)-tetracosapeptide affects protein phosphorylation and polyphosphoinositide metabolism in rat brain. Biochem J 194: 283-291.

Jolles J, Zwiers H, Van Dongen C, Schotman P, Wirtz KWA, Gispen WH (1980) Modulation of brain polyphosphoinositide metabolism by ACTH-sensitive protein phosphorylation. Nature 286: 623-625.

Jork R, De Graan PNE, Van Dongen CJ, Zwiers H, Matthies H, Gispen WH (1984) Dopamine-induced changes in protein phosphorylation and polyphosphoinositide metabolism in rat hippocampus. Brain Res 291: 73-81.

Kristjansson GI, Zwiers H, Oestreicher AB, Gispen WH (1982) Evidence that the synaptic phosphoprotein B-50 is localized exclusively in nerve tissue. J Neurochem 39: 371-378.

Labarca R, Janowsky A, Patel J, Paul SM (1984) Phorbol esters inhibit agonist-induced |³H|-inositol-1-phosphate accumulation in rat hippocampal slices. Biochem Biophys Res Commun 123: 703-709.

Lapetina ET, Watson SP, Cuatrecasas P (1984) Myo-inositol 1, 4, 5-trisphosphate stimulates protein phosphorylation in saponin-permeabilized human platelets. Proc Natl Acad Sci USA 81: 7431-7435.

Litosch I, Wallis C, Fain JN (1985) 5-Hydroxytryptamine stimulates inositol phosphate production in a cell-free system from blowfly salivary glands. Evidence for a role of GTP in coupling receptor activation to phosphoinositide breakdown. J Biol Chem, 260: 5464-5471.

Mahler HR, Kleine LP, Ratner N, Sörensen RG (1982) Identification and topography of synaptic phosphoproteins. Progr Brain Res 56: 27-48.

Nishizuka Y (1984) The role of protein kinase C in cell surface signal transduction and tumor promotion. Nature 308: 693-697.

Oestreicher AB, Dekker LV, Gispen WH (1985) A radioimmunoassay (RIA) for the phosphoprotein B-50: Distribution in rat brain. Neurosci Letters Suppl 22, 8560.

Oestreicher AB, Van Dongen CJ, Zwiers H, Gispen WH (1983) Affinity-purified anti-B-50 protein antibody: Interference with the function of the phosphoprotein B-50 in synaptic plasma membranes, J Neurochem 41:331-340.

Oestreicher AB, Zwiers H, Schotman P, Gispen WH (1981) Immunohistochemical localization of a phosphoprotein (B-50) isolated from rat brain synaptosomal plasma membranes. Brain Res Bull 6: 145-153.

Routtenberg A, Lovinger D, Steward O (1985) Selective increase in the phosphorylation of a 47 kD protein (F1) directly related to long-term potentiation. Behav Neural Biol 43: 3-11.

Schacht J, Agranoff BW (1972) Effects of acetylcholine on labeling of phosphatidate and phosphoinositides by |³²P|-orthophosphate in nerve ending fractions of guinea pig cortex. J Biol Chem 247: 771-777.

Sherman WR, Leavitt AL, Honchar MP, Hallcher LM, Phillips BE (1981) Evidence that lithium alters phosphoinositide metabolism: chronic administration elevates primarily D-myo-inositol-1-phosphate in cerebral cortex of the rat. J Neurochem 36: 1947-1951.

Sörensen RG, Kleine LP, Mahler HR (1981) Presynaptic localization of phosphoprotein B-50. Brain Res Bull 7: 57-61.

Van Dongen CJ (1985) Phosphoprotein B-50 and polyphosphoinositides in rat brain: Target for neuromodulation by ACTH. Ph.D. Thesis, University of Utrecht, Utrecht, The Netherlands.

Van Dongen CJ, Hershkowitz M, Zwiers H, De Laat S, Gispen WH (1983) Lipid fluidity and phosphoinositide metabolism in rat brain membranes of aged rats: Effects of ACTH(1-24). In: Gispen WH, Traber J (eds): Aging of the brain. Elsevier Sci Publ, Amsterdam; pp. 101-104.

Van Dongen CJ, Zwiers H, Gispen WH (1984) Purification and partial characterization of the phosphatidylinositol 4-phosphate kinase from rat brain. Biochem J 223: 197-203.

Van Dongen CJ, Zwiers H, De Graan PNE, Gispen WM (1985) Modulation of the activity of and phosphatidylinositol 4-phosphate kinase by phosphorylated and dephosphorylated B-50 protein. Biochem Biophys Res Commun 128: 1219-1227.

Van Rooijen LAA (1984) Polyphosphoinositide phosphodiesterase: Characterization and physiological significance in brain. Ph.D. Thesis, University of Utrecht, Utrecht, The Netherlands.

Van Rooijen LAA, Seguin EB, Agranoff BW (1983) Phosphodiesteratic breakdown of endogenous polyphosphoinositides in nerve ending membranes. Biochem Biophys Res Commun 112: 919-926.

Verhallen PFJ, Demel RA, Zwiers H, Gispen WH (1984) Adrenocorticotropic hormone (ACTH)-lipid interactions. Implications for involvement of amphipathic helix formation. Biochim Biophys Acta 775: 246-254.

Vicentini LM, Di Virgilio F, Ambrosini A, Pozzan T, Meldolesi J (1985) Tumor promoter phorbol 12-myristate, 13-acetate inhibits phosphoinositide hydrolysis and cytosolic Ca^{2+} rise induced by the activation of muscarinic receptors in PC12 cells. Biochem Biophys Res Commun 127: 310-317.

Weller M (1979) Protein phosphorylation. PION Ltd, London.

Whitman MR, Epstein J, Cantley L (1984) Inositol 1, 4, 5-trisphosphate stimulates phosphorylation of a 62,000-dalton protein in monkey fibroblast and bovine brain cell lysates. J Biol Chem 259: 13652-13655.

Witter A (1979) On the presence of receptors for ACTH-neuropeptides in the brain. In: Pepeu GC, Kuhar M, Enna L (eds): Receptors for neurotransmitters and peptide hormones. Raven Press, New York, pp. 407-414.

Zwiers H, Schotman P, Gispen WH (1980) Purification and some characteristics of an ACTH-sensitive protein kinase and its substrate protein in rat brain membranes. J Neurochem 34: 1689-1699.

Zwiers H, Tonnaer J, Wiegant VM, Schotman P, Gispen WH (1979) ACTH-sensitive protein kinase from rat brain membranes. J Neurochem 33: 247-256.

Zwiers H, Veldhuis D, Schotman P, Gispen WH (1976) ACTH, cyclic nucleotides and brain protein phosphorylation in vitro. Neurochem Res 1: 669-677.

Zwiers H, Verhaagen J, Van Dongen CJ, De Graan PNE, Gispen WH (1985) Resolution of rat brain synaptic phosphoprotein B-50 into multiple forms by two dimensional electrophoresis: evidence for multi-site phosphorylation. J Neurochem 44: 1083-1090.

Zwiers H, Wiegant VM, Schotman P, Gispen WH (1978) ACTH-induced inhibition of endogenous rat brain protein phosphorylation in vitro: structure-activity. Neurochem Res 3: 455-463.

Phospholipid research and the nervous system
Biochemical and molecular pharmacology
L.A. Horrocks, L. Freysz, G. Toffano (eds)
Fidia Research Series, vol. 4.
Liviana Press, Padova. © 1986

POSSIBLE ROLES OF INOSITOL PHOSPHOLIPIDS IN CELL SURFACE SIGNAL TRANSDUCTION IN NEURONAL TISSUES

Y. Nishizuka, U. Kikkawa, A. Kishimoto, H. Nakanishi and K. Nishiyama

Department of Biochemistry, Kobe University School of Medicine, Kobe 650, Japan

Information of various extracellular signals such as a group of neurotransmitters and some peptide hormones flows from the cell surface into the cell interior through two routes, Ca^{2+} mobilization and protein kinase C activation. Except some tissues such as bovine adrenal medullary cells (Swilem and Hawthorne, 1983), both routes become available as the result of a single ligand-receptor interaction as well as of depolarization. Under normal conditions this protein kinase is activated by diacylglycerol in the presence of membrane phospholipids, particularly PtdSer, at a physiologically low concentration of Ca^{2+} (Nishizuka, 1984a, b). The diacylglycerol active in this role is usually absent from the membrane, but is produced from the receptor-mediated or voltage-dependent hydrolysis of inositol phospholipids as shown in Figure 1. Although it is becoming clear that $PtdIns(4,5)P_2$ is the prime target of phosphodiesterase, other inositol phospholipids are probably broken down at different rates. On the other hand, inositol 1,4,5-trisphosphate ($InsP_3$), a water-soluble product of $PtdIns(4,5)P_2$ hydrolysis, has been proposed to serve as an intracellular mediator of Ca^{2+} mobilization from its internal store (Berridge and Irving, 1984).

Under appropriate conditions some synthetic diacylglycerols such as 1-oleoyl-2-acetyl-glycerol are intercalated into the cell membrane and activate protein kinase C directly. Tumour-promoting phorbol esters such as 12-O-tetradecanoylphorbol-13-acetate (TPA) are shown to act as a substitute for diacylglycerol. By using these permeable activators of the enzyme and Ca^{2+}-ionophore it is shown that both routes mentioned above are essential and act synergistically to

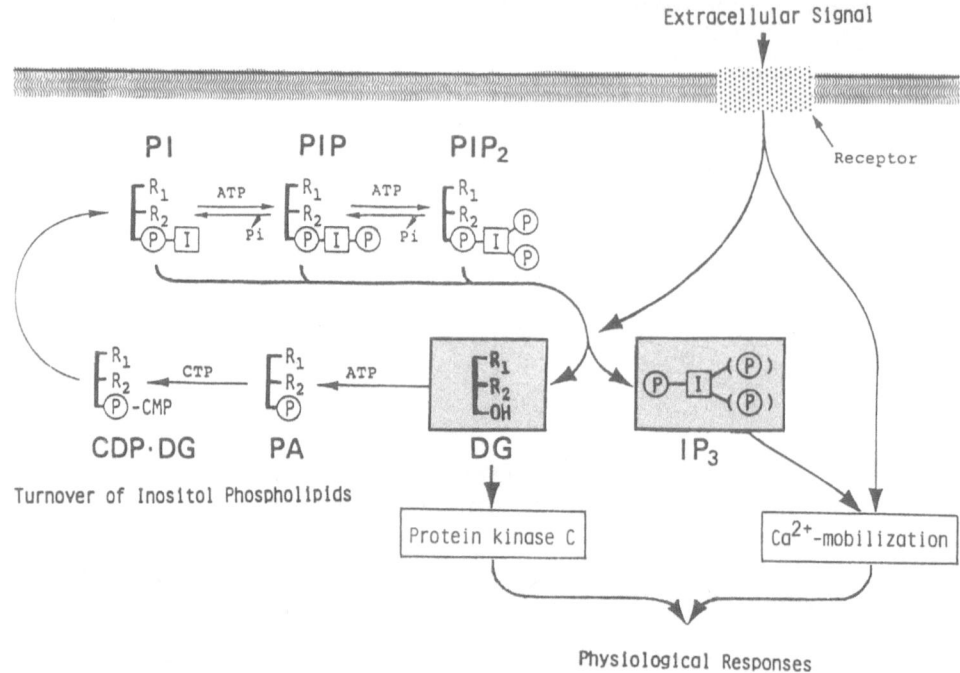

Figure 1. Turnover of inositol phospholipids and signal transduction. PI, PtdIns; PIP, PtdIns4*P*; and PIP$_2$, PtdIns(4,5)P_2. (adapted from Nishizuka, 1984a).

evoke fully many of the subsequent cellular responses. The synergistic role of protein kinase C and Ca^{2+} was first demonstrated for release reactions of platelets (Kaibuchi et al., 1982), and later for a wide variety of cellular processes, including release reactions of neurotransmitters from both peripheral and central nervous tissues, exocytosis of various cell types, smooth muscle contraction, and many other metabolic processes (for reviews, see Nishizuka, 1984b; Kikkawa and Nishizuka, 1986).

Kinetic studies with some cell types such as platelets indicate that, when the receptor is stimulated, diacylglycerol is produced in membrane only transiently, and once produced it disappears very rapidly due to further degradation to arachidonate to generate other messengers such as prostaglandins, and also due to conversion to inositol phospholipids by way of phosphatidate. Similarly, when measured by the aequorin procedure described by Johnson et al. (1985), the appearance of Ca^{2+} is again transient, and it is immediately extruded by some mechanism in a feedback manner as given in Figure 2. In this experiment the amount of aequorin loaded is not limited, since this Ca^{2+}-spike may be repeatedly observed by subsequent addition of thrombin. The biological significance of this feedback control is not fully understood, but it is possible that protein kinase C is also involved in this extrusion process. In microsome fractions from some tissues such as cardiac muscle the activity of Ca^{2+}-transport ATPase is enhanced by the addition of protein kinase C as first described by Limas (1980). In intact cell systems the cytosolic Ca^{2+} concentration is frequently decreased by the

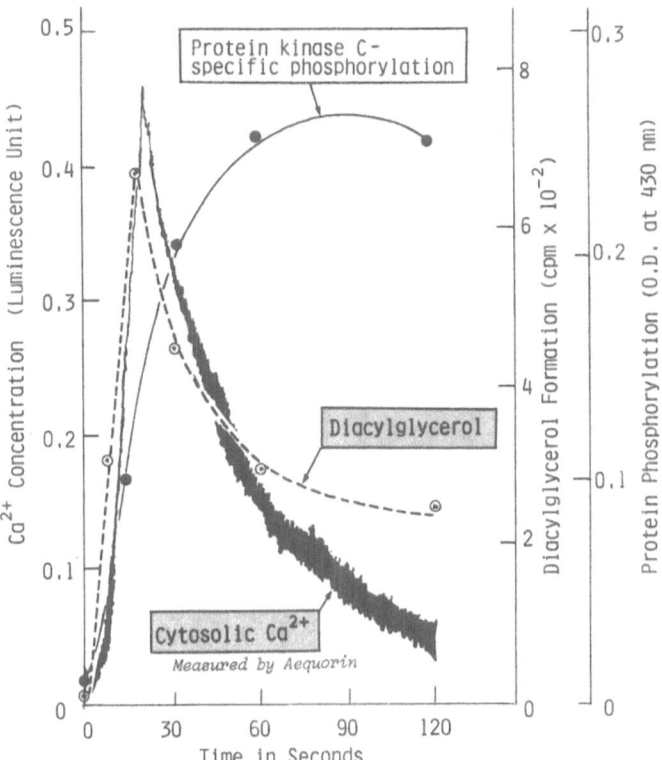

Figure 2. Time courses of diacylglycerol formation, Ca^{2+} mobilization, and protein kinase C-specific phosphorylation of 47K protein in human platelets after thrombin stimulation. Human platelets were isolated, labeled with either [^3H] arachidonate or ^{32}Pi, and stimulated by thrombin at 37°C for various periods of time as indicated. The diacylglycerol produced and radioactive 47K protein were determined as described (Kaibuchi et al., 1982). In a separate set of experiments human platelets were loaded with aequorin under the conditions described by Johnson et al. (1985), and stimulated with thrombin. The detailed experimental procedures will be described elsewhere. Note that the phosphate attached to 47K protein (closed circles) does not necessarily indicate the active state of protein kinase C, but shows the existence of the radioactive phosphoprotein that was phosphorylated by the enzyme.

addition of TPA (Tsien et al., 1982; Rink et al., 1983; Moolenaar et al., 1984; Lagast et al., 1984; Sagi-Eisenberg et al., 1985). In addition, it is attractive to imagine that, as schematically shown in Figure 3, the role of protein kinase C may be extended to the enhancement of Ca^{2+} influx, and protein kinase C and Ca^{2+} act in concert to promote the activation of cellular functions. However, crucial evidence for this mechanism in neuronal tissues is still unavailable. The activation of endogenous protein kinase C by TPA or the micro-injection of this enzyme itself into bag cell neurons enhances the voltage-sensitive Ca^{2+} current (DeRiemer et al., 1985). It is worth noting that the phosphate attached to proteins involved in protein kinase C-specific phosphorylation reactions appears to be often resistant to phosphatases and, thus, the consequence of this phosphorylation may bring a longer persistence in time. Presumably, this may be

46

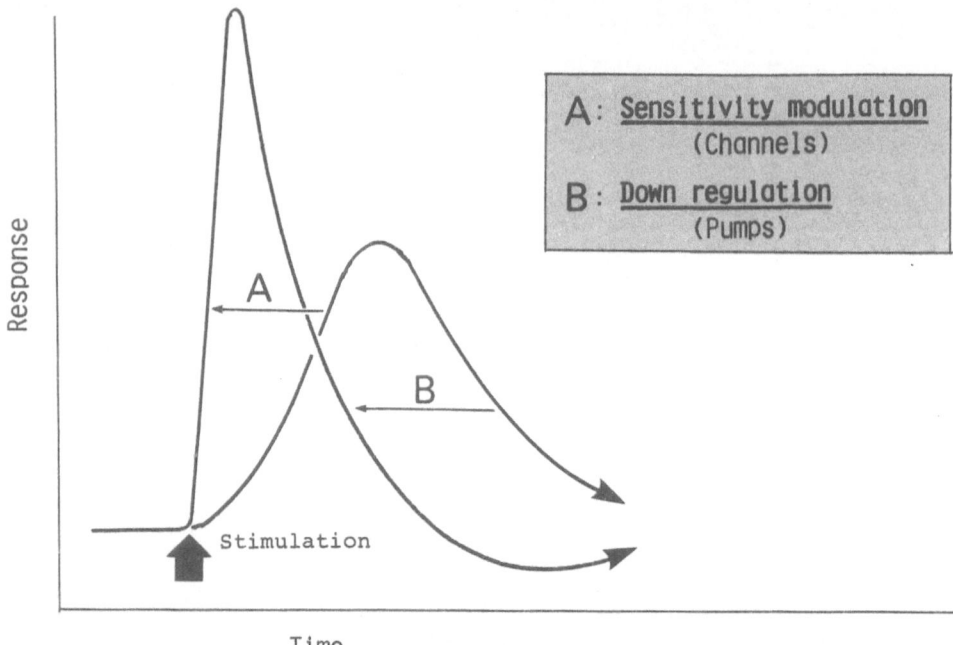

Figure 3. Prospective roles of protein kinase C in membrane conductance in neuronal tissues.

related to neuronal plasticity proposed by Routtenberg (1985). Dual functions of protein kinase C, positive forward action and apparently negative feedback control, are suggested also for some other processes of cellular responses, such as smooth muscle contraction and receptor function of epidermal growth factor (for a review, see Kikkawa and Nishizuka, 1986).

Although protein kinase C and cyclic AMP-dependent protein kinase (protein kinase A) appear to transduce distinctly different pieces of information into the cell, these two signal pathways sometimes cause apparently similar cellular responses and often potentiate each other at the levels of cell surface receptors (Nishizuka, 1984a; Sugden et al., 1985). Recent analysis also indicates that these protein kinases often share the same substrate proteins, even the same seryl and threonyl residues, for phosphorylation. With myelin basic protein as a model substrate, it is shown that, contrary to protein kinase A, protein kinase C reacts with seryl residues that are located at the amino-terminal side close to lysine or arginine as shown in Figure 4. The seryl residues that are commonly phosphorylated by these two enzymes have basic amino acids at both amino- and carboxyl-terminal sides.

Thus, these protein kinases may show different but sometimes similar functions depending on the structure of the target protein (Kishimoto et al., 1985). Although the crucial information on target proteins of protein kinase C in nervous tissues is limited, evidence available at present seems to suggest that the signal transduction through protein kinase C is intimately related to neurotransmitter release, membrane conductance, and interaction between receptors as well as to various metabolic processes such as neurotransmitter biosynthesis. Presumably, the receptors relating to cyclic AMP modulate these processes positively or negatively in the manner briefly described above.

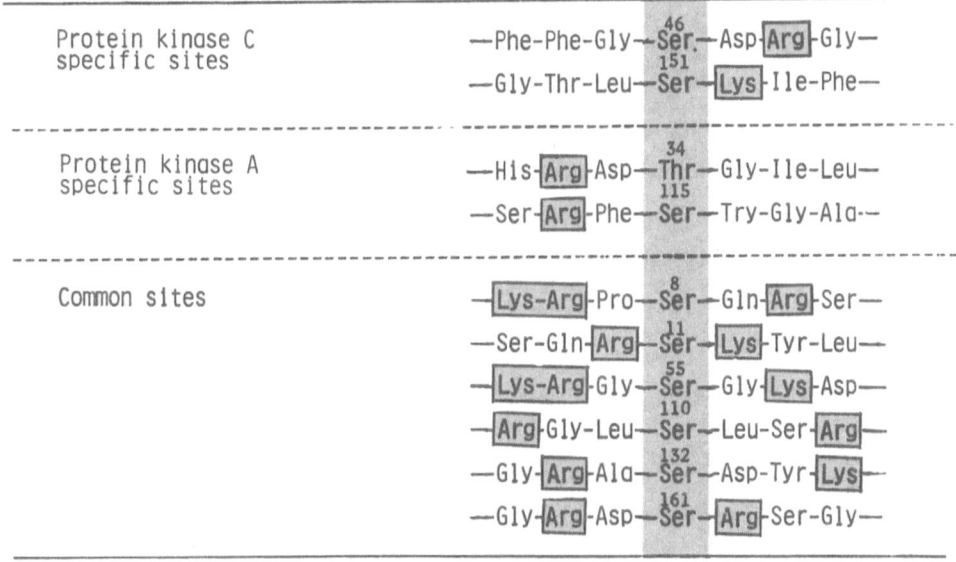

Figure 4. Major phosphorylation sites of myelin basic protein for protein kinases C and A. The detailed experimental conditions are described (Kishimoto et al., 1985).

ACKNOWLEDGMENTS

This investigation has been supported in part by research grants from the Research Fund of the Ministry of Education, Science and Culture, and by the Special Coordination funds from the Ministry of Science and Technology, Japan.

REFERENCES

Berridge MJ, Irvine RF (1984) Inositol trisphosphate, a novel second messenger in cellular signal transduction. Nature 312: 315-321.

DeRiemer SA, Strong JA, Albert KA, Greengard P, Kaczmarek LK (1985) Enhancement of calcium current in *Aplysia* neurons by phorbol ester and protein kinase C. Nature 313: 313-316.

Johnson PC, Ware JA, Cliveden PB, Smith M, Dvorak AM, Salzman EW (1985) Measurement of ionized calcium in blood platelets with photoprotein aequorin. J Biol Chem 260: 2069-2076.

Kaibuchi K, Sano K, Hoshijima M, Takai Y, Nishizuka Y (1982) Phosphatidylinositol turnover in platelet activation: calcium mobilization and protein phosphorylation. Cell Calcium 3: 323-335.

Kikkawa U, Nishizuka Y (1986) Protein kinase C. In: Krebs EG (ed): The enzymes. Academic Press, New York; in press

Kishimoto A, Nishiyama K, Nakanishi H, Uratsuji Y, Nomura H, Takeyama Y, Nishizuka Y (1985) Studies on the phosphorylation of myelin basic protein by protein kinase C and adenosine-3', 5'-monophosphate-dependent protein kinase. J Biol Chem 260: 12492-12499.

Lagast H, Pozzan T, Waldvogel FA, Lew PD (1984) Phorbol myristate acetate stimulates ATP-dependent calcium transport by the plasma membrane of neutrophils. J Clin Invest 73: 878-883.

Limas CJ (1980) Phosphorylation of cardiac sarcoplasmic reticulum by a calcium-activated, phospholipid-dependent protein kinase. Biochem Biophys Res Commun 96: 1378-1383.

Moolenaar WH, Tertoolen LGJ, de Laat SW (1984) Phorbol ester and diacylglycerol mimic growth factor in raising cytoplasmic pH. Nature 312: 371-374.

Nishizuka Y (1984a) The role of protein kinase C in cell surface signal transduction and tumour promotion. Nature 308: 693-697.

Nishizuka Y (1984b) Turnover of inositol phospholipids and signal transduction. Science 225: 1365-1370.

Rink TJ, Sanchez A, Hallam TJ (1983) Diacylglycerol and phorbol ester stimulate secretion without raising cytoplasmic free calcium in human platelets. Nature 305: 317-139.

Routtenberg A (1986) Protein kinase C and substrate protein F1 (47 KD, 4, 5 pI): relation to synaptic plasticity and dendritic spine growth. In: Will B, Schmitt P (eds): Brain Plasticity, Learning and Memory. Plenum, New York; in press

Sagi-Eisenberg R, Lieman H, Pecht I (1985) Protein kinase C regulation of the receptor-coupled calcium signal in histamine-secreting rat basophilic leukaemia cells. Nature 313: 59-60.

Sugden D, Vanecek J, Klein DC, Thomas TP, Anderson WB (1985) Activation of protein kinase C potentiates isoprenaline-induced cyclic AMP accumulation in rat pinealocytes. Nature 314: 359-361.

Swilem AF, Hawthorne JN (1983) Catecholamine secretion by perfused bovine adrenal medulla in response to nicotinic activation is inhibited by muscarinic receptors. Biochem Pharmacol 32: 3873-3874.

Tsien RY, Pozzan T, Rink TJ (1982) T-Cell mitogens cause early changes in cytoplasmic free Ca^{2+} and membrane potential in lymphocytes. Nature 295: 68-71.

Phospholipid research and the nervous system
Biochemical and molecular pharmacology
L.A. Horrocks, L. Freysz, G. Toffano (eds)
Fidia Research Series, vol. 4.
Liviana Press, Padova. © 1986

POSSIBLE GLIAL MODULATION OF NEURONAL ACTIVITY BY EICOSANOIDS AND PHOSPHOINOSITIDE METABOLITES

J.J. DeGeorge, P. Morell and E.G. Lapetina[1]

Biology Science Research Center, Department of Biochemistry and Nutrition, University of North Carolina at Chapel Hill, NC 27514; [1]Department of Molecular Biology, The Wellcome Research Laboratories, Research Triangle Park, NC 27709.

PHOSPHOINOSITIDES AND ARACHIDONIC ACID METABOLISM IN THE NERVOUS SYSTEM

Metabolites of phosphoinositides and of arachidonic acid (AA) may serve central roles as intermediates in signal transmission and regulation of cellular processes. Our understanding of the pathways for the generation of these metabolites and of their biological role is primarily derived from studies in non-neural tissues. An especially detailed body of knowledge has been accumulated concerning their role in the communication between platelets and vascular endothelial cells during platelet activation (Lapetina, 1983; Rittenhouse-Simmons, 1984; Nishizuka, 1984; Zucker and Nacmias, 1985). Similar metabolic pathways are activated in neural tissue in response to a variety of stimuli (for review see: Wolf, 1982; Berridge et al., 1984).

One approach that has been used to examine these pathways is to subject intact neural tissue to a variety of stimuli and observe the generation of either phosphoinositide metabolites, or of free AA and its metabolites. To assess the relevance of the formation of specific products formed to physiological or pathological events, the metabolites themselves or selective inhibitors of the pathways have been tested for the ability to alter the neural responses to stimulation. The use of such techniques has demonstrated that application of neurotransmitters, electrical stimulation, or traumatic insult results in activation of phosphoinositide metabolism (Berridge et al., 1982; Daum et al., 1984; Brown et al., 1984; Briggs et al., 1985), liberation of AA from cellular lipids and accumulation of AA metabolites (Lunt and Rowe, 1971; Bazan, 1971; Hedqvist, 1973; Leslie, 1976; Jonsson and Daniell, 1976; Bernheim et al., 1980). Of particular relevance to the following discussion is the observation that formation of both cyclooxygenase

(Leslie, 1976; Abdel-Halim et al., 1980; Wolfe et al., 1976; Bishai and Coceani, 1981) and lipoxygenase (Spagnuolo et al., 1979; White and Stine, 1984; Lindgren et al., 1984) metabolites of AA occur in CNS tissues. The products of both these pathways alter neuronal activity, regulating release of neurotransmitters and neuronal electrical properties (Bergstrom et al., 1973; Hillier and Templeton, 1980; Palmer et al., 1981; Kimura et al., 1985).

Studies such as those noted above have provided useful information concerning the existence and potential physiological roles of phosphoinositides and AA metabolites in the nervous system. However, in most cases the models used (intact tissues and tissue slices) have consisted of mixed cell types making it difficult to link physiologically relevant biochemical responses to specific cell classes. Various approaches have been used to circumvent this problem and relate the appearance of specific metabolites to specific cell types. Synaptosomes are a useful system to model responses occurring at the nerve terminal. Isolated preparations of brain synaptosomes release AA from phospholipids in response to depolarizing stimuli (Baba et al., 1983; Bradford et al., 1983). The release of AA by synaptosomes, which appears to be phospholipase A_2 dependent, is accompanied by the formation of PGE_2 and $PGF_{2\alpha}$ (Brandfort et al., 1983), metabolites known to alter neurotransmitter release (Hedqvist, 1973) and post synaptic neural responses (Kimura et al., 1985) in some systems. These results have been taken to suggest that electrical activity in vivo may, in addition to being directly coupled to neurotransmitter release, be coupled to altered AA metabolism which can modulate neurotransmitter release and electrical properties of the neuron. Another approach has depended upon the use of neuronal cell lines (Ohsako and Deguchi, 1981; Birkle and Ellis, 1983; Yano et al., 1985; Snider et al., 1984). For example, N1E-115 cells, a cell line widely used as a neuronal model, show stimulation of phosphoinositide metabolism in response to neurotransmitter treatment (Cohen et al., 1983). This same cell line also liberates AA and requires AA metabolism via a lipoxygenase pathway for an increase in cGMP to occur (Snider et al., 1984).

The study of phosphoinositide and AA metabolism by glial cells has received less attention. C6 glioma cells synthesize cyclooxygenase metabolites; these studies involve paradigms meant to model the pathogenesis of edema (Chan and Fishman, 1983). However, few investigations have focused on the question of the glial contribution to neurotransmitter stimulated lipid metabolism. Such a line of investigation would seem promising since a number of studies have demonstrated that cultures of glial cells respond to neurotransmitter stimulation with alterations of cyclic nucleotide metabolism (Clark and Perkins, 1971; McCarthy and de Vellis, 1978; Van Calker et al., 1983; Rougon et al., 1983) and protein phosphorylation (McCarthy et al., 1985; Groppi and Browning, 1980) suggesting that glia in vivo may be capable of monitoring the neurochemical environment. Recently, glial cell lines have been shown to respond to neurotransmitter stimulation with accumulation of inositol phosphate (Masters et al., 1984) and inositol polyphosphates (DeGeorge et al., 1985), as well as with liberation of AA (Mallorga et al., 1980; DeGeorge et al., 1984).

Our studies indicate that stimulated liberation of AA in glial cell lines is accompanied by the accumulation of AA metabolites. Should similar responses occur in glia in vivo, it would suggest the possibility that neuroglia may release AA in response to neurotransmitter stimulation and may contribute to the modulation of neuronal activity exhibited by AA metabolites.

AA METABOLISM BY C62B GLIOMA CELLS

C62B cells, prelabeled with $[1\text{-}^{14}C]AA$ and stimulated with the Ca^{++} ionophore A23187, synthesize several cyclooxygenase and lipoxygenase metabolites of AA (Table 1) (DeGeorge et al., 1984).

Table 1. *Arachidonate liberation* and formation of metabolites following stimulation of C62B glioma cells with ionophore A23187*

Products	Treatments	
	Non-stimulated	A23187
	(CPM/Culture)	
AA	1285 ± 107	4020 ± 643
$PGF_{2\alpha}$	660 ± 135	2050 ± 412
PGE_2	202 ± 60	360 ± 42
PGD_2	123 ± 15	290 ± 15
LM	223 ± 15	490 ± 47
5-HETE	302 ± 14	585 ± 60
HHT	175 ± 13	618 ± 75
12-HETE	300 ± 35	865 ± 73
PtdH	3255 ± 270	6577 ± 885

(*) Cells were prelabeled with 0.25 μCi of $[1\text{-}^{14}C]AA$ for 18 h. After rinsing cultures to remove unincorporated radioactivity, the cultures were treated with culture medium (non-stimulated) or culture medium plus ionophore A23187 (2.5 μM) for 15 min. Lipids were extracted and then separated by TLC. Results are expressed as mean $\pm$ SEM (n = 8). LM, Lipoxygenase metabolite.

In ionophore stimulated cells the cyclooxygenase metabolite $PGF_{2\alpha}$ is the most abundant metabolite, accounting for 39% of the radioactivity in the metabolites formed. Lipoxygenase metabolites of AA are also synthesized, and taken together 12-HETE, 5-HETE, and an unidentified metabolite (LM, believed to be a lipoxygenase metabolite based on its susceptibility to inhibition with lipoxygenase inhibitors, DeGeorge et al., unpublished observation) account for 35% of the radioactivity in metabolites.

NEUROTRANSMITTER STIMULATED LIPID METABOLISM IN C62B GLIOMA CELLS

C62B cells incubated for 18 h incorporate more than 80% of the added label into cellular glycerolipids. About 13% of the radioactivity is extracellular unmetabolized $[1\text{-}^{14}C]AA$ and most of the remaining extracellular radioactivity is in a form that co-chromatographs with the lipoxygenase derivative of AA, 5-HETE (this metabolite does not accumulate in culture medium incubated without cells). Some 0.5% of the cellular radioactivity is present as non-esterified AA and metabolites. It is the increase of the pool of cellular free AA and its metabolites that is monitored in response to neurotransmitter stimulation. When prelabeled C62B cells are treated with 1 mM acetylcholine (ACh), they liberate AA from esterified pools (Table 2). The liberation of AA is accompanied by the accumulation of a lipoxygenase metabolite and of PtdH containing radioactive arachidonate. The accumulation of these products occurs rapidly, with significant accumulations within 30 s after addition of receptor agonist. This rapid return to basal levels probably reflects the stimulated reacylation of AA in phospholipids (DeGeorge et al., 1985b), possibly into lysolipids formed during its liberation. The lipoxygenase metabolite follows a similar time course of accumulation and return to basal levels, but represents less than 10% of the AA liberated. PtdH remains elevated for somewhat longer, but also returns to basal levels in the presence of continued cholinergic stimulation.

Table 2. *Cholinergic stimulation of AA metabolism*.*

Treatment	Incubation Time	AA	PtdH CPM/culture ($x \pm$ SEM, $n = 8$)	LM
Control	zero	1626 ± 315	1957 ± 157	247 ± 26
,,	1min	1369 ± 257	2234 ± 199	225 ± 32
ACh (1mM)	15sec	2377 ± 467	2436 ± 223	287 ± 60
,,	30sec	3553 ± 673	3105 ± 258	377 ± 27
,,	1min	7710 ± 1837	4721 ± 288	528 ± 28
,,	2.5min	2688 ± 463	4557 ± 522	320 ± 39
,,	5min	1731 ± 222	3326 ± 310	258 ± 21
,,	30min	1233 ± 69	1618 ± 116	213 ± 17

(*) Cultures of C62B cells were prelabeled with 0.25 μCi of [1-^{14}C]AA for 18 h. After removing labeling medium, the cultures were treated with medium (control) or medium plus 1 mM ACh for the times indicated. Lipids were extracted and then separated by TLC. LM, lipoxygenase metabolite.

Taken together these data suggest that cells of glial origin are competent to synthesize a variety of AA metabolites, and do synthesize some of these metabolites in response to neurotransmitter stimulation. Furthermore, some of the metabolites produced have been shown in other systems to alter neuronal properties important in neurotransmission.

A MODEL OF NEUROTRANSMITTER COORDINATED GLIAL MODULATION OF NEURONAL ACTIVITY

Based on the observations that 1) glia express neurotransmitter receptors, 2) glial neurotransmitter receptors are functionally coupled to rapid stimulation of cyclic nucleotide metabolism and lipid metabolism (both phosphoinositides and AA) and 3) some of the lipid metabolites synthesized by glia are capable of modulating neuronal activity, we propose a model for one pathway of glial modulation of neuronal activity (Figure 1). In this model the stimulated neuron releases neurotransmitter (T). The release

**A MODEL OF NEUROTRANSMITTER
COORDINATED GLIAL MODULATION
OF NEURONAL ACTIVITY**

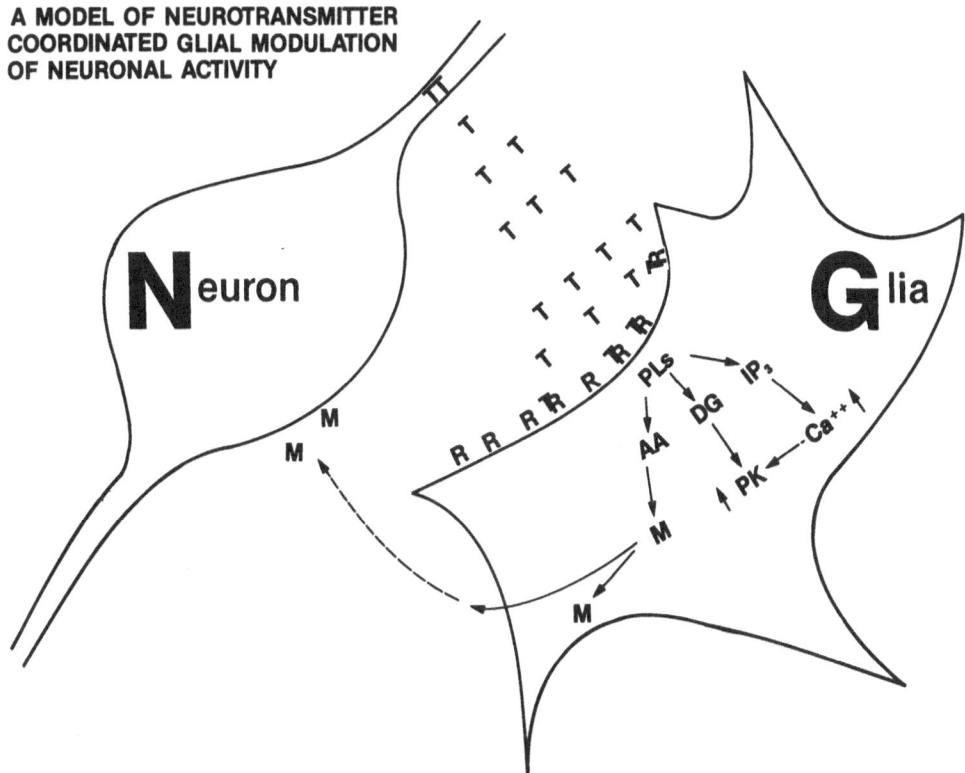

Figure 1. A model of neurotransmitter coordinated glial modulation of neuronal activity. Abbreviations: T, transmitter; R, receptor; M, metabolites of arachidonic acid; DG, DAG; IP_3, InsP_3; PK, protein kinase C.

may occur at neuron-to-neuron synapses where glia form close association with neuronal terminal processes, or possibly at specialized sites of release where glia may function as postsynaptic target cells. The released T binds to receptors (R) on the surface of the glial cell. This binding initiates a transmembrane activation of phospholipase C to begin a cascade of intracellular events resulting in the elevation of intracellular Ca^{++}, activation of protein kinases, and liberation of AA and formation of its metabolites (M; both cyclooxygenase and lipoxygenase derivatives). These metabolites may then act intracellularly to alter glial function (including secretion) or after diffusing to the neuron, transcellularly to complete the communication loop from glia back to neuron. The time course of this feedback modulation of neuronal activity would not be sufficiently rapid to affect the neuronal signal that initiated the response, but could alter neuronal responses to subsequent stimulation.

Most of the data supporting this model are derived from work with glial cell lines, and reports in the literature concerning the effects of AA metabolites on neurons. Preliminary experiments using polygonal astrocytes prepared from neonatal rat brain cerebral cortex have proven promising in extending the evidence for the model to include true glia. In the light of these observations we believe that the model outlined above warrants further consideration as one mechanism by which glia may regulate neuronal activity and influence neurotransmission.

REFERENCES

Abdel-Halim MS, von Holst H, Meyerson B, Sachs C and Änggard E (1980) Prostaglandin profiles in tissue and blood vessels from human brain. J Neurochem 34: 1331-1333.

Baba A, Ohta A and Iwata H (1983) Inhibition of depolarization-induced acetylcholine release and calcium influx in rat brain cortical synaptosomes. J Neurochem 40: 1758-1761.

Bazan NG (1971) Changes in free fatty acids of brain by drug-induced convulsions, electroshock and anesthesia. J Neurochem 18: 1379-1389.

Bergstrom S, Farnebo L-O and Fuxe K (1973) Effect of prostaglandin E_2 on central and peripheral catecholamine neurons. Eur J Pharmacol 21: 362-368.

Bernhein HA, Gilbert TM and Stitt JT (1980) Prostaglandin E levels in third ventricular cerebrospinal fluid of rabbits during fever and changes in body temperature. J Physiol 301: 69-78.

Berridge MJ (1984) Inositol trisphosphate and diacylglycerol as second messengers. Biochem J 220: 245-360.

Berridge MJ, Downes CP and Hanley MR (1982) Lithium amplifies agonist-dependent phosphatidylinositol responses in brain and salivary glands. Biochem J 206: 587-595.

Birkle DL and Ellis EF (1983) Conversion of arachidonic acid to cyclooxygenase and lipoxygenase products, and incorporation into phospholipids in the mouse neuroblastoma clone, neuro-2A. Neurochem Res 8: 319-330.

Bishai I and Coceani F (1980) Metabolism of PGH_2 in feline brain. In: Samuelsson B, Ramwell PW and Paoletti R (eds): Advances in Prostaglandins and Thromboxane Research, Vol. 8. Raven Press, New York, pp. 1221-1223.

Bradford PG, Marinetti GV and Abood LG (1983) Stimulation of phospholipase A_2 and secretion of catecholamines from brain synaptosomes by potassium and A23187. J Neurochem 41: 1684-1693.

Briggs CA, Horwitz J, McAff DA, Tsymbalov S and Perlman RL (1985) Effects of neuronal activity on inositol phospholipid metabolism in the rat autonomic nervous system. J Neurochem 44: 731-739.

Brown E, Kendall DA and Nahorski SR (1984) Inositol phospholipid hydrolysis in rat cerebral cortical slices: I. Receptor characterisation. J Neurochem 42: 1379-1387.

Chan PH and Fishman RA (1982) Alterations of membrane integrity and cellular constituents by arachidonic acid in neuroblastoma and glioma cells. Brain Res 248: 151-157.

Clark RB and Perkins JP (1971) Regulation of adenosine 3':5'-cyclic monophosphate concentration in cultured human astrocytoma cells by catecholamines and histamine. Proc Nat Acad Sci USA 68: 2757-2760.

Daum PR, Downes CP and Yound JM (1984) Histamine stimulation of inositol 1-phosphate accumulation in lithium-treated slices from regions of guinea pig brain. J Neurochem 43: 25-32.

DeGeorge JJ, Morell P, McCarthy K, and Lapetina EG (1984) Cholinergic stimulation of glial cell arachidonic acid and phosphatidic acid metabolism. Soc Neurosci Abs 10: 1181.

DeGeorge JJ, Morell P, McCarthy K, and Lapetina EG (1985) Dexamethasone blocks cholinergic stimulated liberation of arachidonate, but not the rapid increase in inositol phosphates in C62B glioma cells. J Neurochem 44: 551.

Groppi VE Jr, and Browning ET (1980) Norepinephrine-dependent protein phosphorylation in intact C-6 glioma cells. Molec Pharmacol 18: 427-437.

Hedqvist P (1973) In: Ramwell PW (ed): The Prostaglandis, Vol. 1. Plenum Press, New York/London, pp 101-129.

Hillier K and Templeton WW (1980) Regulation of noradrenaline overflow in rat cerebral cortex by prostaglandin E_2. Br J Pharmac 70: 469-473.

Jonsson HT and Daniel HB (1976) Altered levels of PGF in cat spinal cord tissue following traumatic injury. Prostaglandins 11: 51-61.

Kimura H, Okamoto K and Sakai Y (1985) Modulatory Effects of Prostaglandin D_2, E_2 and $F_{2\alpha}$ on the postsynaptic actions of inhibitory and excitatory amino-acids in cerebellar Purkinje cell dendrites in vitro. Brain Res 330: 235-244. ,

Lapetina EG (1983) IV Metabolism of inositides and the activation of platelets. Life Sci 32: 2069-2082.

Leslie CA (1976) Prostaglandin biosynthesis and metabolism in rat brain slices. Res Commun Chem Pathol Pharmacol 14: 455.

Lindgren JA, Hulting A, Dahlén S-E, Hökfelt T, Werner S and Samuelsson B (1984) Evidence for the occurrence of leukotrienes (LT) in the central nervous system and for a neuroendocrine role of LT. Soc Neurosci Abs 10: 1129.

Lunt GG and Rowe CE (1971) The effect of cholinergic substances on the production of unesterified acids in brain. Brain Res 35: 215-220.

Mallorga P, Tallman JF, Henneberry RC, Hirata F, Strittmatter WT and Axelrod J (1980) Mepacrine blocks β-adrenergic agonist-induced desensitization in astrocytoma cells. Proc Natl Acad Sci USA 77: 1341-1345.

Masters SB, Harden TK and Brown JH (1984) Relationships between phosphoinositide and calcium responses to muscarinic agonists in astrocytoma cells. Molec Pharmacol 26: 149-155.

McCarthy KD and de Vellis J (1978) Alpha-adrenergic receptor modulation of beta-adrenergic, adenosine and prostaglandin increased adenosine 3':5'-cyclic monophosphate levels in primary cultures of glia. J Cyclic Nucleotide Res 4: 15-26.

McCarthy KD, Prime J, Harmon T and Pollenz R (1985) Receptor-mediated phosphorylation of astroglial intermediate filament proteins in cultured astroglia. J Neurochem 44: 723-730.

Nishizuka Y (1984) Turnover of inositol phospholipids and signal transduction. Science 225: 1365-1370.

Palmer MR, Mathews WR, Hoffer BJ and Murphy RC (1981) Electrophysiological response of cerebellar Purkinje neurons to leukotriene D_4 and B_4. J Pharmacol Exp Ther 219: 91-96.

Rittenhouse-Simmons S (1984) Activation of human platelet phospholipase C by ionophore A23187 is totally dependent upon cyclo-oxygenase products and ADP. Biochem J 222: 103-110.

Rougon G, Noble M and Mudge AW (1983) Neuropeptides modulate the β-adrenergic response of purified astrocytes in vitro. Nature 305: 715-717.

Snider RM, McKinney M, Forray C and Richelson E (1984) Neurotransmitter receptors mediate cyclic GMP formation by involvement of arachidonic acid and lipoxygenase. Proc Natl Acad Sci USA 81: 3905-3909.

Spagnuolo C, Sautebin L, Galli R, Racagni G, Galli C, Mazzari S and Finesso M (1979) $PGF_{2\alpha}$, thromboxane B_2 and HETE levels in gerbil brain cortex after ligation of common carotid arteries and decapitation. Prostaglandins 18: 53-61.

Van Calker D, Loffler F and Hamprecht B (1983) Corticotropin peptides and melanotropins elevate the level of adenosine 3':5'-cyclic monophosphate in cultured murine brain cells. J Neurochem 40: 418-427.

White, HL and Stine L (1984) Arachidonate lipoxygenase activity in rat brain synaptosomal preparations. Soc Neurosci Abs 10: 1130.

Wolfe LS (1982) Eicosanoids: Prostaglandins, thromboxanes, leukotrienes, and other derivatives of carbon-20 unsaturated fatty acids. J Neurochem 38: 1-14.

Yano K, Higashida H, Hattori H and Nozawa Y (1985) Bradykinin-induced transient accumulation of inositol trisphosphate in neuron-like cell line NG108-15 cells. FEBS Letts. 181: 403-406.

Zucker MB and Nachmias VT (1985) Platelet activation. Arteriosclerosis 5: 2-18.

Phospholipid research and the nervous system
Biochemical and molecular pharmacology
L.A. Horrocks, L. Freysz, G. Toffano (eds)
Fidia Research Series, vol. 4.
Liviana Press, Padova. © 1986

DRUGS AFFECTING MEMBRANE LIPID CATABOLISM: THE BRAIN FREE FATTY ACID EFFECT

Elena B. Rodriguez de Turco *

Instituto de Investigaciones Bioquimicas, Universidad Nacional del Sur (INIBIBB), Consejo Nacional de Investigaciones Cientificas y Tecnicas (CONICET), Bahia Blanca, Argentina

INTRODUCTION

In the past 15 years, the brain "free fatty acid effect" has become a well-recognized phenomenon, elicited by a variety of stimuli, including: electroconvulsive shock (ECS) (Bazan 1970; Bazan and Rakowski, 1970; Aveldano and Bazan, 1979), drug-induced convulsions (Bazan, 1971; Marion and Wolfe, 1978; Siesjö et al., 1982; Rodriguez de Turco et al., 1983), anoxia and ischemia (Bazan, 1970; Cenedella et al., 1975; Aveldano and Bazan, 1975a; Rehncrona et al., 1982; Shiu et al., 1983; Yoshida et al., 1983), and cryogenic brain injury (Chan et al., 1983; Bazan et al., 1984). An increase in membrane phospholipid catabolism is the earliest response to brain injuries (Bazan, 1976). Furthermore, diacylglycerols (DAG) are simultaneously accumulated during ischemia (Aveldano and Bazan, 1975b) and convulsions (Rodriguez de Turco et al., 1983). The largest contribution to increases in both free fatty acids (FFA) and DAG is the accumulation of stearic (18:0) and arachidonic (20:4) acids. Based upon this observation, it has been suggested that the breakdown of phospholipids enriched in these fatty acids (i.e. inositol lipids) is involved (Aveldano and Bazan, 1975b).

The cellular and/or subcellular level where these lipid effects occur and the detailed events leading to enzyme activation remain poorly defined. Moreover, the simultaneous accumulation of FFA and DAG with similar fatty acid profiles, have led

Abbreviations used: FFA, free fatty acids; ECS, electroconvulsive shock; 20:4, arachidonic acid; 18:0, stearic acid

* Present Address: Louisiana State University Medical Center, Department of Physiology, 1901 Perdido Street, New Orleans, LA 70112-3098 U.S.A.

to the suggestion that "...the produced DAG and FFA originate in separate sources or are related through sequential reactions involving DAG-lipases..." (Aveldano and Bazan, 1975b). The stimulation of phospholipases A_2 and C may be coupled to receptor activation during convulsions. A decreased activity of acylation systems due to energy failure may also contribute to an increase in FFA (Pediconi et al., 1985). However, FFA release (prior to detection of a measurable ATP decrease) has been reported during bicuculline-induced convulsions (Siesjö et al., 1982) and severe hypoxia (Gardiner et al., 1981). This suggests that the phospholipases are very important, at least at the earliest times of insult, to the accumulation of FFA in brain.

Over the past several years, there has been an increased interest in these lipid changes, because of the involvement of 20:4 and DAG in cell signaling mechanisms. The release of 20:4, through phospholipase activation, is the rate-limiting step in eicosanoid synthesis (Flower and Blackwell, 1976). DAG can activate a specific protein kinase, protein kinase C, following the receptor-mediated, enhanced turnover of inositol lipids (Nishizuka, 1984).

This chapter describes some pharmacological and biochemical aspects of FFA and DAG accumulation in the central nervous system during convulsions. Recent studies to be presented are: 1) comparative changes in the accumulation of FFA and DAG triggered by ECS and bicuculline-induced convulsions, 2) dexamethasone-induced retention of lipid changes due to ECS, 3) involvement of muscarinic receptor activation in brain FFA and DAG accumulation after ECS, and 4) lateralization with respect to cerebral hemispheres in the FFA response after ECS.

ACCUMULATION OF FREE FATTY ACIDS AND DIACYLGLYCEROLS TRIGGERED BY CONVULSIONS

Electroconvulsive shock or a single intraperitoneal injection of bicuculline induce a rapid accumulation of FFA and DAG at the onset of tonic-clonic convulsions (Bazan and Rodriguez de Turco, 1983). Despite differences in the mechanism involved in neuronal discharge, the amount and composition of fatty acid release is quite similar between the two stimuli (Table 1). Arachidonic and stearic acids represent approximately 70% of the total FFA accumulated in a 1:1 molar relationship. Palmitic and oleic acids are present in a similar ratio, although the amount released in both cases is less. Although DAG accumulation is 2.5-fold higher after bicuculline-induced convulsions than after seizures induced by ECS, the fatty acid profiles are similar. DAG containing 18:0 and 20:4 are the principal molecular species accumulated and represent 65-70% of the total DAG.

DEXAMETHASONE DECREASES ECS-INDUCED FREE FATTY ACID AND DIACYLGLYCEROL ACCUMULATION

To gain insight into the enzymatic mechanism involved in membrane phospholipid breakdown, the effect of dexamethasone on the ECS-induced release of FFA and DAG was explored. Glucocorticoids induce the synthesis of a phospholipase-inhibitory pro-

Table 1. *Accumulation of free fatty acids and diacylglycerols in rat brain after bicuculline - or electroconvulsive shock-induced convulsions*

Fatty acid	Free fatty acids		Diacylglycerols	
	Bicuculline* (6)	ECS (4)	Bicuculline* (6)	ECS (4)
16:0	32.9 ± 12.9	24.5 ± 6.3	34.6 ± 15.1	16.9 ± 1.8
18:1	25.4 ± 7.0	20.1 ± 6.1	18.4 ± 6.4	6.8 ± 0.3
18:0	108.9 ± 14.0	88.2 ± 5.3	62.5 ± 17.8	23.7 ± 2.1
20:4	101.7 ± 3.4	77.4 ± 1.8	66.4 ± 15.5	25.0 ± 2.6
22:6	2.5 ± 0.9	1.2 ± 0.2	5.1 ± 1.5	1.7 ± 0.2
TOTAL	266.3 ± 32.6	221.1 ± 10.1	187.3 ± 51.7	76.3 ± 4.6
16:0/18:1	1.3	1.2	1.9	2.5
18:0/20:4	1.1	1.1	0.9	0.9

Data are expressed as nmol fatty acids (experimental values minus basal values) per 100 mg protein; mean values ± S.D. Number of samples are given in parentheses. Bicuculline was injected intraperitoneally (10 mg/kg body weight). ECS was applied using platinum needle electrodes placed under the skin on the temporal bone. Unidirectional rectangular pulses (0.5 msec at 45 V) were applied at 7 msec intervals over a 1 sec period.
*Data taken from Rodriguez de Turco et al., 1983.

tein (lipocortin) in non-neural tissue (Flower and Blackwell, 1979). Dexamethasone pretreatment exerts an inhibitory action on ECS-induced lipid changes (Fig. 1). The accumulation of FFA and DAG is inhibited by 40% and 30%, respectively. In both FFA and DAG, stearic and arachidonic acids show the greatest inhibition. In contrast, 24 hr after cryogenic brain injury dexamethasone inhibits free 20: 4 accumulation by 100%, while the accumulation of 20:4 in DAG is affected only minimally (Politi et al., 1985). The difference in effects of dexamethasone may be due to the different cellular focus of the two insults (cryogenic injury vs. ECS), which could result in a variety of effects to the different cell types.

BLOCKAGE OF MUSCARINIC CHOLINERGIC RECEPTORS BY ATROPINE INHIBITS THE ECS-INDUCED DEGRADATION OF LIPIDS CONTAINING STEARIC AND ARACHIDONIC ACIDS

Stimulation of muscarinic cholinergic receptors in brain synaptosomes has been linked to the phosphodiesteratic cleavage of polyphosphoinositides (Fisher and Agranoff, 1981). An increased turnover of these lipids results in the transient accumulation of DAG. To evaluate the possible involvement of muscarinic receptors in the ECS-induced accumulation of 20:4-DAG, rats were pretreated with atropine [160 mg/kg body weight, injected intraperitoneally (Friedel and Schanberg, 1972)] 30 min prior to ECS. The blockage of muscarinic receptors results in the selective inhibition (30%) of DAG containing 18:0 and 20:4 (Fig. 2). Also, the accumulation of free 18:0 and 20:4 is inhibited to a similar degree. These results suggest that after ECS muscarinic receptor activation is coupled to the selective breakdown of membrane lipids containing 18:0 and 20:4 (e.g. inositol lipids). A similar atropine-mediated inhibition of bicuculline induced inositol lipid degradation, has been reported to occur in rat brain in vivo (Vadnal et al., 1985).

60

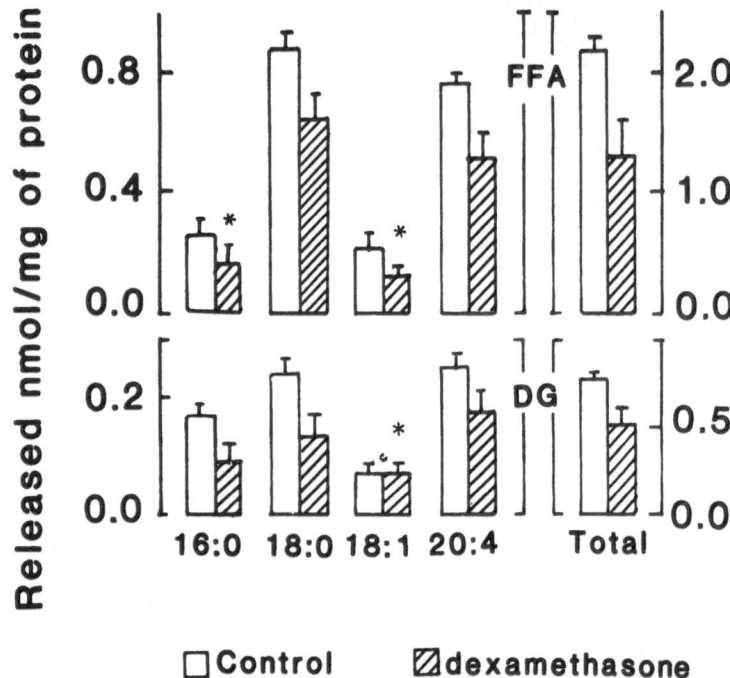

Figure 1. The inhibitory effect of dexamethasone on the electroconvulsive shock-induced accumulation of free fatty acids (FFA) and diacylglycerols (DG) in rat brain. Values were calculated by subtracting the basal from the experimental total contents. Asterisk denotes differences that are NOT statistically significant from untreated animals. Animals received an intraperitoneal injection of dexamethasone (1.25 mg per kg body weight) once every 12 hr for 48 hr prior to electroconvulsive shock.

LATERALIZATION OF FREE FATTY ACID CONTENT AND RELEASE AFTER ELECTROCONVULSIVE SHOCK IN RAT BRAIN HEMISPHERES

The endogenous content of FFA exhibits a differential distribution between the left and right cerebral hemispheres in the mouse (Pediconi and Rodriguez de Turco, 1984) and rat (Ginobile de Martinez et al., 1985). The right hemisphere displays a lower basal content and a higher capacity to release fatty acids, during ischemia (Pediconi and Rodriguez de Turco, 1984) and after ECS (Table 2). In addition, the right hemisphere has a greater capacity for re-establishing basal FFA levels. This is reflected in the recovery phase (300 sec) after ECS (Table 2). The stimulation of phospholipases A_2 and A_1 during the tonic convulsion may be involved in the fast release of FFA (10 sec after ECS), which is followed by a much slower recovery of pre-stimulation levels (300 sec after ECS) mediated by acyl CoA synthetase and transferases. A long-lasting disturbance in the equilibrium of these enzymatic systems occurs after repetitive ECS treatments (Table 2). Twenty-four hr after application of four ECS (at a rate of one ECS per 24 hr), FFA levels are increased 30% in both hemispheres with 20:4 showing the greatest change (260% of control).

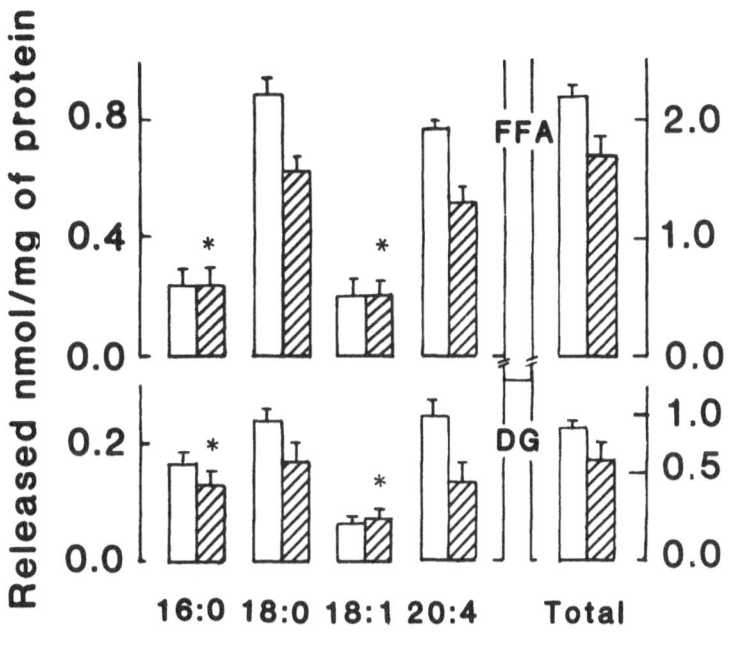

□ Control   ▨ Atropine treated

Figure 2. Lower accumulation of free fatty acids and diacylglycerols in the brain of rats pretreated with atropine. Data represent the difference between experimental and basal values. Asterisk denotes differences that are NOT statistically significant from untreated animals. Rats received one intraperitoneal injection of atropine (160 mg per kg body weight) 30 min prior to ECS.

Table 2. *Release and removal of free fatty acids after electroconvulsive shock in the right and left cerebral hemispheres of the rat brain*

Condition	Time Between Last ECS and Killing	Right Hemisphere		Left Hemisphere	
		20:4	Total FFA	20:4	Total FFA
Control	--	5.3± 1.6	110.8± 2.4	5.7±0.5	149.3± 6.7[a]
1 ECS Treatment	10 sec	94.1±10.8	348.7±19.3	68.7±5.2[a]	312.4±19.5
	300 sec	11.8± 2.2	137.4±13.0	15.8±1.2[a]	242.9± 6.7[a]
4 ECS Treatments	24 hr	13.6± 1.5[c]	151.4± 2.2[c]	14.3±1.4[c]	190.3± 4.2[ac]
5 ECS Treatments	10 sec	76.2± 3.9[b]	334.7±•18.5	69.7±4.5	322.9± 6.6
	300 sec	14.6± 0.5[b]	153.5±19.0	7.9±1.1[ab]	288.6±43.5[a]

Data represent nmol free fatty acid per 100 mg protein ± S.D. from four samples. Controls for the 5-ECS group were subjected to four successive daily shocks and killed 24 hr after receiving the last shock (4-ECS). Statistically significant differences: a = between right and left hemispheres (p-values from <0.001 to <0.01); b = between 1-ECS and 5-ECS (p-values from <0.025 to <0.001); and c = 24 hr after 4 ECS and control values ($p<0.001$). Data taken from Ginobile de Martinez et al., 1985. Other details as in Table 1.

Regardless of the differences in basal levels when the first or fifth ECS treatments are given, a similar final level of FFA is seen in each hemisphere during the tonic convulsion. Hence, a lower basal level results in a higher ECS-induced FFA release and vice versa. Arachidonic acid is the only FFA showing a 30% higher concentration in the right hemisphere than the left after the first ECS, and the lateralized response is abolished after successive ECS treatments (Table 2).

CONCLUSIONS

Increased turnover of acyl groups in membrane phospholipids is the earliest response triggered by stimulation of the central nervous system. This is reflected in a rapid accumulation of FFA and DAG, enriched in 20:4. This is a transient phenomenon, compatible with life (Bazan and Rakowski, 1970). A sustained lipid breakdown occurs during repeated convulsions or ischemia and may play a central role in biochemical alterations leading to brain damage.

Under steady-state conditions a very low FFA and DAG concentration is maintained (about 0.1% of the total acyl groups of phospholipids). During convulsions the accumulation of FFA and DAG is equivalent to the degradation of about 0.2% of the total phospholipids (Fig. 3). When considering the accumulation of 20:4 and 20:4-DAG,

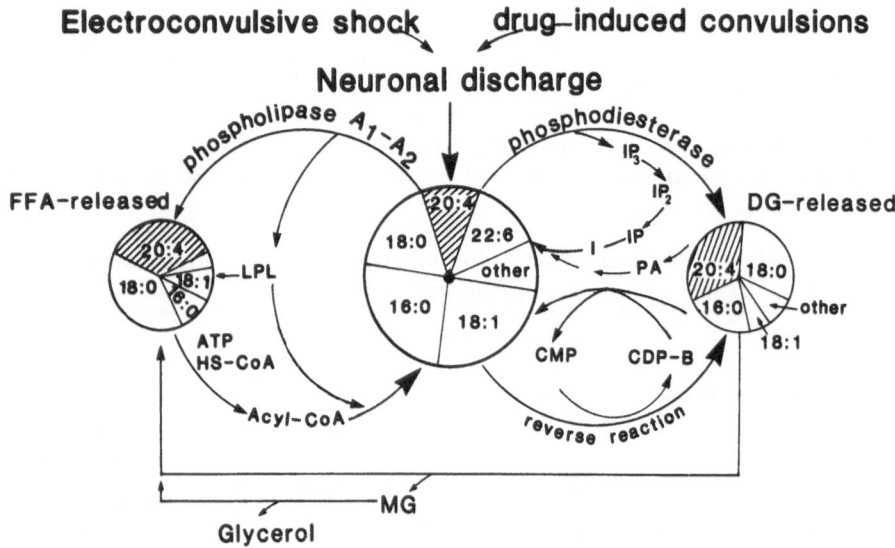

Figure 3. Enzymatic pathways involved in the selective degradation of membrane phospholipids enriched in arachidonic acid upon neurotransmitter discharge. Central circle: percent distribution of esterified acyl groups in total phospholipids from rat brain. Black dot in middle of central circle represents the labile phospholipid pool susceptible to degradation during convulsions. Lateral circles: percent composition of FFA and DAG-acyl groups released upon stimulation. CDP-B = CDP base; I = Ins; PA = PtdH; DG = DAG.

with respect to the total esterified 20:4, the magnitude of change is greater (0.8%). The preferential accumulation of 20:4 and 20:4-DAG reflects a selective degradation of 20:4-containing lipids. This hypothesis is supported by the convulsion-induced increase in the eicosanoid synthesis (Zatz et al., 1975; Folco et al., 1977; Marion and Wolfe, 1978; Berchtold-Kanz et al., 1981), which requires the selective release of 20:4.

Arachidonic acid release may be a consequence of the sequential activation of phospholipase C and DAG and monoacylglycerol lipases or the activation of phospholipase A_2 (Fig. 3). A muscarinic receptor-mediated activation of phospholipase A_2 has been shown in neuroblastoma cells in culture (Snider et al., 1984) and in non-neuronal cells (Blackwell et al., 1977; van de Bosch, 1980; Susuki et al., 1980; Billah and Lapetina, 1980; Bormann et al., 1984). The atropine-induced decrease in accumulation of 20:4 and 20:4-DAG in ECS (Fig. 2) supports the suggestion that accumulation of both FFA and DAG may occur simultaneously and may be linked to muscarinic receptor activation. Muscarinic stimulation of nerve endings results in a minimal contribution from the inositide cycle intermediates to free 20:4 accumulation (van Roijen et al., 1985); however, 20:4-DAG accumulates. The recovery to basal FFA and DAG levels after ECS suggests that there is no precursor-product relationship between DAG and FFA. Moreover, while DAG levels approach basal levels within 1 min after ECS (unpublished observations), FFA levels recover much more slowly, reaching pre-stimulation levels in 5 min (Table 2). All these observations add support to the contention that FFA and DAG accumulation are related lipid effects occurring simultaneously.

The close relationship between FFA and DAG accumulation also is apparent in dexamethasone pretreated animals. This drug induced a similar inhibition in both FFA and DAG accumulation induced by ECS. This suggests that either the drug can directly inhibit phospholipase A_2 and C and/or that the inhibition of one (e.g. phospholipase A_2) results directly or indirectly in a low activity of the other. This is the first evidence showing a glucocorticoid effect on phospholipases in the central nervous system.

The differential response of FFA between cerebral hemispheres in response to ischemia (Pediconi and Rodriguez de Turco, 1984) and ECS (Ginobile de Martinez et al., 1985) and the normal differences in distribution of FFA between cerebral hemispheres may be a manifestation of asymmetry in the neurochemical mechanisms involved in fatty acid metabolism. Differences in the activity of enzymes committed to the regulation of FFA turnover could result in a lower basal content and a higher FFA release in the right, as opposed to the left, hemisphere. In addition, lateralization in deacylation-reacylation reactions of complex lipids could be secondary to differences in the neurotransmitter systems, known to be distributed asymmetrically in the brain (Zimmerberg et al., 1974; Glick and Jerussi, 1974).

Chronic stimulation of the brain results in a higher steady-state level of FFA and a lower response to ECS in both hemispheres. This observation raises the possibility of long-lasting metabolic perturbations, affecting lipids in close proximity to membrane receptors. The physiological consequences of a new metabolic equilibrium towards a higher FFA level could be the first event, at the membrane level, leading to receptor down-regulation. A decreased density in β-adrenergic receptors (Bergstrom and Kellar, 1979; Pandley et al., 1979) and muscarinic cholinergic receptors (Lerer et al., 1983) has been demonstrated after repeated ECS. Phospholipase A_2 has been implicated in down-regulation of muscarinic and other receptors (Haigler et al., 1980; Higuchi et al.,

1983; Mallorga et al., 1980; Torda et al., 1981). Inasmuch as a net increase in phospholipase A_2 activity seems to be unnecessary (Higuchi et al., 1983), the possibility remains that a decrease in the acylation of fatty acids may be involved. This would result in an imbalance in phospholipid acyl group turnover, with a consequent increase in FFA and lysophospholipids.

Although these studies have been done in complex structures, the entire cerebrum or brain hemispheres, our present results suggest an active participation of specific phospholipids enriched in arachidonic acid. This appears to be the earliest response of the brain to stimulation. In addition to short-term lipid effects, the sustained alteration in the metabolism of glycerolipids following chronic stimulation could affect signaling mechanisms and physiological responses.

ACKNOWLEDGMENTS

I would like to express my sincere thanks and appreciation to Dr. Nicolas G. Bazan for his 11 years of support and encouragement with INIBIBB. I also would like to thank the many individuals who have collaborated with us over the past 15 years at INIBIBB.

Special thanks to Celia Carucci and Carlos Nugenser for their excellent technical assistance.

REFERENCES

Aveldano MI, Bazan NG (1975a) Differential lipid deacylation during brain ischemia in a homeotherm and a poikilotherm. Content and composition of free fatty acids and triacylglycerols. Brain Res 100: 99-110.

Aveldano MI, Bazan NG (1975b) Rapid production of diacylglycerols enriched in arachidonate during early brain ischemia. J Neurochem 25: 919-920.

Aveldano MI, Bazan NG (1979) Alpha-methyl-p-tyrosine inhibits the production of free arachidonic acid and diacylglycerols in brain after a single electroconvulsive shock. Neurochem Res 4: 213-221.

Bazan NG (1970) Effect of ischemia and electroshock on free fatty acid pool in the brain. Biochim Biophys Acta 218: 1-10.

Bazan NG (1971) Changes in free fatty acids of brain by drug-induced convulsions, electroshock and anesthesia. J Neurochem 18: 1379-1385.

Bazan NG (1976) Free arachidonic acid and other lipids in the nervous system during early ischemia and after electroshock. In: Porcellati G, Amaducci L, Galli C (eds): Function and Metabolism of Phospholipids in the Central and Peripheral Nervous System. Plenum Press, New York, pp. 317-335.

Bazan NG, Politi LE, Rodriguez de Turco EB (1984) Endogenous pools of arachidonic acid-enriched membrane lipids in cryogenic brain edema. In: Go KG, Baethmann A (eds): Recent Progress in the Study and Therapy of Brain Edema. Plenum Publishing Corp, New York, pp. 203-212.

Bazan NG, Rodriguez de Turco EB (1983) Seizures promote breakdown of membrane phospholipids in the brain. In: Caputto R, Ajmone Marsan C (eds): Neuronal Transmission, Learning and Memory. Raven Press, New York, pp. 187-194.

Bazan NG, Rakowski H (1970) Increased levels of free fatty acids after electroconvulsive shock. Life Sci 9: 501-507.

Berchtold-Kanz E, Anhut H, Heldt R, Neufang B, Hertting G (1981) Regional distribution of arachidonic acid metabolites in rat brain following convulsive stimuli. Prostaglandins 22: 65-79.

Bergstrom DA, Kellar KJ (1979) Effect of electroconvulsive shock on monoaminergic receptor binding sites in rat brain. Nature 278: 464-466.

Billah MM, Lapetina EG (1982) Formation of lysophosphatidylinositol in platelets stimulated with thrombin or Ionophore A-23187. J Biol Chem 257: 5196-5200.

Blackwell GJ, Duncombe WG, Flower RJ, Parsons MF, Vane JR (1977) The distribution and metabolism of arachidonic acid in rabbit platelets during aggregation and its modification by drugs. Br J Pharmacol 59: 353-366.

Bormann BJ, Huang C-K, Mackin WM, Becker EL (1984) Receptor-mediated activation of a phospholipase A_2 in rabbit neutrophil plasma membrane. Proc Natl Acad Sci 81: 767-770.

Cenedella RJ, Galli C, Paoletti R (1975) Brain free fatty acid levels in rats sacrificed by decapitation versus focused microwave irradiation. Lipids 10: 290-293

Chan PH, Longar S, Fishman RA (1983) Phospholipid degradation and edema development in cold-injured rat brain. Brain Res 277: 329-337.

Fisher SK, Agranoff BW (1981) Enhancement of the muscarinic synaptosomal phospholipid labeling effect by the ionophore A-23187. J Neurochem 37: 968-977.

Flower RJ, Blackwell GJ (1976) The importance of phospholipase A_2 in prostaglandin biosynthesis. Biochem Pharmacol 25: 285-291.

Flower RJ, Blackwell GJ (1979) Anti-inflammatory steroids induce biosynthesis of a phospholipase A_2 inhibitor which prevents prostaglandin generation. Nature 278: 456-459.

Folco GC, Longiave D, Bosisio E (1977) Relations between prostaglandin E_2, $F_{2\alpha}$ and cyclic nucleotide levels in rat brain and induction of convulsions. Prostaglandins 13: 893-900.

Friedel RO, Schanberg SM (1972) Effects of carbamylcholine and atropine on incorporation in vivo of intracisternally injected ^{33}Pi into phospholipids of rat brain. J Pharmacol Exp Ther 183: 326-332.

Gardiner M, Nilsson B, Rehncrona S, Siesjö B (1981) Free fatty acids in the rat brain in moderate and severe hypoxia. J Neurochem 36: 1500-1505.

Ginobile de Martinez MS, Rodriguez de Turco EB, Barrantes FJ (1985) Asymmetric distribution and differential effects of electroconvulsive shock on lipids of rat cerebral hemispheres. Dominance of the right hemisphere in free fatty acid metabolism. Brain Res 339: 315-322.

Glick SD, Crane AM, Jerussi TP, Fleisher LN, Green JP (1975) Functional and neurochemical correlates of potentiation of striatal asymmetry by callosal section. Nature 254: 616-617.

Haigler HT, Willingham MC, Pastan I (1980) Inhibitors of ^{125}I-epidermal growth factor internalization. Biochem Biophys Res Comm 94: 630-637.

Higuchi H, Uchida S, Matsumoto K, Yoshida H (1983) Inhibition of agonist-induced degradation of muscarinic cholinergic receptor by quinacrine and tetracaine - Possible involvement of phospholipase A_2 in receptor degradation. Eur J Pharmacol 94: 229-239.

Lerer B, Stanley M, Demetriou S, Gershon S (1983) Effect of electroconvulsive shock on muscarinic cholinergic receptors in rat cerebral cortex and hippocampus. J Neurochem 41: 1680-1683.

Mallorga P, Tallman JF, Henneberry RC, Hirata F, Strittmatter WT, Axelrod J (1980) Mepacrine blocks beta-adrenergic agonist-induced desensitization in astrocytoma cells. Proc Natl Acad Sci USA 77: 1341-1345.

Marion J, Wolfe LS (1978) Increase in vivo of unesterified fatty acids, prostaglandin $F_{2\alpha}$ but not thromboxane B_2 in rat brain during drug-induced convulsions. Prostaglandins 16: 99-110.

Nishizuka Y (1984) The role of protein kinase C in cell surface signal transduction and tumor promotion. Nature 308: 693-698.

Pandley GN, Heinze WJ, Brown BD, Davis JM (1979) Electroconvulsive shock treatment decreases beta-adrenergic receptor sensitivity in rat brain. Nature 280: 234-235.

Pediconi MF, Rodriguez de Turco EB (1984) Free fatty acid content and release kinetics as manifestation of cerebral lateralization in mouse brain. J Neurochem 43: 1-7.

Pediconi MF, Rodriguez de Turco EB, Bazan NG (1985) Reduced labeling of brain phosphatidylinositol, triacylglycerols and diacylglycerols by [1-^{14}C]arachidonic acid after electroconvulsive shock. Potentiation of the effect by adrenergic drugs and comparison with palmitic acid labeling. Neurochem Res 11: 217-230.

Politi LE, Rodriguez de Turco EB, Bazan NG (1985) Dexamethasone effect on free fatty acid and diacylglycerol accumulation during experimentally-induced vasogenic brain edema Neurochem Pathol 3: 249-269.

Rehncrona S, Westerberg E, Åkesson B, Siesjö B (1982) Brain cortical fatty acids and phospholipids during and following complete and severe incomplete ischemia. J Neurochem 38: 84-93.

Rodriguez de Turco EB, Morelli de Liberti SA, Bazan NG (1983) Stimulation of free fatty acid and diacylglycerol accumulation in cerebrum and cerebellum during bicuculline-induced status epilepticus. Effect of pretreatment with α-methyl-p-tyrosine and p-clorophenylalanine. J Neurochem 40: 252-259.

Schiu GK, Nemmer JP, Nemoto EM (1983) Reassessment of brain free fatty acid liberation during global ischemia and its attenuation by barbiturate anesthesia. J Neurochem 40: 880-884.

Siesjö BK, Ingvar M, Westerberg E (1982) The influence of bicuculline-induced seizures on free fatty acid concentration in cerebral cortex, hippocampus and cerebellum. J Neurochem 39: 796-802.

Snider RM, McKinney M, Forray C, Richelson E (1984) Neurotransmitter receptors mediate cyclic GMP formation by involvement of arachidonic acid and lipoxygenase. Proc Natl Acad Sci USA 81: 3905-3909.

Torda T, Yamaguchi I, Hirata F, Kopin IJ, Axelrod J (1981) Quinacrine-blocked desensitization of adrenoceptors after immobilization stress or repeated injection of isoproterenol in rats. J Pharmacol Exp Ther 216: 334-338.

Vadnal R, Van Rooijen LAA, Bazan NG (1985) Effect of atropine on bicuculline-induced changes in inositol lipid metabolism in rat brain in vivo. J Neurochem, in press

Van den Bosch H (1980) Intracellular phospholipase A$_2$. Biochim Biophys Acta 604: 191-246.

Van Rooijen LAA, Hajra AK, Agranoff BW (1985) Tetraenoic species are conserved in muscarinically enhanced inositide turnover. J Neurochem 44: 540-543.

Yoshida S, Inoh S, Asano T, Sano K, Shimasaki H, Ueta N (1983) Brain free fatty acids, edema, and mortality in gerbils subjected to transient, bilateral ischemia, and the effect of barbiturate anesthesia. J Neurochem 40: 1278-1286.

Zatz M, Roth RH (1975) Electroconvulsive shock raises prostaglandins F in rat cerebral cortex. Biochem Pharmacol 24: 2101-2103.

Zimmerberg B, Glick SD, Jerussi TP (1974) Neurochemical correlates of a spatial preference in rat. Science 185: 623-625.

Phospholipid research and the nervous system
Biochemical and molecular pharmacology
L.A. Horrocks, L. Freysz, G. Toffano (eds)
Fidia Research Series, vol. 4.
Liviana Press, Padova. © 1986

UNIQUE METABOLIC FEATURES OF DOCOSAHEXAENOATE METABOLISM RELATED TO FUNCTIONAL ROLES IN BRAIN AND RETINA

Haydee E.P. Bazan*, Brock Ridenour, Dale L. Birkle and Nicolas G. Bazan

Louisiana State University Medical School, LSU Eye Center
and Eye, Ear Nose and Throat Hospital,
New Orleans, Louisiana

INTRODUCTION

To define the physiological significance of the highly unsaturated phospholipids in photoreceptor and synaptic membranes, a better understanding of the metabolic pathways involved is required. Some of the aspects to be defined include the regulation of fatty acid metabolism by physiological stimuli (e.g. light), pathological alterations that occur in diseases (e.g. retinal degenerations, seizures, ischemia, neuronal degenerations, etc.) and precursor-product relationships (e.g. the role of the dietary supply of essential fatty acids).

In the retina, molecular species of glycerolipids with 65% of their total acyl groups being docosahexaenoic acid (22:6) have been isolated (Aveldano and Bazan, 1977, 1983; Aveldano et al., 1983). Moreover, a proportion of these molecular species seems to take the form of phospholipids with two 22:6 molecules acylated to the same glycerol backbone (termed supraenoic glycerolipids). These species could be synthesized by deacylation-reacylation or by de novo routes. There are several experimental findings that point to the synthesis de novo of 22:6-containing glycerolipids. The supraenoic species have a high rate of synthesis as measured by the incorporation of radiolabeled glycerol (Aveldano de Caldironi and Bazan, 1980). In bovine retinal microsomes, a key intermediate in synthesis de novo, phosphatidic acid (PtdH), contains 21% of the total esterified 22:6 (Giusto and Bazan, 1979) and cationic amphiphilic drugs, such as propranolol, increase the rate of synthesis of 22:6-PtdH (Bazan et al., 1976b, 1981, 1982;

* Address reprint requests and correspondence to Dr. H.E.P. Bazan, LSU Eye Center, 136 S. Roman St., New Orleans, LA 70112.

Bazan 1982a, 1983; Bazan and Giusto, 1980; Giusto et al., 1983; Ilincheta de Boschero and Bazan, 1982, 1983). The first metabolic step for acylation of fatty acids into glycerolipids, synthesis of coenzyme A (CoA) derivatives (e.g. docosahexaenoyl-CoA), is very active in the retina and brain (Reddy and Bazan, 1984a, 1984b, 1985a, 1985b; Reddy et al., 1984). In bovine retinal microsomes, 22:6 is acylated during the de novo synthesis of PtdH, followed by channeling to other lipids (Bazan et al., 1984a). Furthermore, [1-14C]eicosapentaenoic acid (20:5, n-3) is elongated and desaturated to 22:6, and incorporated into PtdH in the rat retina in vivo (Bazan et al., 1982a).

This chapter will describe some of the metabolic pathways for 22:6 in the retina and brain, with emphasis on the possible physiological significance of the active metabolism and avid retention of this fatty acid in photoreceptor and synaptic membranes. A summary scheme of the possible routes for the metabolism of 22:6 is shown in Figure 1.

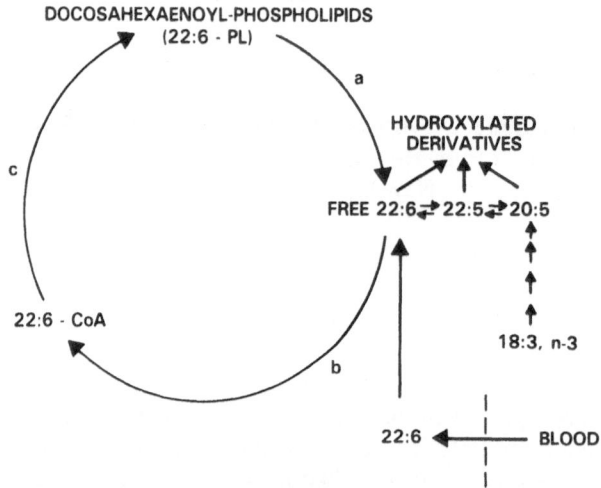

Figure 1. Docosahexaenoic acid (22:6) metabolism, a = phospholipase A₂, giving rise to lysophospholipid or monoacylphosphoglyceride (not shown) and free 22:6; b = docosahexaenoyl-coenzyme A synthetase; c = acyltransferase. 18:3, n-3, the precursor of docosahexaenoic acid, is connected to 20:5 with arrows that represent elongation and desaturation reactions. Hydroxylated derivatives are products of lipoxygenase activity.

ACYLATION OF 22:6 INTO RAT BRAIN LIPIDS IN VIVO

Anesthetized rats were injected intraventricularly and bilaterally with 1 μCi radiolabeled 22:6 (prepared as a complex with bovine serum albumin) and sacrificed by decapitation into liquid nitrogen at 15, 30, 60 or 120 min after injection. The frozen brains were pulverized and the lipids extracted with chloroform:methanol (2:1 by vol.). Approximately 20% of the injected radiolabel was recovered in the brain. Figure 2 shows

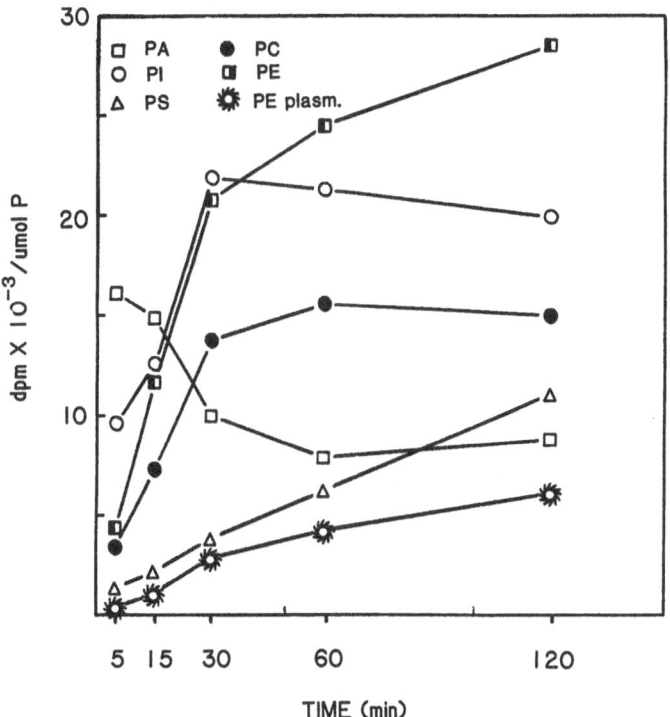

Figure 2. Incorporation of [1-14C]22:6 into rat brain phospholipids. Rats were injected intraventricularly with radiolabeled fatty acid and sacrificed at the times indicated. Date are calculated as specific activity (dpm/μmol phosphorus in each individual phospholipid). Values are the mean of three or more separate experiments (standard error less than 10% of the mean).

the incorporation of 22:6 into rat brain phospholipids. Phosphatidylethanolamine (PtdEtn) and phosphatidylcholine (PtdCho) incorporated the largest percentage of the label. The labeling of PtdCho and phosphatidylinositol (PtdIns) increased from 15 min to one hour and then leveled off. The specific activity of PtdH and diacylglycerols (data not shown) peaked at 15 min, indicating rapid acylation of these key intermediates in glycerolipid biosynthesis. Triacylglycerol labeling remained constant throughout the time period studied (data not shown), while the specific activity of PtdEtn and phosphatidylserine (PtdSer) increased. Further analysis of the PtdEtn fraction revealed that about 30% of this lipid was ethanolamine plasmalogen (PlasEtn). The specific activity of PlasEtn also increased over the time course. Total methanolysis of the lipid extract and analysis of the fatty acid methyl ester by high performance liquid chromatography (Fig. 3) showed that more that 95% of the radioactivity was recovered as 22:6, indicating no major retroconversion of the fatty acid to docosapentaenoic acid (22:5) or 20:5 during the time course studied. The incorporation of radiolabeled 22:6 followed basically the same pattern as the endogenous distribution of this fatty acid in brain glycerolipids.

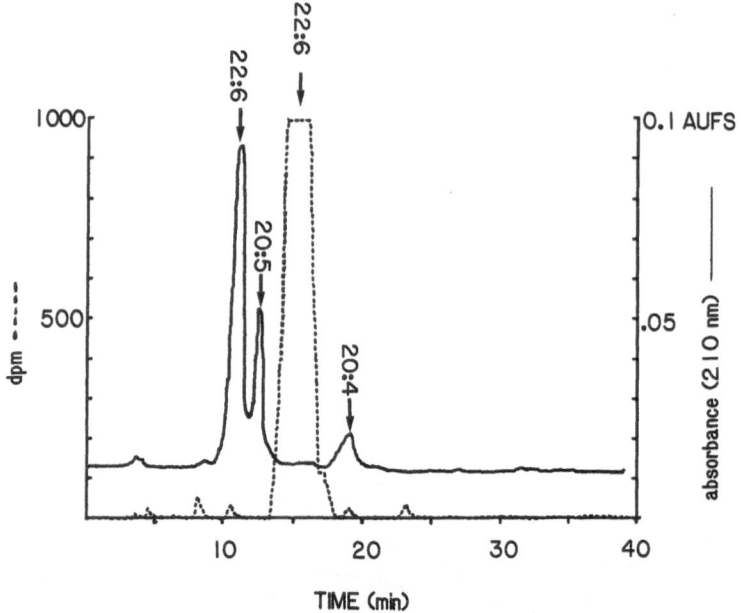

Figure 3. High performance liquid chromatographic analysis of total fatty acid methyl esters from rat brain prelabeled in vivo with [1-¹⁴C]22:6. Methyl esters were separated by reverse phase chromatography. Endogenous 22:6, 20:5 and arachidonic acid (20:4) were seen by ultraviolet detection (solid line), while the only detected radiolabeled fatty acid was 22:6 (dashed line).

IN VIVO ACYLATION OF DOCOSAHEXAENOIC ACID IN THE EYE

Twenty-four hours after injection of [1-¹⁴C]22:6 into the vitreous of the frog, 49% of the total radioactivity was found in the retina (Fig. 4). Inner segment microsomes are a subcellular fraction obtained from the P_1 pellet of the retina after hypoosmotic shock and differential centrifugation. This fraction, which is derived from the visual cell bodies, displayed the most incorporation of 22:6.

When rat retinas were labeled in vivo with [1-¹⁴C]22:6 (Table 1), most of the radio-label was incorporated into PtdEtn, PtdCho, and PlasEtn. This in vivo labeling pattern agrees with the distribution of endogenous 22:6 in these phospholipids. Several years ago, Dorman et al. (1977) reported that PlasEtn accounts for 26% of the ethanolamine fraction in the calf retina. The acyl group composition of these plasmalogens has not been studied in retina. The composition of the phospholipid fraction shown in Table 1 includes the composition of the plasmalogen fraction, because in this particular study the two lipids were not separated. PtdSer did not show a pattern of incorporation similar to the endogenous content of docosahexaenoic acid. We have observed previously (Bazan et al., 1984a) that retinal microsomes have a very low acylation activity towards PtdSer, possibly due to the lack of some required soluble factor in the assay.

When the time course of [1-¹⁴C]22:6 incorporation into individual lipid classes was

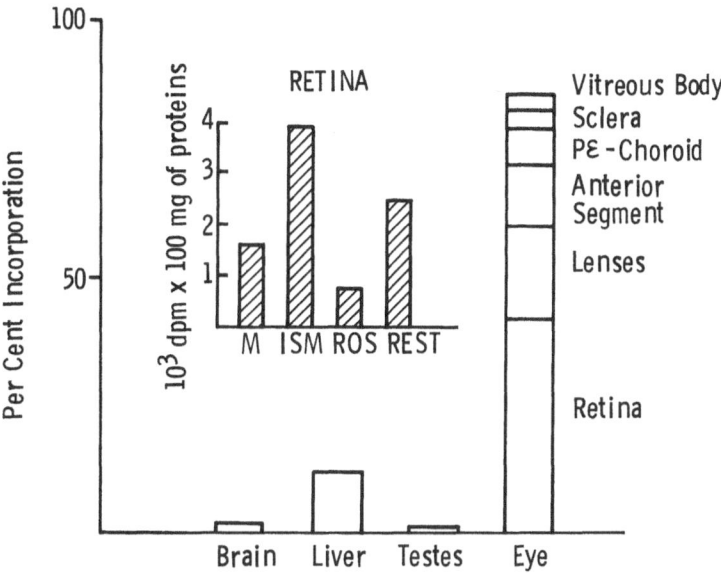

DISTRIBUTION OF LIPID LABELING
24 hrs. AFTER AN INTRAVITREAL
INJECTION OF ^{14}C-22.6 IN THE FROG

Figure 4. Distribution of the labeling of total lipids after intravitreal injection of [^{14}C]22:6 in the frog. Frogs were adapted for 15 days on a dark-light cycle (14 hr dark, 10 hr light) and kept at 27°C. Frogs were injected intravitreally with 0.5 μCi of [U-^{14}C]docosahexaenoate-Na$^+$ salt (sp. act. 0.1 μCi/19.4 μmol). Frogs were kept on the dark-light cycle for 24 hr prior to decapitation. Tissues were removed and homogenized in chloroform:methanol (2:1 by vol), lipids were extracted and radioactivity was determined. M = neural microsomes; ISM = inner segment microsomes; ROS = rod outer segments; REST = all other retinal subcellular fractions; PE = pigment epithelium.

followed in the rat eye in vivo, there was a very rapid incorporation in PtdH (Fig. 5A); the early peak of labeling of PtdH preceeded the appearance of this fatty acid in other phospholipids. At longer time points (Fig. 5B), the highest incorporation was found in PtdEtn.

It has been shown previously (Anderson et al., 1980) that in frog retina there is an active synthesis of PtdEtn via decarboxylation of PtdSer. In the rat retina, despite the high concentration of hexaenoic molecular species in PtdSer, there was not an active acylation of 22:6 in this phospholipid and at longer times PtdEtn showed the highest acylation activity for this fatty acid. In previous experiments in vivo, we have observed a different relationship in the acylation of PtdEtn to PtdSer in different frog retina subcellular fractions. The ratio was 8 for rod outer segments, 1.3 for microsomes and 0.32 for pigment epithelium (unpublished results).

These results were in agreement with previous studies (Bazan et al., 1984a; Bazan and Bazan, 1985), indicating that 22:6 may be introduced during biosynthesis de novo of PtdH. This pathway yields docosahexaenoyl-PtdH, which is channeled to other phospholipids

Table 1. *Content of 22:6 and incorporation of [1-¹⁴C]-22:6 in rat retinal lipids in vivo*

Lipid	Incorporation (%)	Levels of 22:6 composition
PtdH	0.5 ± 0.08	15**
PtdSer	1.4 ± 0.03	42.4*
PtdIns	2.7 ± 0.1	4
PtdCho	30 ± 3	16.7*
PtdEtn	33 ± 2.9	42.2*
PlasEtn	11 ± 0.5	nd
PlasCho	2.8 ± 0.4	nd
DG	7.5 ± 1.1	7.1***
TG	9.2 ± 1.9	17.2***

Rat retinas were labeled in vivo by injecting 1 μCi of [1-¹⁴C]-22:6 intravitreally in both eyes. The animals were killed 90 minutes after the injection. The values are expressed as mole %.
nd: values not determined
*Data taken from Anderson and Maude, 1972.
**Bovine retinal lipids (Bazan et al., 1981b)
***Bovine retinal lipids (Aveldano and Bazan, 1974)
PlasEtn: ethanolamine plasmalogens, PlasCho: choline plasmalogens; DG: diacylglycerol; TG: triacylglycerol.

through a 22:6-diacylglycerol intermediate. A small portion, if any, of the docosahexaenoyl-PtdH pool could be expected to be channeled through CDP-diacylglycerol; however this route is likely to be minor because the phosphoinositides and disphosphatidylglycerol (cardiolipin) contain very low quantities of 22:6. This new proposed route is an alternative to the deacylation-reacylation cycle that removes a fatty acyl chain from the second carbon of the glycerol backbone by phospholipase A_2, and reacylates a polyunsaturated fatty acylchain (e.g. docosahexaenoyl or arachidonoyl).

DESATURATION OF DOCOSAPENTAENOIC ACID (22:5, n-3)

There is also an active desaturation of polyunsaturated fatty acids in the rat retina in vivo after intravitreal injection of [1-¹⁴C]22:5. Retinal lipid extracts were methanolyzed and radiolabeled fatty acid methyl esters were analyzed by high performance liquid chromatography with flow scintillation detection. Seven days after injection, more than 30% of the precursor was converted to 22:6 (Table 2).

THE FREE 22:6 POOL

The free fatty acid pool is very small in the retina and brain (Aveldano and Bazan, 1972, 1973, 1974b, 1981, Bazan et al., 1981b; Bazan 1979, 1982b, Bazan et al., 1976a), but increases rapidly in the brain after ischemia (Bazan, 1970; Aveldano and Bazan, 1974a, 1975), electroconvulsive shock (Bazan, 1970, 1971) and drug-induced convulsions (Bazan et al., 1982b), and in the retina during anoxia (Giusto and Bazan, 1983). One of the principal free fatty acids released is 22:6 and, as a possible source for the released fatty acid, it has been proposed that membrane phospholipids could be deacylated through the activation of phospholipase A_2 (Bazan, 1970).

Recently, a new pathway for the oxygenation of free 22:6 has been described in canine retinas incubated in vitro with [U-14C]22:6 (Table 3). Several products were isolated by high performance liquid chromatography (Bazan et al., 1984b, 1985a) with the primary product identified as 11-hydroxydocosahexaenoic acid (Peak 1, Table 3). The synthesis of these compounds was inhibited 50% by 10 μM nordihydroguaiaretic acid, a lipoxygenase inhibitor and antioxidant, and stimulated in neuronal degenerative

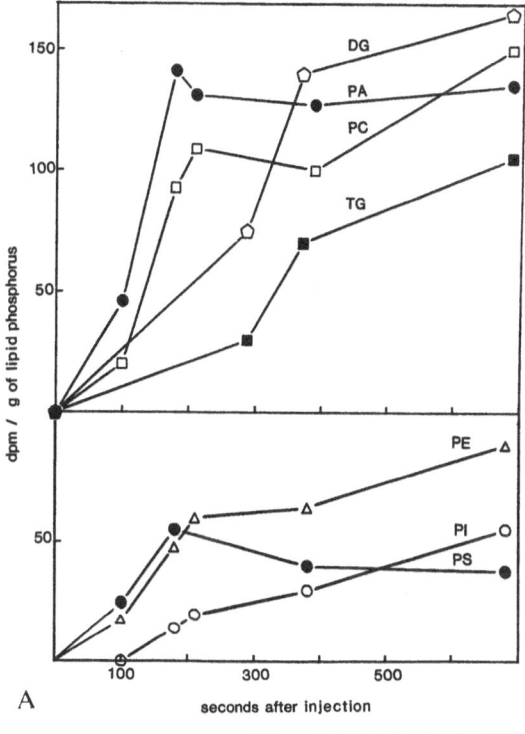

Figure 5. Time course of [1-14C]22:6 in phospholipids of the rat retina in vivo. A: Early time course in sec; B: Later time course in min. DG = diacylglycerol; PA = phosphatidic acid; PC = phosphatidylcholine; TG = triacylglycerol; PE = phosphatidylethanolamine; PI = phosphatidylcholine; PS = phosphatidylserine.

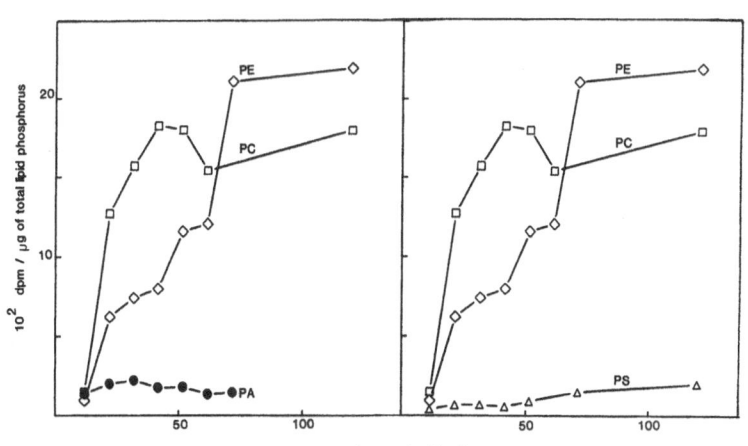

Table 2. *Elongation of 22:5 in the rat retina in vivo*

%o of Conversion		
90 min	3 days	7 days
0.91 ± 0.09	11.73 ± 0.93	34.38 ± 7.11

The values represent the mean ± SE of four individual samples.
Rat retinas were labeled in vivo by intravitreal injection of [1-14C]-22:5 in both eyes.

Table 3. *Oxidative metabolites of [1-14C]-22:6 in canine retina*

Compound	Retention Time (min)	%o of total products
PEAK I	27	61
PEAK II	29	16
PEAK III	35	10
PEAK IV	45	12

Dog retinas were incubated in Ames buffer with 1 μCi of [1-14C]22:6 for one hr at 37°C and the incubation medium was extracted and analyzed by high pressure liquid chromatography. This table was recalculated from data given in Bazan et al., 1984b.

disease, canine ceroid lipofuscinosis (Reddy et al., 1985). Very little is known about the oxidative metabolites of 22:6. Lipid peroxidation in rat retina produces a decreased amplitude in the electroretinogram (Doly et al., 1984). Recently, enzymatic oxidation of 22:6 also has been shown in human platelets (Aveldano and Sprecher, 1983) and in rat liver microsomes (Van Rollins et al., 1984), although the function of these metabolites remains to be elucidated.

CONCLUSION

We have discussed in this chapter some new findings about the metabolism of 22:6 in vivo in the retina and brain. The activation-acylation system is very active in these tissues, and 22:6 is acylated to a significant extent during the de novo synthesis of PtdH. The retina also possesses a highly active elongation-desaturation system to convert 20:5 to 22:6. This implies that the retina is able to synthesize 22:6 from blood-borne precursors, or precursors provided by other retinal cells, such as the retinal glia (Mueller cells). It is not known if glial cells in the nervous system support neurons through the metabolism of essential fatty acids. The presence of the elongation-desaturation pathway also implies retroconversion of 22:6 to 20:5, which is the precursor of prostaglandins of the 3-series (e.g. PGE_3). These eicosanoids, which have been isolated from platelets, and vascular endothelial cells, antagonize the aggregatory, pro-thrombotic activity of the 2-series eicosanoids formed from 20:4 (e.g. thromboxane A_2). In retina and brain, the conversion of 20:5 to eicosanoids has not been investigated.

Docosahexaenoic acid is derived from α-linolenic acid (18:3, n-3), which animals are unable to synthesize. Little is known about the function of 22:6 except the fact that phospholipids of photoreceptor and synaptic membranes are enriched in this acyl group (Fliesler and Anderson, 1983; Bazan and Reddy, 1984). The electroretinogram (Wheeler et al., 1975), visual acuity (Neuringer et al., 1984), and learning ability (Lamptey and Walker, 1976) are impaired when there is a decrease in docosahexaenoate in retina and brain due to a dietary deficiency of the linolenic acid series. Prolonged dietary deprivation of 18:3 is needed to decrease the tissue level of 22:6 (Futterman et al., 1971; Anderson and Maude, 1972; Tinoco et al., 1977), because retina and brain tenaciously retain this fatty acid. The mechanism of this conservation is not known. It has been proposed that thiol ester formation, i.e. conversion of lipid-soluble free 22:6 to water-soluble docosahexaenoyl-CoA, may be important in trapping the fatty acid inside the cell, in a fashion analogous to the intracellular trapping of glucose by phosphorylation (Reddy and Bazan, 1985a, b). It is likely that 22:6 or some structurally similar fatty acid plays an important role in the function of photoreceptor membranes, because depletion of 22:6 is attained only when 18:3-deficient diets are supplemented with linoleic acid (18:2, n-6) (Tinoco et al., 1977). With this dietary regimen, docosapentaenoate (22:5, n-6), the most similar fatty acid to 22:6 that the retina can produce, increases in glycerolipids.

Although active membrane biogenesis takes place in the visual cells to support the ontological development (Papermaster and Schneider, 1982) and the continous renewal of outer segments (Papermaster and Schneider, 1982; Basinger et al., 1976), no biochemical or autoradiographic information is available about how 22:6 is utilized in this process. These data are important to understand the pathways in the retina for the synthesis of docosahexaenoyl-phospholipids and to understand why this fatty acid is so avidly retained in photoreceptor and synaptic membranes. During the renewal of rod outer segments, 22:6 may be taken up by the phagolysosomal system of the pigment epithelium. How 22:6 is recycled back to the visual cell is not known. In this regard, an interphotoreceptor retinol binding protein containing noncovalently bound 22:6 has been described recently. The suggestion was made that this protein may mediate the retention of 22:6 in the retina by shuttling the fatty acid from the pigment epithelium to the visual cell (Bazan et al., 1985b).

The discovery of the further metabolism of 22:6 (Bazan et al., 1984b), resulting in the formation of lipoxygenase-reaction products, opens up the search for physiological roles for these metabolites. The importance of these pathways in the retina is not known, but one of the possibilites is that these metabolites (docosanoids) act as intracellular mediators between the retina and pigment epithelium, perhaps to signal the onset of rod outer segment shedding.

ACKNOWLEDGMENTS

The authors wish to acknowledge the support of the National Institutes of Health, National Eye Institute grants EY04428 and EY07073, and the Ernest C. and Yvette C. Villere Chair for Research in Retinal Degeneration.

REFERENCES

Anderson RE, Kelleher PA, Maude MB (1980) Metabolism of phosphatidylethanolamine in the frog retina. Biochim Biophys Acta 620: 227-235.

Anderson RE, Maude MB (1972) Lipids in ocular tissues VIII. The effects of essential fatty acid deficiency on the phospholipids of the photoreceptor membranes of rat retina. Arch Biochem Biophys 15: 270-276.

Aveldano MI, Bazan NG (1972) High content of docosahexaenoate and of total diacylglycerol in retina. Biochem Biophys Res Comm 48: 689-693.

Aveldano MI, Bazan NG (1973) Fatty acid composition and level of diacylglycerols and phosphoglycerides in brain and retina. Biochim Biophys Acta 296: 1-9.

Aveldano MI, Bazan NG (1974a) Displacement into incubation medium by albumin of highly unsaturated retina free fatty acids arising from membrane lipids. FEBS Lett 40: 53-56.

Aveldano MI, Bazan NG (1974b) Free fatty acids, diacyl- and triacylglycerols and total phospholipids in vertebrate retina: Comparison with brain, choroid and plasma. J Neurochem 23: 1127-1135.

Aveldano MI, Bazan NG (1975) Different lipid deacylation during brain ischemia in a homeotherm and a poikilotherm. Content and composition of free fatty acids and triacylglycerols. Brain Res 100: 99-110.

Aveldano de Caldironi MI, Bazan NG (1977) Acyl groups, molecular species and labeling by ^{14}C-glycerol and ^{3}H-arachidonic acid of vertebrate retina glycerolipids. Adv Exp Med Biol 83: 397-404.

Aveldano de Caldironi MI, Bazan NG (1980) Composition and biosynthesis of molecular species of retina phosphoglycerides. Neurochem Internat 1: 381-392.

Aveldano MI, Bazan NG (1983) Molecular species of phosphatidylcholine,-ethanolamine, -serine and -inositol in microsomal and photoreceptor membranes of bovine retina. J Lipid Res 24: 620-627.

Aveldano MI, Sprecher H (1983) Synthesis of hydroxy fatty acids from 4,7,10,13,16,19-[1-^{14}C] docosahexaenoic acid by human plateletes. J Biol Chem 258: 9339-9343.

Aveldano de Caldironi MI, Giusto NM, Bazan NG (1981) Polyunsaturated fatty acids of the retina. Prog Lipid Res 20: 49-57.

Aveldano MI, Pasquare de Garcia SJ, Bazan NG (1983) Biosynthesis of molecular species of inositol, choline, serine, and ethanolamine glycerophospholipids in the bovine retina. J Lipid Res 24: 628-638.

Basinger S, Bok D, Hall M (1976) Rhodopsin in the rod outer segment plasma membrane. J Cell Biol 69: 29-31.

Bazan HEP, Bazan NG (1985) Metabolism of docosahexaenoyl groups in phosphatidic acid, and in other phospholipids of the retina. In: Horrocks L, Kanfer J, Porcellati G (eds): Phospholipids in the Nervous System, Vol 2: Physiological Roles. Raven Press, New York, pp 209-217.

Bazan HEP, Careaga MM, Bazan NG (1981a) Propranolol increases the biosynthesis of phosphatidic acid, phosphatidylinositol and phosphatidylserine in the toad retina. Studies in the entire retina and subcellular fractions. Biochim Biophys Acta 666: 63-71.

Bazan HEP, Marcheselli VL, Careaga MM, Bazan NG (1981b) Biosynthesis and metabolism of essential and acidic phospholipids in the central nervous system. In: Bazan NG, Paoletti R, Iacono J (eds): New Trends in Nutrition, Lipid Research and Cardiovascular Diseases. Alan R Liss, New York, pp 101-1120.

Bazan HEP, Careaga MM, Sprecher H, Bazan NG (1982a) Chain elongation and desaturation of eicosapentaenoate to docosahexaenoate and phospholipid labeling in the rat retina in vivo. Biochim Biophys Acta 712: 123-128.

Bazan HEP, Sprecher H, Bazan NG (1984a) De novo biosynthesis of docosahexaenoyl phosphatidic acid in bovine retinal microsomes. Biochim Biophys Acta 796: 11-19.

Bazan NG (1970) Effects of ischemia and electroconvulsive shock on free fatty acid pool in the brain. Biochim Biophys Acta 218: 1-10.

Bazan NG (1971) Changes in free fatty acids of brain by drug-induced convulsions, electroshock and anesthesia. J Neurochem 18: 1379-1385.

Bazan NG (1982a) Biosynthesis of phosphatidic acid and polyenoic phospholipids in the central nervous system. In: Horrocks LA, Ansell GB, Porcellati G (eds): Phospholipids in the Nervous System Vol. I: Metabolism. Raven Press, New York, pp 49-62.

Bazan NG (1982b) Metabolism of phospholipids in the retina. Vision Res 22: 1539-1548.

Bazan NG (1983) Metabolism of phosphatidic acid. In: Lajtha A (ed): Handbook of Neurochemistry, Vol. 3. Plenum Pub, New York, pp 17-39.

Bazan NG, Giusto NM (1980) Docosahexaenoyl chains are introduced in phosphatidic acid during de novo synthesis in retinal microsomes. In: Kates M, Kuksís A (eds): Control of Membrane Fluidity. Humana Press, New Jersey, pp 223-236.

Bazan NG, Reddy TS (1985) Retina. In: Lajtha A (ed): Handbook of Neurochemistry, Vol. 8. Plenum Press, New York, pp. 507-575.

Bazan NG, Aveldano MI, Bazan HEP, Giusto MN (1976a) Metabolism of retina acylglycerides and arachidonic acid. In: Paoletti R, Porcellati G, Jacini G (eds): Lipids, Vol. I. Raven Press, New York, pp 89-97.

Bazan NG, Ilincheta de Boschero MG, Giusto NM, Bazan HEP (1976b) De novo glycerolipid biosynthesis in the toad and cattle retina. Redirecting of the pathway by propranolol and phentolamine. Adv Exp Med Biol 72: 139-149.

Bazan NG, Aveldano de Caldironi MI, Giusto NM, Rodriguez de Turco EB (1981) Phosphatidic acid in the central nervous system. Prog Lipid Res 20: 307-313.

Bazan NG, di Fazio de Escalante MS, Careaga MM, Bazan HEP, Giusto NM (1982) High content of 22:6 (docosahexaenoate) and active [2-^{3}H]glycerol metabolism of phosphatidic acid from photoreceptor membranes. Biochim Biophys Acta 712: 702-706.

Bazan NG, Rodriguez de Turco EB, Morelli de Liberti SA (1982b) Free arachidonic acid and membrane lipids in the central nervous system during bicuculline-induced status epilepticus. Adv Neurology 34: 305-310.

Bazan NG, Birkle DL, Reddy TS (1984b) Docosahexaenoic acid (22:6, n-3) is metabolized to lipoxygenase reaction products in the retina. Biochem Biophys Res Comm 125: 741-747.

Bazan NG, Birkle DL, Reddy TS (1985a) Biochemical and nutritional aspects of the metabolism of polyunsaturated fatty acids and phospholipids in experimental models of retinal degeneration. In: LaVail MM, Anderson G, Hollyfield J (eds): Retinal Degeneration: Contemporary Experimental and Clinical Studies. Alan R. Liss, Inc., New York, pp 159-187.

Bazan NG, Reddy TS, Redmond TM, Wiggert B, Chader GJ (1985b) Endogenous fatty acids are covalently and non-covalently bound to interphotoreceptor retinoid-binding protein in the monkey retina. J Biol Chem 260: 13677-13680.

Birkle DL, Bazan NG (1986) The arachidonic acid cascade and phospholipid and decosahexaenoic acid metabolism in the retina. In: Osborne NO, Chader G (eds): Progress in Retinal Research. Pergamon Press, London, pp. 309-335.

Doly M, Braguet P, Bonhomme B, Meyniel G (1984) Effects of lipid peroxidation on the isolated rat retina. Ophthalmic Res 16: 292-296.

Dorman RV, Dreyfus H, Freysz L, Horrocks LA (1977) Ether lipid content of phosphoglycerides from the retina and brain of chicken and calf. Biochim Biophys Acta 486: 55-59.

Fliesler SY, Anderson RE (1983) Chemistry and metabolism of lipids in the vertebrate retina. Prog Lipid Res 22: 79-131.

Futterman S, Downer JL, Hendrickson A (1971) Effect of essential fatty acid deficiency on the fatty acid composition, morphology and electroretinographic response of the retina. Inv Ophthalmol 10: 151-154.

Giusto NM, Bazan NG (1979) Phosphatidic acid of retinal microsomes contains a high proportion of docosahexaenoate. Biochem Biophys Res Comm 91: 791-794.

Giusto NM, Bazan NG (1983) Anoxia-induced production of methylated and free fatty acids in retina, cerebral cortex and white matter. Comparison with triglycerides and with other tissues. Neurochem Pathol 1: 17-41.

Giusto NM, Ilincheta de Boschero MG, Bazan NG (1983) Accumulation of phosphatidic acid in microsomes from propranolol-treated retinas during short-term incubations. J Neurochem 40: 563-568.

Ilincheta de Boschero MG, Bazan NG (1982) Selective modification in the de novo biosynthesis of retinal phospholipids and glycerides by propranolol or phentolamine. Biochem Pharmacol 31: 1049-1055.

Ilincheta de Boschero MG, Bazan NG (1983) Reversibility of propranolol-induced changes in the biosynthesis of monoacylglycerol, diacylglycerol, triacylglycerol, and phospholipids in the retina. J Neurochem 40: 260-266.

Lamptey MA, Walker BL (1976) A possible essential role for dietary linolenic acid in the development of the young rat. J Nutr 106: 86-96.

Neuringer M, Connor WE, Van Petten C, Bastard L (1984) Dietary omega-3 fatty acid deficiency and visual loss in infant rhesus monkeys. J Clin Invest 73: 272-275.

Papermaster DS, Schneider BG (1982) Biosynthesis and morphogenesis of outer segment membranes in vertebrate cell. In: McDewitt DS (ed): Cell Biology of the Eye. Academic Press, Inc, New York; pp. 475-531.

Reddy TS, Bazan NG (1984a) Activation of polyunsaturated fatty acids by rat tissues in vitro. Lipids 19: 987-989.

Reddy TS, Bazan NG (1984b) Synthesis of arachidonoyl coenzyme A and docosahexaenoyl coenzyme A in retina. Curr Eye Res 3: 1225-1232.

Reddy TS, Bazan NG (1985a) Synthesis of arachidonoyl coenzyme A and docosahexaenoyl coenzyme A in synaptic plasma membranes of cerebrum, cerebellum and brain stem of rat brain. J Neurosci Res 13: 381-390.

Reddy TS, Bazan NG (1985b) Synthesis of docosahexaenoyl-, arachidonoyl- and palmitoyl-coenzyme A in ocular tissues. Exp Eye Res 41: 87-95.

Reddy TS, Sprecher H, Bazan NG (1984) Long chain acyl coenzyme A synthetase from rat brain microsomes: Kinetic studies using [1-^{14}C]docosahexaenoic acid substrate. Eur J Biochem 145: 21-29.

Reddy TS, Birkle D, Armstrong D, Bazan NG (1985) Change in content, incorporation and lipoxygenation of docosahexaenoic acid in retina and retinal pigment epithelium in canine ceroid lipofuscinosis. Neurosci Lett 59: 67-72.

Tinoco J, Miljanich P, Medwadowski B (1977) Depletion of docosahexaenoic acid in retinal lipids of rats fed a linolenic acid-deficient, linoleic acid-containing diet. Biochim Biophys Acta 486: 575-580.

Van Rollins M, Baker R, Sprecher HW, Murphy RC (1984) Oxidation of docosahexaenoic acid by rat liver microsomes. J Biol Chem 259: 5779-5783.

Wheeler TG, Benolken RM, Anderson RE (1975) Visual membranes, Specifics of fatty acid precursors for the electrical response to illumination. Science 188: 1312-1314.

Phospholipid research and the nervous system
Biochemical and molecular pharmacology
L.A. Horrocks, L. Freysz, G. Toffano (eds)
Fidia Research Series, vol. 4.
Liviana Press, Padova. © 1986

PHARMACOLOGICAL AGENTS AND THE RELEASE OF ARACHIDONIC ACID AND PROSTAGLANDIN BIOSYNTHESIS BY RABBIT IRIS MUSCLE

Ata A. Abdel-Latif, Shokofeh Naderi, Sardar Y.K. Yousufzai

Department of Cell and Molecular Biology Medical College of Georgia
Augusta, GA 30912, USA

INTRODUCTION

Arachidonic acid (AA), the precursor of prostaglandins (PGs) and other biological active substances, is found mainly esterified to membrane phospholipids, almost exclusively in the 2-acyl position, from which it could be released by the action of lipases (Lands and Samuelson, 1968; Vogt, 1978) and reincorporated by a deacylation-reacylation cycle (for reviews see Hill and Lands, 1970; Holub and Kuksis, 1978; Irvine, 1982).

In the unstimulated iris, most of the AA is esterified to PtdIns, PtdCho, and PtdEtn (Yousufzai and Abdel-Latif, 1985a), and exogenously added radioactive AA is taken up by the iris and rapidly incorporated into the arachidonate-containing glycerolipids (Abdel-Latif and Smith, 1982). Addition of various agonists, such as NE (Yousufzai and Abdel-Latif, 1983; Abdel-Latif et al., 1983a; Yousufzai and Abdel-Latif, 1985b) to the iris increased the release of PGE_2, measured by radioimmunoassay, and [^{14}C]AA from prelabelled irides, measured by radiochromatography, in a dose-dependent manner. While the precise mechanism for AA release from membrane phospholipids by various pharmacological agents is still unclear, a number of pathways have been proposed. These include: (a) degradation of PtdIns by a specific phospholipase C to Ins P and DG; AA is then released from the latter by the DG/monoacylglycerol pathway (Bell et al., 1979; Okazaki et al., 1981), and (b) direct deacylation by phospholipase A_2 (Flower and Blackwell, 1979; McKean et al., 1981; Walsh et al., 1981). Although several investigators have proposed a metabolic relationship between AA release and

---

[1] Abbreviations: PG, prostaglandin; PAF, platelets-activating factor; NE, norepinephrine; DG, 1, 2-diacylglycerol.

enhanced phosphoinositide turnover, the precise nature of this relationship has not yet been elucidated. This paper summarizes our recent studies on the effects of NE, ACh, PAF, and other pharmacological agents on AA release and PG synthesis in the rabbit iris.

EFFECTS OF ADRENERGIC AGENTS ON AA RELEASE AND PG SYNTHESIS

Catecholamines have been reported to stimulate PG biosynthesis in a wide variety of tissues including the rabbit iris (for review see Abdel-Latif et al., 1983b). The mechanism underlying the stimulatory effect of these amines on PG synthesis is poorly understood. It has been suggested that: (a) They act as cofactors for the cyclooxygenation of AA (b) They stimulate AA release and consequently PG synthesis through receptor-mediated mechanisms.

Structure-Activity Studies on the Cyclooxygenation of AA

Previously, we have suggested that the structural requirement for maximal catecholamine stimulation of PG synthesis by the iris is a catechol nucleus and ethylamine polar side chain (Abdel-Latif et al., 1983a). Thus the deaminated metabolites of NE had little effect on PG synthesis by the iris and iris microsomes; in constrast normetanephrine and catechol exerted significant stimulatory and inhibitory effects, respectively (Abdel-Latif et al., 1983a, 1983b). In these studies dopamine, but not DOPA, had little effect on PG synthesis in the iris. This is probably due to the fact that this catecholamine is impermeable to the plasma membrane of the smooth muscle.

The above studies demonstrate that NE and other catecholamines are taken up by the tissue where they could act as cofactors for the cyclooxygenation of AA.

Involvement of Adrenoceptors in the Release of AA and PG Synthesis

To answer the question whether adrenoceptors are involved in the NE-stimulation of PG release in the rabbit iris, we have investigated the effects of α-adrenergic blockers on NE-stimulated release of PGE_2 and AA (Tables 1 and 2, respectively). As can be

Table 1. *Effects of α-adrenergic blockers on NE-stimulated release of PGE_2 by rabbit iris*

Additions	PGE_2 (μg/g tissue)	Effect of Drug (% of control)
Control	3.06	100
50 μM NE	5.30	173
50 μM NE + 0.1 nM Prazosin	4.32	141
50 μM NE + 1 nM Prazosin	3.24	106
50 μM NE + 20 μM Yohimbine	5.18	169

In this experiment one iris from each pair was incubated in 1 ml of Krebs-Ringer buffer in the absence and presence of the pharmacological agents as indicated at 37°C for 30 min. At the end of incubation, PGE_2 in the medium was quantitated by RIA. Results are means of three different experiments and each experiment was run in triplicate.

Table 2. *Effects of prazosin on NE-stimulated release of AA from PtdIns of iris muscle prelabelled with [¹⁴C]AA and its conversion into PGE₂*

Additions	Radioactivity (dpm/iris)					
	Tissue		Medium			
	PtdIns	% control	AA	% control	PGE₂	% control
Control	945	100	4505	100	1216	100
100 μM NE	643	68	5467	121	1541	127
100 μM NE + 10 μM Prazosin	926	98	4370	97	1143	94

Two irides from the same rabbit were preincubated in 2 ml Krebs-Ringer buffer that contained 0.2 μCi [¹⁴C]AA/ml for 1 h. The prelabelled irides were washed three times with nonradioactive medium, then incubated (of the pair one was used as control) in absence and presence of the pharmacological agents as indicated for 30 min at 37° C. At the end of incubation the medium was analyzed for PGE₂ and the tissue for glycerolipids as described previously (Yousufzai and Abdel-Latif, 1984). Results are means of three different experiments.

seen from Table 1, the stimulatory effect of NE on PGE₂ release by the iris was blocked by prazosin (1 nM), but not by yohimbine. Under the same experimental conditions prazosin alone had no effect on PGE₂ release (data not shown). These data suggest that the NE-stimulated PGE₂ release by the iris is mediated through α_1-adrenoceptors.

The first step in the synthesis of PGs is the liberation of AA from membrane phospholipids. In the stimulated iris, phosphoinositides are rapidly degraded by a Ca^{2+}-activated phospholipase C (Akhtar and Abdel-Latif, 1978; Abdel-Latif et al., 1980 and for review see Abdel-Latif et al., 1985). In order to demonstrate a relationship between the phosphoinositide response and the release of AA for PG synthesis we have investigated the effects of NE and its antagonists on irides prelabelled with [¹⁴C]AA. NE increased the breakdown of PtdIns by 32%, and this was accompanied by an increase in the release of radioactive AA and consequently PGE₂ (Table 2). The stimulatory effect of NE was blocked by prazosin, thus suggesting that the release of AA from membrane phosphoinositides is controlled by α_1-adrenoceptors. Under the same experimental conditions prazosin alone had no effect on the release of AA from PtdIns (data not shown). These data indicate that activation of α_1-adrenergic receptors in the iris leads to the release of AA from membrane phosphoinositides.

Since there are two major pathways for the release of AA from PtdIns, namely a direct release by phospholipase A_2 and an indirect release by PtdIns-specific phospholipase A_2 (see Introduction above), it was of interest to determine which of the two pathways is active in the iris. To answer this question we have investigated the two possibilities by determining the effect of NE on the formation of [³H] lysoPtdIns and on the metabolism of [¹⁴C] DG in irides prelabelled with [³H]myo-inositol and [¹⁴C]AA with time. As can be seen from Fig. 1, in 10 s the NE-induced increase in the radioactivity of [³H]lysoPtdIns and increase in the metabolism of that of [¹⁴C]DG were 20 and 15%, respectively, and in 60 s the changes in these lipids were 49 and 35%, respectively. These data suggest the presence of both pathways in the iris.

EFFECTS OF CHOLINERGIC AGENTS ON AA RELEASE AND PG SYNTHESIS

To further demonstrate a relationship between the phosphoinositide response and the release of AA for PG synthesis we have investigated the effects of ACh, and its

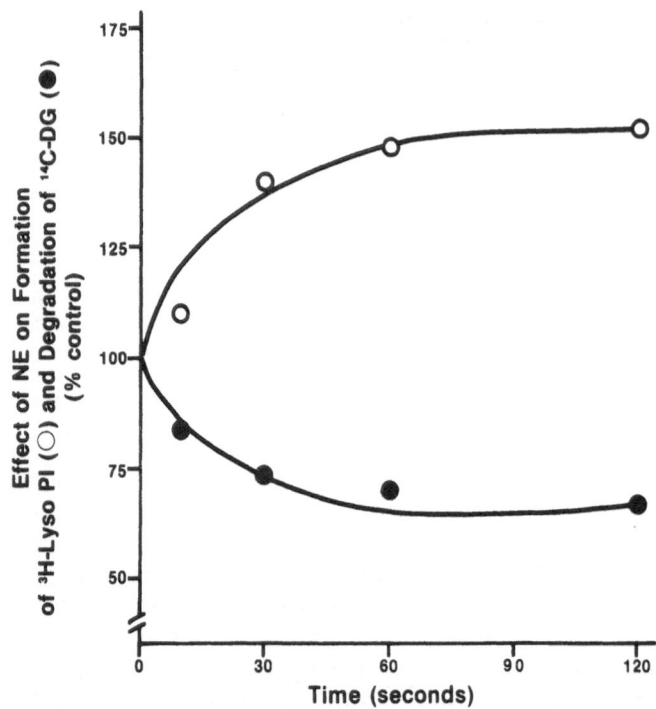

Figure 1. Time course of NE-induced formation of [³H]Lyso PI ([³H] lysoPtdIns) and metabolism of [¹⁴C]DG in iris muscle prelabelled with [³H]myo-inositol and [¹⁴C]AA. Two irides (one from each pair of eyes) were preincubated in 2 ml of Krebs-Ringer bicarbonate buffer that contained 0.2 μCi of [¹⁴C]AA and 5.7 μCi of [³H]myo-inositol for 90 min at 37°C. At the end of incubation the prelabelled irides were washed three times with nonradioactive Krebs-Ringer buffer, then incubated in 2 ml buffer in the absence and presence of NE (0.1 mM) for various time intervals as indicated. Phospholipids were extracted and lysoPtdIns and DG were separated by TLC and their radioactive contents determined as described previously (Yousufzai and Abdel-Latif, 1984). The control values (zero time) for [³H]lysoPtdIns and [¹⁴C]DG were (dpm/2 irides) 21740 ± 892 and 10506 ± 213, respectively. Values are the mean of two separate experiments, and each experiment was run in duplicate. (Taken from Yousufzai and Abdel-Latif, 1984, with permission).

antagonists in contrast to NE this cholinergic agonist does not act as a cofactor in the cyclooxygenation of AA (Yousufzai and Abdel-Latif, 1984), on the release of AA from PtdIns in irides prelabelled with [¹⁴C]AA. ACh increased the breakdown of PtdIns by 22%, and this was accompanied by an increase in the release of radioactive AA and consequently PGE_2 (Table 3). The stimulatory effect of ACh was blocked by atropine, thus suggesting that the release of AA from membrane PtdIns is controlled by muscarinic cholinergic receptors. Atropine alone had no effect on the release of AA from PtdIns (data not shown).

Again these data provide further evidence that in the iris the stimulated release of AA from PtdIns for PG synthesis is coupled to the activation of muscarinic cholinergic receptors.

Table 3. *Effects of atropine on ACh-stimulated release of radioactive AA from PtdIns of rabbit iris prelabelled with [¹⁴C]AA and its conversion into PGE₂*

Additions	Radioactivity (dpm/iris)					
	Tissue		Medium			
	PtdIns	% control	AA	% control	PGE₂	% control
Control	945	100	4505	100	1216	100
100 μM ACh	737	78	5271	117	1484	122
100 μM ACh + 10 μM Atropine	1021	108	4280	95	1082	89

Conditions of incubation were the same as those described under Table 2.

EFFECTS OF Ca^{2+}, IONOPHORE A23187, NE, AND ACh ON PGE₂ RELEASE

Several neurotransmitters and hormones have been proposed to stimulate PG synthesis by causing activation of a Ca^{2+}-dependent phospholipase (for review see Irvine, 1982). However, the precise role played by Ca^{2+} in the stimulatory action of these agents on PG synthesis is still unknown. In the present studies, the concentration of intracellular Ca^{2+} was increased by enhancing the entry rate of Ca^{2+} from incubation medium with A23187. As can be seen from Table 4, when A23187 or Ca^{2+} was added alone the increase in PGE₂ release was 24-36% of that of the control; however, when they were added together the release of PGE₂ increased by 53%. Similarly, when NE and ACh were added without Ca^{2+} there was a 37 and 22% increase in PGE₂ release, respectively; however, when the neurotransmitters and the cation were added together there was a 90 and 51% increase in PGE₂ release, respectively. These data suggest that Ca $^{2+}$ is required for maximal neurotransmitter-induced release of PGE₂ by the rabbit iris.

Table 4. *Effects of Ca^{2+}, ionophore A23187, NE, and ACh on PGE₂ release by rabbit iris*

Additions	Effects on PGE₂ Release (% of control)
Control (Ca^{2+}-free medium)	100
A23187 (20 μM)	124
Ca^{2+} (2.5 mM)	136
Ca^{2+} + A23187	153
NE (50 μM)	137
NE + Ca^{2+}	191
ACh (50 μM)	122
ACh + Ca^{2+}	151

In this experiment pairs of irides were first washed with Ca^{2+}-free Krebs-Ringer containing EGTA (0.25 mM), then with Ca^{2+}-free Krebs-Ringer buffer in the absence and presence of Ca^{2+} and the pharmacological agents added as indicated at 37°C for 30 min. At the end of incubation PGE₂ in the medium was quantitated by RIA. The control value for PGE₂ was 4.87 ± 0.376 μg/g tissue. Results are means of three different experiments.

EFFECTS OF PLATELET-ACTIVATING FACTOR
ON THE RELEASE OF AA AND PGs

Platelet-activating factor (PAF) has been reported to possess inflammatory and hypotensive properties and to provoke cellular responses in a wide variety of tissues (for reviews see Vargaftig et al., 1981; Pinckard et al., 1982). PAF induces contraction in guinea pig ileal smooth muscle (Findley et al., 1981), promotes the degradation of phosphoinositides in platelets (Shukla and Hanahan, 1982; Ieyasu et al., 1982; Mauco et al., 1983; Billah and Lapetina, 1983; MacIntyre and Pollock, 1983; Kloprogge and Akkerman, 1984), and stimulates the release of AA in platelets (Shaw et al., 1981). In platelets there is experimental evidence which suggests that PAF could exert its biological effects by increasing the permeability of membranes to Ca^{2+} (Lee et al., 1981). However, while there is a considerable amount of information about the effects of PAF on several metabolic pathways in a variety of tissues, the mechanisms of action of this bioactive lipid are largely uninvestigated.

Previously, we have investigated the properties and the mechanism of action of PAF on AA release and PG synthesis in the rabbit iris (Yousufzai and Abdel-Latif, 1985b). We have shown (a) that PAF stimulates the release of arachidonate and consequently PG synthesis in a time- and dose-dependent manner; (b) that arachidonate release by PAF is probably mediated through phospholipase A_2; and (c) that PAF-induced release of arachidonate is Ca^{2+}-dependent, is inhibited by Ca^{2+}-antagonistic drugs, and that it is not mediated through metabolites generated by the cyclooxygenase and lipoxygenase pathways.

Effect of PAF on ^{32}P Labelling of Phosphoinositides

As can be seen from Table 5 both NE and carbachol appreciably increased phosphoinositide turnover; in contrast PAF, at concentrations which ranged from 1×10^{-12} to 5×10^{-5}M, had no effect on ^{32}P labelling of phospholipids in this tissue. This is in contrast to the rapid accumulation of [^3H]myoinositol phosphates and the en-

Table 5. *Effects of NE, carbachol, and PAF on ^{32}P labelling of phosphoinositides in rabbit iris.*

Phospholipid	Effect of Agonist on ^{32}P Incorporation (% of control)		
	NE	Carbachol	PAF
PtdH	228	296	104
PtdIns	171	182	95
PtdInsP_2	74	76	102

Irides (in pairs) were preincubated for 50 min in 2 ml of modified Krebs-Ringer bicarbonate buffer (pH 7.4) containing 20 μCi of ^{32}Pi at 37°. The irides (one of the pair was used as control and the other as experimental) were transferred to separate tubes containing 1 ml of fresh ^{32}P-containing buffer and the agonist (50 μM) was added as indicated, then incubated for 10 min. At the end of incubation the phospholipids were extracted and analyzed by two-dimensional and one-dimensional TLC as previously described (Akhtar and Abdel-Latif, 1984). The control values for PtdH, PtdIns and PtdInsP_2 were (cpm/iris): 4860, 16026, and 102414, respectively. The data are means of 8 determinations.

hanced ^{32}P labelling of phosphoinositides and PtdH by carbachol and other Ca^{2+}-mobilizing agonists previously reported in this tissue (Abdel-Latif et al., 1977; Akhtar and Abdel-Latif, 1980; Akhtar and Abdel-Latif, 1984; Abdel-Latif et al., 1985).

Dose-Response to PAF for Release of $[^{14}C]AA$ and PGE_2

There is general agreement now that PG formation must be preceded by a lipolytic process to release free AA from the tissue phospholipids (for review see Irvine, 1982). To investigate the mechanism of action of PAF on AA release and PG synthesis, irides were prelabelled with $[^{14}C]AA$ and the effect of various concentrations of PAF on the release of radioactivity was determined by radiochromatography. As can be seen from Fig. 2, PAF increased the release of labelled arachidonate and PGE_2 in a dose-

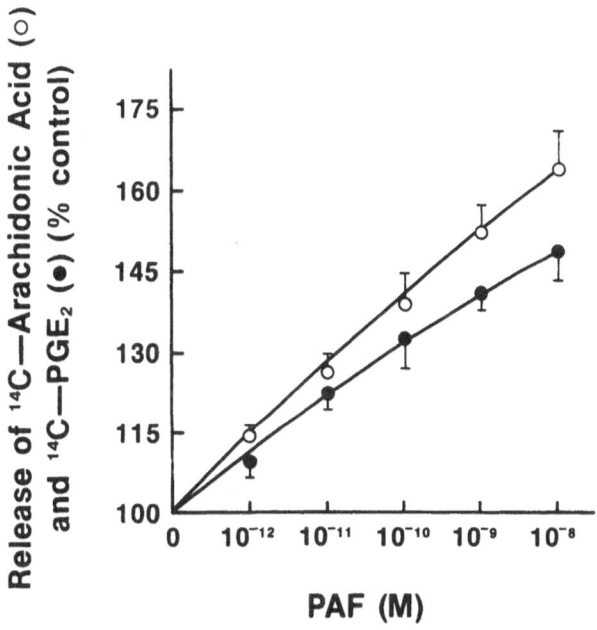

Figure 2. Dose-response to PAF for release of ^{14}C-labelled AA and PGE_2 in iris muscle. Irides (in pairs) were preincubated for 1 h in 2 ml of Krebs-Ringer bicarbonate buffer (pH 7.4) containing 0.2 μCi of $[^{14}C]AA/ml$. The prelabelled irides were washed twice with non-radioactive buffer, then incubated in the absence and presence of various concentrations of PAF as indicated for 10 min at 37°C. Extraction of AA and PGs from the medium and analysis of radioactivity were as described under Table 2. Typical control values (in absence of PAF) for $[^{14}C]AA$ and PGE_2 were (d.p.m./iris) 13257 ± 985 and 2298 ± 137, respectively. The effects of PAF on AA and PGE_2 release are expressed as percentages of their respective controls. Each point is the mean of values from three separate experiments run in triplicate. (Taken from Yousufzai and Abdel-Latif, 1985b with permission.)

dependent manner. Thus, PAF at 10^{-12} and 10^{-9}M increased arachidonate release by 15 and 53%, respectively. These data indicate that release of arachidonate by PAF from membrane phospholipids is dose dependent.

Time Course for the Effect of PAF on the Release of [^{14}C]AA

The time course for the action of 10^{-9}M PAF on the release of AA from irides prelabelled with [^{14}C]AA in the presence of indomethacin is given in Fig. 3. Significant increase in the release of labelled AA was observed within 1 min after addition of 10^{-9}M PAF and after 10 min this increased by more than 150%. These data demonstrate that the metabolites of AA are not involved in the PAF-induced release of AA in the iris.

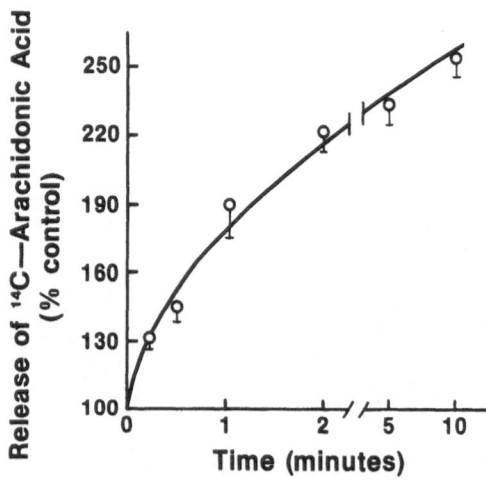

Figure 3. Time course for the effect of PAF on the release of [^{14}C]AA in the iris. Conditions of incubation were the same as described under Fig. 2 except that the experimental media contained 10^{-9}M PAF and the incubations were carried out in the presence of 1.6 μM indomethacin and for various time intervals as indicated. Control value (after 10 min) for [^{14}C]AA release was about (d.p.m./iris) 16165. Each point is the mean of values from two separate experiments run in duplicate.

Time Course for the Effect of PAF on the Breakdown of Doubly Labelled PtdIns and Release of AA

To answer the possibility raised above, namely that PAF could act by stimulating the phospholipase A_2 pathway to increase the release of free AA and consequently PG synthesis, we have investigated the time course for the effect of PAF on the breakdown of PtdIns in iris muscle prelabelled with [^{3}H]AA and [^{14}C]myo-inositol. Within 15 s PAF (10^{-9}M) increased the release of [^{3}H]AA and formation of [^{14}C]lysoPtdIns from doubly labelled PtdIns by 15% and 10%, respectively, and after 10 min of incubation the increases were 58 and 38%, respectively (Fig. 4). Under the same experimental conditions, we were able to observe only slight changes in the release of [^{14}C]myo-inositol

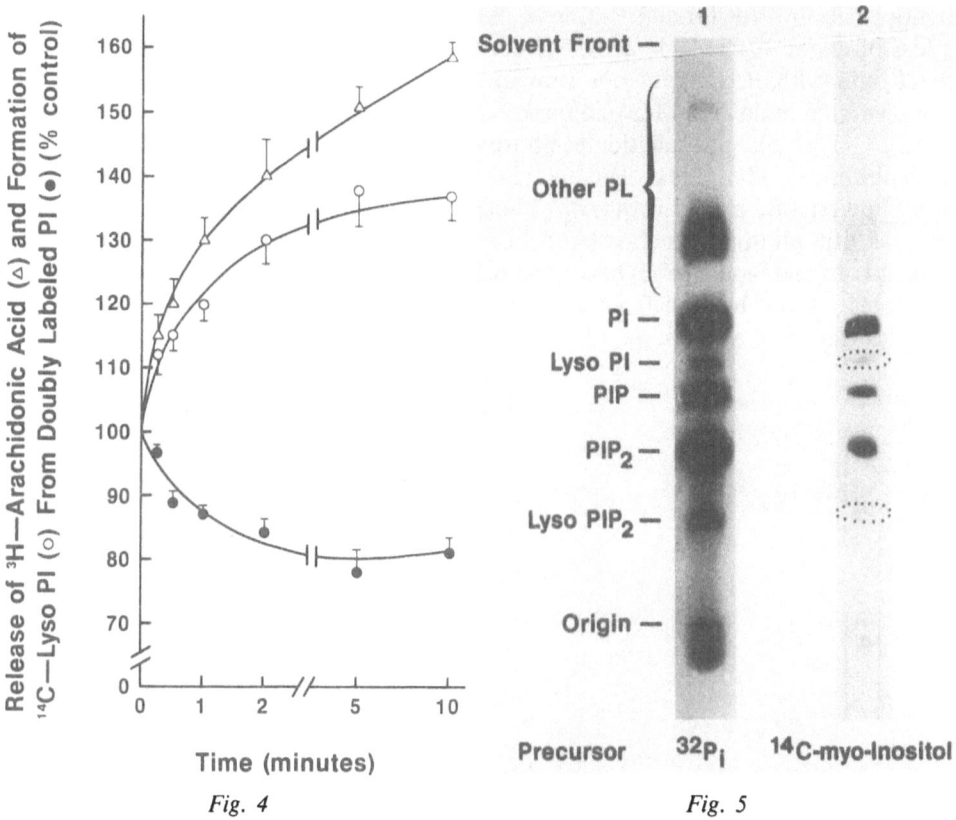

Fig. 4

Fig. 5

Figure 4. Time course for the effect of PAF on the release of ³H-labelled AA and formation of ¹⁴C-labelled lysoPtdIns (lysoPI) from doubly labelled PtdIns. Irides (in pairs) were preincubated for 2 h in 2 ml Krebs-Ringer bicarbonate buffer (pH 7.4) containing 2 μCi [³H]AA and 0.3 μCi [¹⁴C]myo-inositol. The doubly prelabelled irides were washed twice with non-radioactive buffer, then incubated in the absence and presence of PAF (10⁻⁹M) for various time intervals as indicated. Extractions of [³H]AA from the medium and phospholipids from the tissue and analysis of radioactivity were as described previously (Yousufzai and Abdel-Latif, 1985b). The control values (zero time) for ¹⁴C radioactivities in lysoPtdIns and in doubly labelled PtdIns were (d.p.m./iris) 5451 ± 165 and 20395 ± 726, respectively. The effects of PAF on AA release and on the formation of lysoPtdIns from PtdIns are expressed as percentages of their respective controls. Each point is the mean of values from two separate experiments run in triplicate. (Taken from Yousufzai and Abdel-Latif, 1985b, with permission.)

Figure 5. Autoradiographs of thin-layer chromatograms obtained after incubation of iris muscle in a medium containing either 20 μCi ³²Pi/ml (lane 1) or 0.2 μCi [¹⁴C]myo-inositol/ml (lane 2) for 1 h at 37°C. Total lipids were extracted and phospholipids were separated by one-dimensional TLC as previously described (Akhtar et al., 1983).
PI, PtdIns; PIP, PtdIns4*P*; PIP₂, PtdIns(4,5)*P*₂.

88

phosphates by PAF (data not shown). These data suggest that PAF acts to release AA from PtdIns by stimulating the phospholipase A_2 pathway. Analysis of ^{14}C-radioactivity in PtdCho and PtdIns in iris muscle prelabelled with [14C]AA revealed a loss of about 3% of radioactivity from these phospholipids upon incubation for 1 to 10 min with PAF (data not shown). The following phospholipases have been demonstrated in the iris: phospholipase A_2 against PtdIns(4,5)P_2, PtdIns (Fig. 5), and PtdCho (Fig. 6); phosphatidate phosphohydrolase against PtdH (Fig. 6), and phospholipase C against PtdIns (Fig. 6). The properties of PtdIns(4,5)P_2 phosphodiesterase and PtdIns(4,5)P_2 phosphomonoesterase (Akhtar and Abdel-Latif, 1978), PtdIns phosphodiesterase (Abdel-Latif et al., 1980), and PtdH phosphohydrolase (Abdel-Latif and Smith, 1984) have already been described in iris microsomal and soluble fractions. As can be seen from Figures 5 and 6, PtdIns(4,5)P_2, PtdIns, PtdH, and PtdCho can serve as a source of AA for PG synthesis in this tissue.

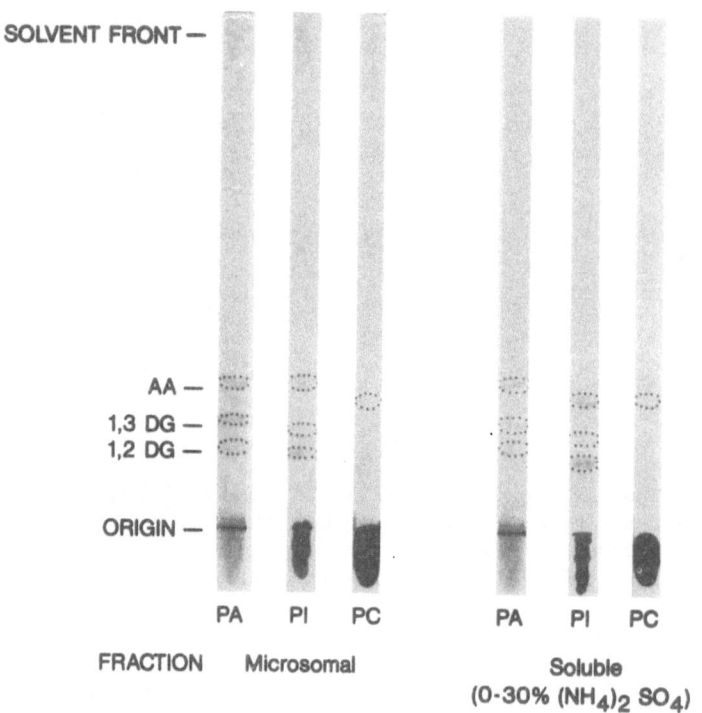

Figure 6. Autoradiographs of thin-layer chromatograms obtained after incubation of 2-[1-^{14}C]arachidonyl-PtdH, 2-[1-^{14}C]arachidonyl-PtdIns, or 2-[1-^{14}C]arachidonyl-PtdCho with iris microsomal and 110,000 × g soluble (0-30% (NH$_4$)$_2$SO$_4$ precipitate) fractions. Phospholipase activities in microsomal and soluble fractions were determined in a reaction mixture containing 120 nmol of labelled phospholipids (about 60,000 cpm) sonicated in 100 mM Hepes/NaOH buffer (pH 7.5), 2 mM CaCl$_2$, 50 mM KCl, and 200 μg protein in a final volume of 200 μl. Incubations were carried out at 37°C for 10 min and terminated by lipid extraction with chloroform-methanol (2:1). The neutral lipids were separated from the total lipid extract by means of one-dimensional TLC as previously described (Abdel-Latif and Smith, 1982).
Abbreviations: PA, PtdH; PI, PtdIns; PC, PtdCho.

Effect of PAF, in Absence and Presence of Ca^{2+}, Trifluoperazine, and Other Ca^{2+}-Antagonists, on Release of AA

In the light of our observations on the effects of neurotransmitters, other pharmacological agents, and Ca^{2+} on AA release and PG synthesis in the iris (see above), and the present findings with PAF (Fig. 4), both of which provided evidence that a phospholipase A_2-hydrolyzing PtdIns is activated when the iris is stimulated for PG synthesis, we have investigated the role of Ca^{2+} in PAF-induced release of AA and PG synthesis in this tissue (Yousufzai and Abdel-Latif, 1985b). During these studies it was found that in the absence of Ca^{2+}, PAF had no effect on the release of PGE_2 and labelled arachidonate. However, when Ca^{2+} was added to the incubation medium, the release of PGE_2 and AA by the agonist increased by 105 and 85%, respectively (Yousufzai and Abdel-Latif, 1985b). These data indicate that Ca^{2+} is required for the PAF-induced AA release in this tissue.

In view of the finding that Ca^{2+} is required for PAF-induced release of AA and PGE_2, it was of interest to determine the effects of trifluoperazine and other Ca^{2+}-channel antagonists on their release in the iris. As can be seen from Fig. 7 trifluoperazine inhibited PAF-induced release of $[^{14}C]AA$ in a dose-dependent manner. Thus at $10^{-10}M$ the drug inhibited AA release by 15%, and at $10^{-6}M$ it inhibited the release by 49%. Trifluoperazine alone had no effect on the release of ^{14}C-radioactivity (data not shown). Similar effects were observed with other Ca^{2+}-channel antagonists such as nifedipine, verapamil, diltiazem, and manganese (Yousufzai and Abdel-Latif, 1985b). These studies indicate that in the iris muscle, PAF actions on the release of AA and PGs are inhibited by Ca^{2+}-channel antagonists.

CONCLUSIONS

We have investigated the effects of NE, ACh, and PAF on PG production by rabbit iris, measured by RIA, and the type of phospholipase activated by these agonists in irides in which phosphoinositides were doubly prelabelled with $[^{14}C]$myo-inositol and $[^3H]AA$, quantitated by radiometric and chromatographic methods. (a) The studies on the properties and mechanism of AA release from PtdIns in prelabelled irides by NE and ACh (10-100 μM) suggest the involvement of both phospholipase C and phospholipase A_2. (b) The studies on the properties and mechanism of AA release from PtdIns in prelabelled irides by PAF (10^{-12}-$10^{-8}M$) suggest the involvement of phospholipase A_2. (c) PtdIns(4,5)P_2, PtdCho and PtdEtn could serve as a source of AA for PG synthesis. (d) In contrast to the neurotransmitters NE and ACh, PAF does not behave as a Ca^{2+}-mobilizing agonist in this tissue. (e) The data presented demonstrate that as with PtdIns(4,5)P_2 phosphodiesterase, phospholipase A_2 is also, although indirectly, under neurotransmitter and hormonal control.

ACKNOWLEDGMENTS

This work was supported by USPHS Grants EY-04387 and EY-04171 from the National Eye Institute. This is contribution no. 0900 from the Department of Cell and Molecular Biology. We are grateful to Ronald Gracy and Robert Reps for technical assistance.

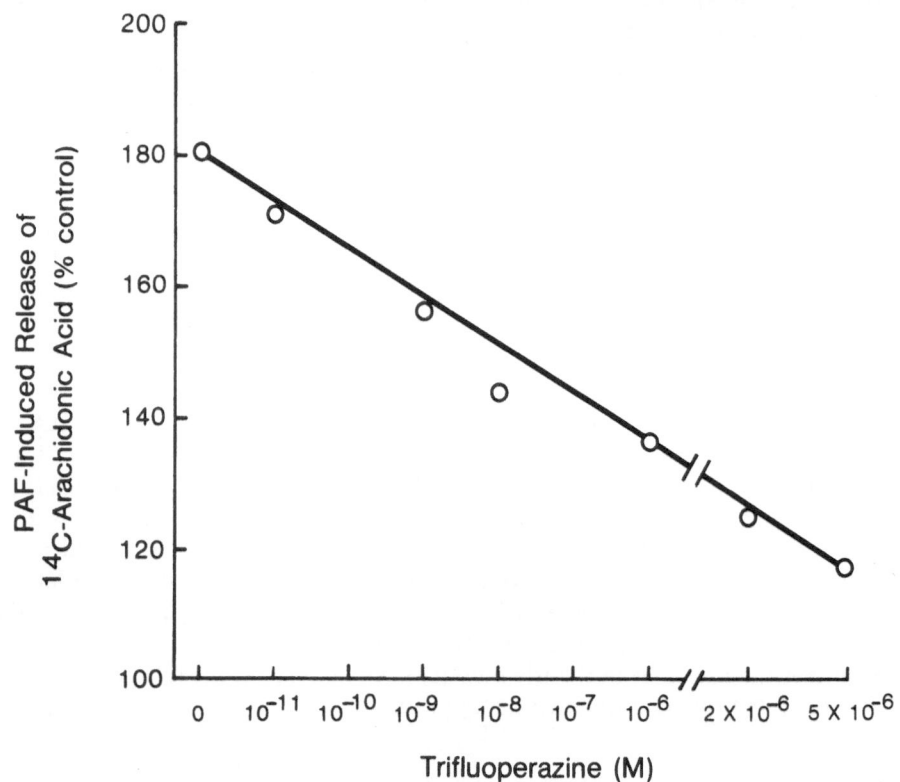

Figure 7. Dose-response curve for inhibition of PAF-induced release of [¹⁴C]AA by trifluoperazine in iris muscle. Irides (in pairs) were preincubated for 1 h in 2 ml of Krebs-Ringer bicarbonate buffer (pH 7.4) containing 0.2 μCi of [¹⁴C]AA/ml. The prelabelled irides were washed twice with nonradioactive buffer, then incubated in the absence and presence of PAF (10^{-9}M) and various concentrations of trifluoperazine for 10 min at 37°C. Extraction of AA from the medium and analysis of radioactivity were as described above. The inhibitory effects of trifluoperazine on PAF-induced release of AA are expressed as percentage of control. Each point is the mean of values from two separate experiments run in triplicate.

REFERENCES

Abdel-Latif AA, Akhtar RA and Hawthorne JN (1977) Biochem J 162: 61-73.

Abdel-Latif AA, Luke B and Smith JP (1980) Biochim Biophys Acta 614: 425-434.

Abdel-Latif AA and Smith JP (1982) Biochim Biophys Acta 711: 478-489.

Abdel-Latif AA and Smith JP (1984) Can J Biochem Cell Biol 62: 170-177.

Abdel-Latif AA, Smith JP and Dover RK (1983a) Biochem Pharmacol 32: 729-732.

Abdel-Latif AA, Smith JP, Yousufzai SYK and Dover RK (1983b) In: Sun GY, Bazan N, Wu J-Y, Porcellati G and Sun AY (eds): Neural Membranes. The Humana Press, Clifton, New Jersey pp 97-122.

Abdel-Latif AA, Smith JP and Akhtar RA (1985) In: Bleasdale JE, Eichberg J and Hauser G (eds): Inosital and phosphoinositides: Metabolism and Regulation. The Humana Press, Clifton, New Jersey, pp. 275-298.

Akhtar RA and Abdel-Latif AA (1978) Biochim Biophys Acta 527: 159-170.

Akhtar RA and Abdel-Latif AA (1980) Biochem J 192: 783-791.

Akhtar RA and Abdel-Latif AA (1984) Biochem J 224: 291-300.

Akhtar RA, Taft WC and Abdel-Latif AA (1983) J Neurochem 41: 1460-1468.

Bell RL, Kennerly DA, Stanford N and Majerus PW (1979) Proc Natl Acad Sci USA 76: 3238-3241.

Billah MM and Lapetina EG (1982) J Biol Chem 257: 5196-5200.

Findley SR, Lichtenstein LM, Hanahan DJ and Pinckard RN (1981) Am J Physiol 241: C130-134.

Flower RJ and Blackwell GJ (1979) Nature (London) 278: 456-459.

Hill EE and Lands WEM (1970) In: Wakil SJ (ed): Lipid Metabolism. Academic Press New York pp. 185-279.

Holub BJ and Kuksis A (1978) Adv Lipid Res 16: 1-125.

Ieyasu H, Takai Y, Kaibuchi K, Sawamura M and Nishizuka Y (1982) Biochem Biophys Res Commun 108: 1701-1708.

Irvine RF (1982) Biochem J 204: 3-16.

Kloprogge E and Akkerman JWN (1984) Biochem J 223: 901-909.

Lands WEM and Samuelsson B (1968) Biochim Biophys Acta 164: 426-429.

Lee TC, Malone B, Blank ML and Snyder F (1981) Biochem Biophys Res Commun 102: 1262-1268.

MacIntyre DE and Pollock WK (1983) Biochem J 212: 433-437.

Mauco G, Chap H and Douste-Blazy L (1983) FEBS Lett 153: 361-365.

McKean ML, Smith JB and Silver MJ (1981) J Biol Chem 256: 1522-1524.

Okazaki T, Sagawa N, Okita JR, Bleasdale JE, MacDonald PC and Johnston JM (1981) J Biol Chem 256: 7316-7321.

Pinckard RN, McManus LM and Hanahan DJ (1982) Adv Inflam Res 4: 147-180.

Shaw JO, Klusick SJ and Hanahan DJ (1981) Biochim Biophys Acta 663: 222-229.

Shukla SD and Hanahan DJ (1982) Biochem Biophys Res Commun 106: 697-703.

Vargaftig BB, Chignard M, Benveniste J, Lefort J and Wal F (1981) Ann NY Acad Sci 370: 119-137.

Vogt W (1978) Adv Prostaglandin and Thromboxane Res 3: 89-95.

Walsh CE, Waite BM, Thomas MJ and DeChatlet LR (1981) J Biol Chem 256: 7228-7234.

Yousufzai SYK and Abdel-Latif AA (1983) Exp Eye Res 37: 279-292.

Yousufzai SYK and Abdel-Latif AA (1984) Prostaglandins 28: 399-415.

Yousufzai SYK and Abdel-Latif AA (1985a) Biochem Pharmacol 34: 539-544.

Yousufzai SYK and Abdel-Latif AA (1985b) Biochem J 228: 697-706.

Phospholipid research and the nervous system
Biochemical and molecular pharmacology
L.A. Horrocks, L. Freysz, G. Toffano (eds)
Fidia Research Series, vol. 4.
Liviana Press, Padova. © 1986

PHOSPHOLIPID METABOLISM IN NERVOUS TISSUES: MODIFICATION OF PRECURSOR INCORPORATION AND ENZYME ACTIVITIES BY CATIONIC AMPHIPHILIC DRUGS

George Hauser, Omanand Koul and Ubaldo Leli

Ralph Lowell Laboratories, McLean Hospital, Harvard Medical School, Belmont, MA 02178, USA

INTRODUCTION

Although the effects of individual drugs on the incorporation of precursors into lipids in vitro have been sporadically examined for thirty years or so, it is primarily during the last ten years that a substantial body of information has been accumulated in this field. During this time, it has come to be recognized that numerous drugs with varied therapeutic actions share the capacity to redirect phospholipid metabolism. The compounds involved exhibit certain common physical-chemical features of their structures, namely a bulky lipophilic ring system and an aliphatic side-chain containing a quaternary, positively charged nitrogen which bestows hydrophilic character upon the molecule. The typical shift in incorporation of $^{32}P_i$ into phospholipids, elicited by drugs of this type, manifests itself in enhanced labeling of acidic and reduced labeling of neutral phospholipids. This has been demonstrated in numerous biological systems (Hauser and Pappu, 1982; Abdel-Latif, 1983; Bazan et al., 1985).

Certain of these substances have also been studied as agents which when administered to animals or humans can lead to an accumulation of phospholipids in tissues. Such accumulations consist primarily of polar lipids and tend to be localized in lysosomes. As a result specific and characteristic pathological features appear which have been extensively described (Lüllmann et al., 1975, 1978; Drenckhahn and Lüllmann-Rauch, 1979; Lüllmann-Rauch, 1979). In the nervous system neuronal cell bodies are the main site of pathological changes which consist of enlarged lysosomes filled with what appears from several lines of evidence to be phospholipid in an arrangement of either lamellar or crystalloid subunits. This may represent a means for storing phospholipids which cannot be metabolized owing to inhibition of the degradative enzymes. Whether a particular cationic amphiphilic drug has in fact the capacity to induce these accumulations and morphological alterations depends primarily on the rate with which it can be degraded and removed from the body.

The cationic amphiphilic drugs used in the studies discussed in this paper are relatively rapidly metabolized and do not give rise to lipidosis. They can nevertheless have significant influence on membrane events, in particular reactions concerned with lipid metabolism, in the short term. Three drugs have been used in this work, one of which, propranolol (PRO), has previously been shown in our laboratory to be very effective in perturbing phospholipid metabolism in pineal gland (Eichberg et al., 1973, 1979; Hauser and Eichberg, 1975) and brain mince (Pappu and Hauser, 1981a, b; Hauser and Pappu, 1982). PRO has also been used for similar experiments in other systems by several other investigators (Abdel-Latif and Smith, 1976; Giusto et al., 1983). Furthermore, after PRO administration to rats, tissues continued to exhibit disturbed incorporation of precursors into phospholipids for some time, presumably owing to retention of the drug (Pappu and Hauser, 1982). The other two drugs, chlorpromazine (CPZ) and desmethylimipramine (DMI), both psychotropic agents, were next to PRO most potent in the system used (Pappu and Hauser, 1981a) and for this reason a more extensive examination of their effects was undertaken. This paper primarily describes experiments to date on (a) incorporation of precursors into C6 glioma cells and modification of the drug effects, (b) inhibition of PtdH phosphohydrolase of rat brain and (c) characterization and modification of lysophospholipase activity.

DRUG EFFECTS ON PHOSPHOLIPID METABOLISM IN C6 CELLS

The effects of cationic amphiphilic drugs on phospholipid metabolism were studied in C6 glioma cells in culture, in order to characterize their actions in a system in which parameters could be easily varied. The purpose was to establish cationic amphiphilic drugs as potential tools for dissecting different events of phospholipid biosynthesis and degradation, and perhaps to gain insight into the molecular mechanism of action of CPZ, DMI, and PRO. Data on the specific characteristics of glial phospholipid metabolism could also be produced by using this system. Lipids in C6 glioma cells were labeled by replacement of the cell culture medium with buffer containing radioactive precursors and the effects of the inclusion of CPZ, DMI, and PRO were examined. The incorporation of $^{32}P_i$ into total lipids was increased by each of the drugs in a dose-dependent manner, reaching more than double the level of incorporation seen in the absence of cationic amphiphilic drugs at concentrations above 200 μM.

Cationic amphiphilic drugs have been shown to associate with phospholipid bilayers (Conrad and Singer, 1981), and this is thought to modify them and interfere with active and passive membrane transport phenomena. If such an altered membrane permeability were responsible for increased uptake of $^{32}P_i$, resulting in higher specific activity of ATP, this could account for the elevation in total labeling. However, if this were the sole effect, one would expect a parallel increase in labeling in all phospholipid classes.

When the labeling pattern of phospholipids was determined after 1 hr exposure to CPZ (Fig. 1), changes in $^{32}P_i$ incorporation into individual phospholipid classes were

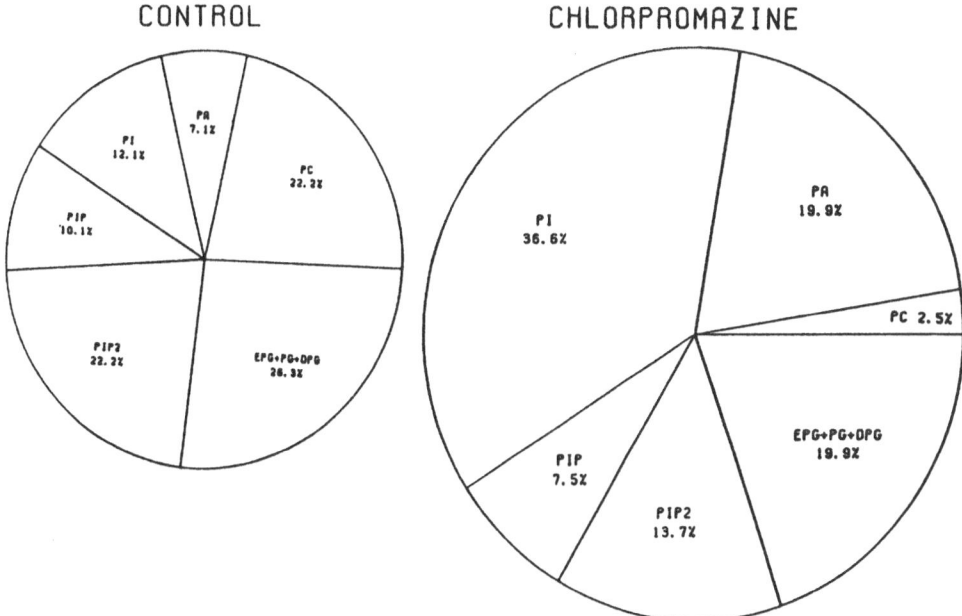

CONTROL CHLORPROMAZINE

Figure 1. Pattern of ³²Pi incorporation into C6 glioma cell phospholipids. C6 cells were grown for four days, the last one without serum. After removal of growth medium, Krebs-Ringer-HEPES-glucose medium, containing about 5 μCi ³²P$_i$ and drugs when desired, was added and the incubation was allowed to proceed for 60 min at 37°C. CPZ, 100 μM. Cells were harvested and extracted with neutral and acidic CHCl$_3$-CH$_3$OH (2:1). After TLC of the washed combined lipid extracts and radioautography, bands corresponding to individual phospholipid classes were scraped off and counted. Values are percentages of the sum of the radioactivity in recovered phospholipids. The area of the circles represents total incorporation. EPG = ethanolamine phosphoglycerides; DPG = diphosphatidylglycerol, PA = PtdH, PC = PtdCho, PG = PtdGro, PI = PtdIns; PIP = PtdIns4P, PIP$_2$ = PtdIns(4,5)P_2.

qualitatively very similar for the three drugs. Increased labeling of PtdH, PtdIns, and PtdGro and decreased labeling of PtdCho and PtdEtn was observed as also seen, for example, in brain slices (Hokin-Neaverson, 1980). Effects on Ptd-CMP labeling which are dramatic in pineal gland (Hauser and Eichberg, 1975; Eichberg et al., 1979) were not examined in these cells. Changes were dose-dependent as shown for DMI in Figure 2. In contrast to DMI, 100 μM CPZ inhibited PtdCho labeling completely. When PtdIns and PtdIns(4,5)P_2 were expressed as percentages of total labeling, reciprocal effects by CPZ were seen during the first hour (Fig. 3). The relative incorporation of ³²P$_i$ into PtdIns(4,5)P_2 was considerably higher than that into PtdIns at the earliest time point, suggesting either an extremely rapid conversion of PtdIns to PtdIns(4,5)P_2 or a very active replacement of monoesterified phosphate groups in pre-existing PtdIns(4,5)P_2. The shape of the curves favors the latter interpretation.

A special feature of C6 glioma cells is the high labeling rate of ethanolamine phosphoglycerides and the high plasmalogen content of this phospholipid class. The

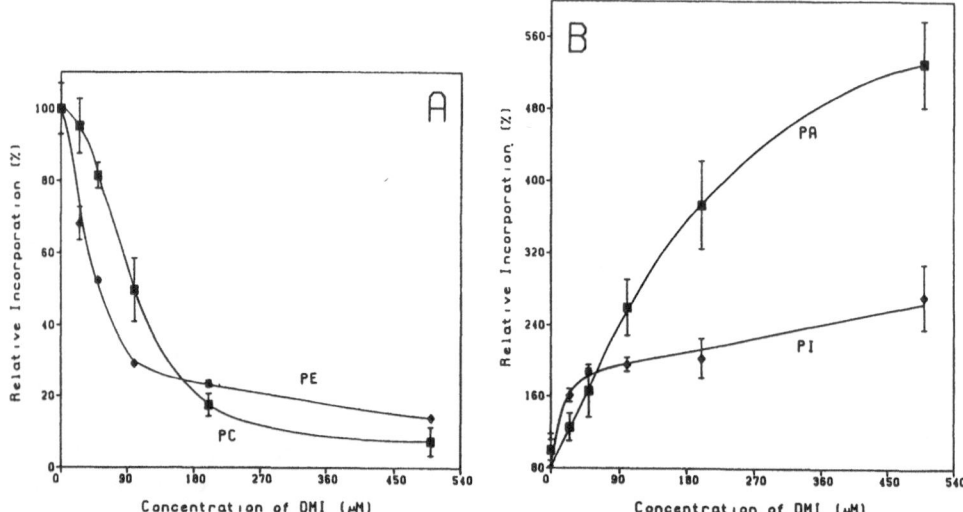

Figure 2. Dose-response curves for $^{32}P_i$ incorporation into acidic and neutral phospholipids of C6 glioma cells in presence of DMI. Lipids were labeled, extracted, and counted as in Figure 1. Values are percentages of the sum of the radioactivity recovered in phospholipids. A: neutral phospholipids; B: acidic phospholipids (See Fig. 1 for abbreviations).

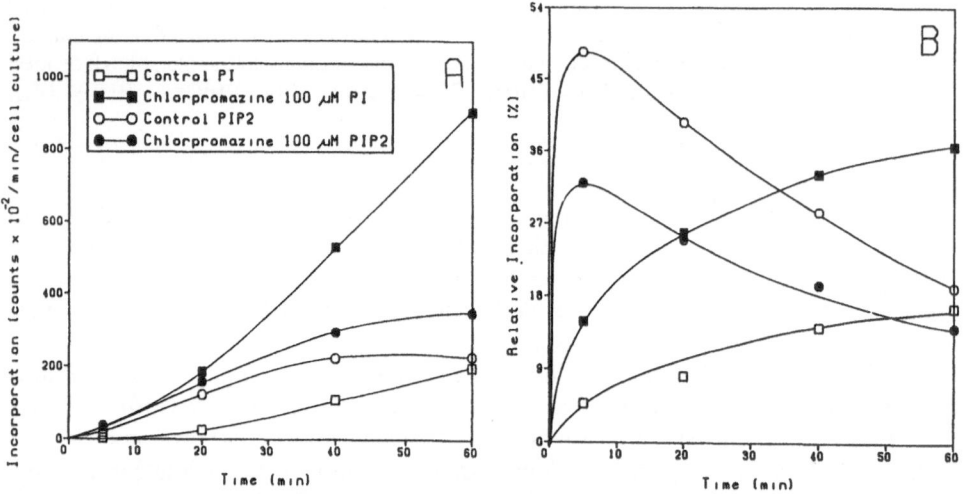

Figure 3. Time-course of incorporation of $^{32}P_i$ into C6 glioma cell PtdIns and PtdIns(4,5)P_2 - effect of chlorpromazine. Lipids were labeled, extracted and counted as in Figure 1. A: Counts per minute incorporated per cell culture after incubation with 5 μCi of ^{32}Pi. B: Relative incorporation is the incorporation expressed as percentage of the sum of radioactivity recovered in phospholipids. (See Fig. 1 for abbreviations).

drugs influence the labeling of the plasmalogen component in parallel with that of PtdEtn, since radioactivity in these two components as a percentage of the label in total ethanolamine phosphoglycerides remains unchanged. All of these changes can be attributed to a series of effects on different enzymatic steps that are affected by the drugs.

One of these is catalyzed by PtdH phosphohydrolase which has been shown to be inhibited by cationic amphiphilic drugs in many systems (e.g. Bowley et al., 1977; Sturton and Brindley, 1977; Pappu and Hauser, 1983). This could account for a reduction of DAG synthesis and reduced availability of this substrate for PtdCho and PtdEtn formation. The accumulation of the PtdH resulting from such a block could explain the increased metabolic flow towards PtdIns and PtdGro, which are synthesized via PtdCMP. Studies on PtdH phosphohydrolase are discussed below. Another enzyme is phosphocholine: CTP cytidylyltransferase, the regulatory enzyme for PtdCho synthesis, that is also inhibited by CPZ (Pelech and Vance, 1984) and can therefore control the shift of incorporation away from PtdCho. This enzyme is stimulated by fatty acids and DAG and the amount of CDPcholine produced is thought to be the critical regulatory variable for the synthesis of PtdCho.

The time-course of [³H]glycerol incorporation into DAG in the presence of 100 μM CPZ shows a marked depression in labeling of this lipid (Fig. 4). This suggests that the effects of cationic amphiphilic drugs might in fact be mediated to some extent by a decrease in DAG synthesis. To test this hypothesis further we added diolein to the cell culture medium in an attempt to increase the substrate for PtdCho synthesis and

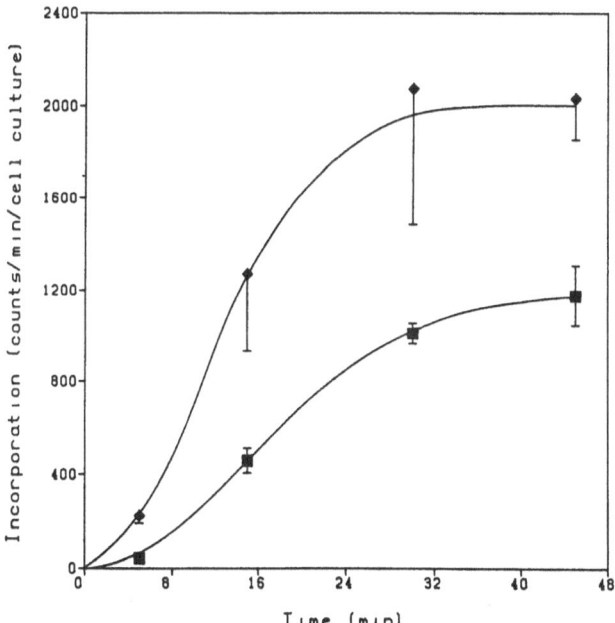

Figure 4. Time-course of [³H]glycerol incorporation into C6 glioma cell DAG. Cells were treated as in Fig. 1. Incubation medium contained about 10 μCi of [³H]-glycerol. Values are cpm incorporated into DAG per cell culture. Control, ◆; 100 μM CPZ, ■.

counteract the effect of the drugs on PtdCho and PtdEtn labeling. However, the reduced labeling was unaffected, perhaps owing to the fact that an apolar compound such as diolein cannot effectively penetrate the cell membrane (results not shown). To overcome this difficulty 1-oleoyl, 2-acetyl glycerol (OAG) was tested since it more easily enters into cells. PtdCho labeling was restored by 1 mM OAG to the level seen in the absence of drug but the effects of CPZ on PtdH and PtdIns were not altered (Fig. 5). It is uncertain from these findings whether added OAG is a substrate for PtdCho synthesis or acts in a different manner to prevent the expression of the drug effect on PtdCho labeling.

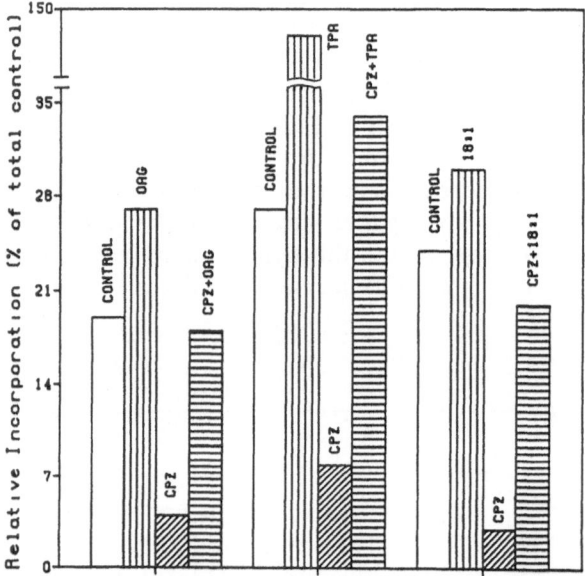

Figure 5. Modification of chlorpromazine effects on $^{32}P_i$ incorporation into phosphatidylcholine of C6 glioma cells by 1-oleoyl-2-acylglycerol (OAG), 12-O-tetradecanoylphorbol-13-acetate (TPA) and oleic acid (18:1). Cells were treated as in Figure 1. Incubations contained 100 mM CPZ and/or 1 mM OAG, 10 μM TPA or 100 μM 18:1.
Values are percentages of the sum of the radioactivity recovered in phospholipids from the controls.

Bimodally distributed enzymes like PtdH phosphohydrolase and phosphocholine: CTP cytidylyltransferase are thought to be regulated by association with and dissociation from cell membranes. DAG and fatty acids have been shown to promote the transfer of enzymes to the particulate fraction which constitutes the more active form (Choy et al., 1979; Pelech et al., 1983, 1984a, b; Cascales et al., 1984; Martin-Sanz et al., 1985; Hall et al., 1985); the decrease in DAG presumably produced by the influence of cationic amphiphilic drugs could thus result in lower cytidylyltransferase activity with a consequent decrease in PtdCho labeling. To check this possibility we tested the capacity of an analog of DAG that is not a substrate for PtdCho synthesis to counteract the effects of CPZ. When 12-O-tetradecanoylphorbol-13-acetate was added to the culture medium, stimulation of PtdCho labeling by $^{32}P_i$ was observed both in the absence and in the presence of 100 μM CPZ (Fig. 5). This suggests that the decrease in DAG labeling is due to the inhibition of PtdH phosphohydrolase and may be related to the changes observed in the labeling of PtdCho and PtdEtn in a dual manner, reflecting both inadequate availability of DAG as substrate and lowered activity of the rate-determining, DAG-dependent enzyme. This would be in agreement with the fact that the regulatory step in PtdCho biosynthesis is the formation of CDPcholine via phosphocholine cytidylyltransferase and that the amount of this substance is critical, regardless of the availability of DAG as substrate. The results obtained by adding 100 μM oleate to the incubation medium also support this hypothesis. As seen in Fig. 5, PtdCho labeling, which is reduced 8-fold by CPZ, is elevated to levels very close to control if oleate is also present, possibly owing to a similar activation of cytidylyltransferase by translocation from the cytosolic to the membrane-bound form.

STUDIES ON PHOSPHATIDATE PHOSPHOHYDROLASE

PtdH phosphohydrolase in brain is localized in cytosolic and microsomal fractions utilizing free or membrane-bound PtdH at different rates to give DAG and inorganic phosphate. Pappu and Hauser (1983) showed that brain microsomal enzyme is 5-6 times as active against free as against membrane-bound substrate (Fig. 6). On the other hand, the cytosolic enzyme is 2-3 times as active against the membrane-bound substrate. PRO

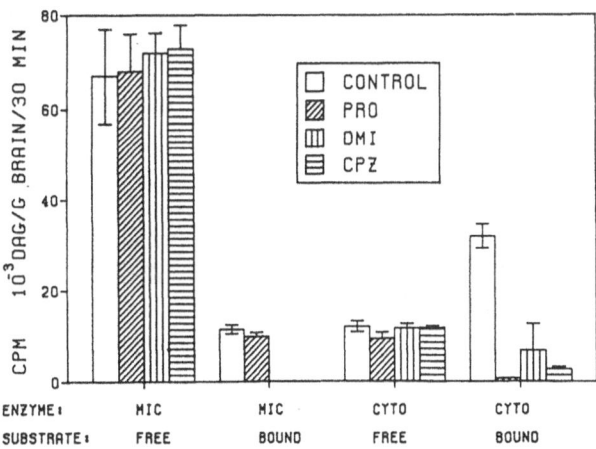

Figure 6. Effect of cationic amphiphilic drugs on rat brain PtdH phosphohydrolase. Activity of cytosolic (CYT) and microsomal (MIC) PtdH phosphohydrolase was determined with free or membrane-bound [^{14}C]oleoyl PtdH as substrate. The enzyme was assayed in Tris-HCl buffer, pH 7.4, at 37°C for 30 min with about 150 μg of CYT or MIC protein. Labeled free PtdH was sonicated in the assay buffer before use. Radioactivity in DAG was measured after separation by TLC. In some experiments [^{32}P]PtdH was used as substrate and the $^{32}P_i$ counts were converted to DAG equivalents for the purpose of this graph.

only inhibited the cytosolic enzyme acting on bound substrate whereas other combinations of enzyme and substrate were not affected to a significant extent. We have repeated these experiments and confirmed the findings using DMI and CPZ (Fig. 6). Dose-response curves for PRO, CPZ, and DMI show that the activity of the cytosolic enzyme acting on membrane-bound substrate is inhibited by all three drugs in a dose-dependent manner, with similar potencies (Fig. 7).

In view of our findings with C6 glioma cells reported above we examined the phosphohydrolase from these cells to see if the same effects of cationic amphiphilic drugs could be demonstrated. In this system using free PtdH, the microsomal enzyme was about half as active as the cytosolic enzyme. It hydrolyzed only about 9% of the substrate in 30 min, less than the enzyme from brain cytosol. However, both free and bound substrates were hydrolyzed to the same extent (about 18%) by the cytosolic enzyme. In these cells, as in rat brain studied earlier (Pappu and Hauser, 1983), cationic amphiphilic drugs inhibited only the cytosolic enzyme acting on membrane-bound substrate (Table 1). At 200 μM drug concentration, the enzyme was inhibited more than 95% which may partially account for the phospholipid changes seen after incubation of these cells with drugs.

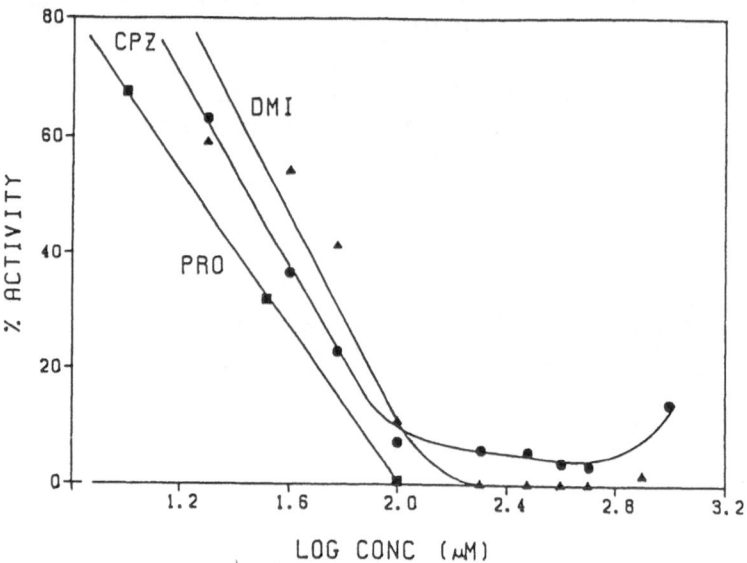

Figure 7. Dose response curves for cationic amphiphilic drugs inhibition of cytosolic PtdH phosphohydrolase. Rat brain cytosolic activity against membrane-bound substrate was assayed in presence of the drugs indicated. Assay conditions were as given in Figure 6, except that 960 μg of enzyme protein were used in a total volume of 400 μl.

Cytosolic Enzyme and Free PtdH: no Inhibition by Cationic Amphiphilic Drugs

We prepared membrane vesicles to study the effect of the drugs on cytosolic PtdH phosphohydrolase in order to examine why no inhibition is observed with the free substrate. Free PtdH ([^{14}C]oleoyl) was sonicated in buffer alone, with total liver lipids

Table 1. *Phosphatidate phosphohydrolase of C6 cells*

Conditions	Addition	Relative Activity (%)
Cytosolic enzyme and membrane-bound substrate	None	100
	PRO	3.9
	CPZ	2.2
	DMI	4.0
Cytosolic enzyme and free substrate	None	100
	PRO	112
	P_2 Fraction	107
	P_2 + PRO	112
Microsomal enzyme and free substrate	None	100
	DMI	101

Microsomes or supernatant of C6 cells (150 μg of protein) were used as enzyme sources. Assay conditions were as given in Figure 7. All drugs were 200 μM. The P_2 fraction from C6 cells (14,000 × g pellets, 33 μg protein per assay) was used after heat denaturation.

Table 2. *Effect of addition of membranes and lipids on phosphatidate phosphohydrolase activity with soluble substrate*

Addition	Relative Activity (%)	Hydrolysis (% of added substrate)
None	100	19.3
P_3 fraction	71.0	13.8
P_3 + PRO	55.9	10.8
Lipids (liver)	107	20.6
Lipids + PRO	98.4	19.0

The enzyme was assayed in duplicate according to the conditions given in Figure 7. Heat-denatured P_3 fraction (liver microsomes, 500 μg protein per assay) or total liver lipids (2 mg per assay) were mixed with phosphatidate and sonicated before use.

or with heat-denatured membranes (subcellular P_2 fraction of C6 cells or P_3 fraction of rat liver) and used for assaying the enzyme in presence of the drugs (Tables 1 and 2). Addition of P_3 membranes from liver reduced the activity compared to the controls. In the presence of P_3 membranes and 200 μM PRO appreciable activity remained, although this concentration of the drug inhibits the enzyme almost completely when it acts on membrane-bound substrate. Addition of liver lipids produced a slight stimulation of the basal activity, but no inhibition in the presence of PRO (Table 2), nor did PRO inhibit the cytosolic enzyme acting on PtdH mixed and sonicated with P_2 fraction from C6 cells (Table 1). These studies indicate the necessity of an additional factor for mediating the inhibition of the cytosolic enzyme by cationic amphiphilic drugs.

Several authors have speculated on the mechanism of action of cationic amphiphilic drugs on this enzyme. Brindley and Bowley (1975) studied PtdH phosphohydrolase in liver and postulated that the inhibition is a result of direct binding of the drug to the substrate, rendering it unavailable for enzyme action. In apparent support of this hypothesis Pappu and Hauser (1983) found that membrane-bound PtdH substrate, isolated from PRO-pretreated tissue, inhibited the enzyme, whereas the enzyme from the same source was fully active.

If the drug were to bind lipids directly and thereby inhibit the enzyme, one would expect the inhibitory capacity of a drug to be related only to the acidity of the substrate used. In analogous experiments IC_{50} values for several cationic amphiphilic drugs acting on liver phospholipase A with PtdCho, PtdEtn, and PtdIns as substrates were 5-7 times as high for PtdIns as for PtdEtn and PtdCho (Pappu and Hostetler, 1984) suggesting that the inhibition occurs by a complex mechanism rather than the direct binding of drug to substrate.

Mg^{2+} Requirement for Enzyme Activity and for Inhibition by Cationic Amphiphilic Drugs

Phosphohydrolase activity is dependent not only on whether the substrate is free or bound but also on which metal is associated with it (Smith et al., 1967; Bowley et al., 1977). Surface structure and charge density produced by association with proteins also determine the activity of this enzyme (Ide and Nakazawa, 1985). Both Mg^{2+}-dependent and Mg^{2+}-independent fractions exist in rat adipocyte cytosol and microsomes (Lamb and Fallon, 1974). The Mg^{2+}-dependent activity in rat hepatocyte cytosol can be regulated by translocation by oleic acid (Cascales et al., 1984). In their studies to elucidate the failure of cationic amphiphilic drugs to exert an inhibitory effect when free PtdH is the substrate for cytosolic enzyme, Bowley et al. (1977) found that not only was Mg^{2+} needed for optimal activity, it was also required for inhibition. Without Mg^{2+}, CPZ at 100-200 μM had no significant effect on the activity, but as concentrations of CPZ were increased the enzyme was first stimulated and then inhibited. Our data with free sonicated substrate show that there is no significant effect of cationic amphiphilic drugs on this enzyme in the absence of Mg^{2+} (Fig. 6 and Table 1). It is important to note that the membrane-bound substrate used by Brindley and Bowley (1975), Pappu and Hauser (1983), and in this study to demonstrate inhibition of cytosolic enzyme activity by cationic amphiphilic drugs is prepared in presence of Mg^{2+}. Although the membranes are washed, some Mg^{2+} remains, part of it presumably bound to PtdH (Bowley et al., 1977). Depletion of Mg^{2+} by EDTA has been shown to be inhibitory (Mitchell et al., 1971). Mg^{2+} involvement is also demonstrated by the inhibition of brain cytosolic enzyme activity by CPZ when free PtdH is used as the Mg-salt (Table 3). It can, therefore, be inferred that the inhibition of cytosolic phosphohydrolase acting on the free or membrane-bound substrate occurs in presence of Mg^{2+}. The experiments done to demonstrate direct binding of the drugs to PtdH have also been conducted in presence of Mg^{2+} (Bowley et al., 1977). Because Mg^{2+} is needed for optimal activity and for inhibition by cationic amphiphilic drugs the potency of the drug must be related to the binding constant of the drug and cation, the concentration of Mg^{2+} in the medium, the ionization of Mg-PtdH and the prevailing pH, and hence on an equilibrium between the ionized and non-ionized species. Therefore, a complex mechanism has to be invoked to explain drug action.

Most of the studies on the effects of cationic amphiphilic drugs have been on enzymes concerned with lipid degradation. We have studied the effects of these drugs on two synthetic enzymes as well. Neither DAG kinase nor PtdIns4P kinase in the cytosolic fraction of rat brain was inhibited by PRO, CPZ, or DMI (Table 4). Selective metabolic steps must be affected to explain the observed shift in precursor incorporation into phospholipids.

Table 3. *Effects of EDTA on cytosolic phosphatidate phosphohydrolase acting on free* Mg^{2+}-*phosphatidate*

Addition	Relative Activity (%)
None	100
CPZ	19.1
EDTA	11.9
EDTA + CPZ	2.6

Labeled [^{14}C]phosphatidate from rat liver microsomes was dispersed in Tris·HCl buffer, pH 7.4. Assay was run for 20 min at 37° in a total volume of 210 μl with 960 μg of cytosolic protein as enzyme source. CPZ, 360 μM, and EDTA, 10 mM, were added where indicated.

Table 4. *Effect of cationic amphiphilic drugs on diacylglycerol kinase and* *phosphatidylinositolphosphate kinase of brain*

Addition	Phosphatidic Acid Formed	PIP$_2$ Formed
	(cpm in product/min)	
None	7737 ± 670	2554 ± 328
PRO	7591 ± 384	2503 ± 235
CPZ	8500 ± 197	2864 ± 194
DMI	7567 ± 557	2825 ± 120

100,000 × g rat brain supernatant was used as the enzyme source and [γ-^{32}P]ATP was used as substrate. Assays were run in Tris-HCl buffer pH 7.4 according to Kai et al. (1968) for PtdIns4P kinase, and Kanoh et al. (1983) for diacylglycerol kinase. About 6% and 0.13% of the added [^{32}P]ATP radioactivity was recovered in phosphatidic acid and PtdIns(4,5)P_2 (PIP$_2$), respectively. The final concentration of the drugs was 100 μM; 100-200 μg of the supernatant protein was used in each assay.

STUDIES ON LYSOPHOSPHOLIPASE

Lysophospholipase A catalyzes the hydrolysis of lysophospholipids to free fatty acids and glycero-phospho-base derivatives. The enzyme is widely distributed and its function is to eliminate toxic lysocompounds formed in membranes by phospholipase A activities. Lysophospholipids have strong detergent properties and have effects on a variety of enzyme activities. Pertinent for this discussion, CTP: phosphocholine cytidylyltransferase, the regulatory step for PtdCho synthesis, is activated by these compounds in liver (Fiscus and Schneider, 1966; Choy et al., 1977; Choy and Vance, 1978). Beef brain Na$^+$, K$^+$-ATPase and Ca^{2+}-ATPase are also activated by lysoPtdCho (Tanaka and Strickland, 1965; Martonosi et al., 1968; The and Hasselbach, 1972) and modification of ATPase would alter incorporation results. Consequently, the capacity of cationic amphiphilic drugs to interfere with the enzymatic steps that control the amount of lysocompounds in cell membranes was explored.

Lysophospholipase A has been described in liver (Dawson, 1956; Lands, 1960; Erbland and Marinetti, 1962; van den Bosch et al., 1968), in pancreas (Shapiro, 1953; van den Bosch et al., 1973; De Jong et al., 1973), in rat brain (Leibowitz and Gatt, 1968), in lung tissue (Brumley and van den Bosch, 1977; Vianen and van den Bosch, 1978; van Heusden et al., 1981; Casals et al., 1982), in heart (Gross and Sobel, 1982;

Reddy et al., 1983) and in amniotic membranes (Jarvis et al., 1984). Rat brain enzymes are active toward lysoPtdCho and lysoPtdEtn with a pH optimum of 7.2-8.6 (Leibowitz and Gatt, 1968). The enzyme is bimodally distributed between cytosolic and particulate subcellular fractions. CPZ and local anesthetics have been reported to inhibit lysophospholipase from rat liver directly (Kunze et al., 1976). Also treatment of mice with chloroquine resulted in inhibition of liver lysosomal lysophospholipase (Pakalapati and Debuch, 1982).

When [^{32}P]lysoPtdCho and [^{32}P]lysoPtdEtn and a brain microsomal enzyme preparation were used, the drugs displayed a qualitatively similar biphasic effect on the activity with a rank order of potency CPZ > DMI > PRO. Stimulation was seen at low drug concentration but inhibition occurred above 200 μM (Fig. 8A, B). By comparison, liver lysosomal phospholipases A and C were inhibited with IC$_{50}$ values of 30-70 μM for CPZ and 200-400 μM for imipramine and PRO (Hostetler and Matsuzawa, 1981). The stimulatory effect of cationic amphiphilic drugs on lysophospholipase was also observed with the soluble enzyme. With [^{32}P]lysoPtdCho as substrate it was 25% for 100 μM PRO, 20% for 50 μM CPZ and 30% for 50 μM DMI.

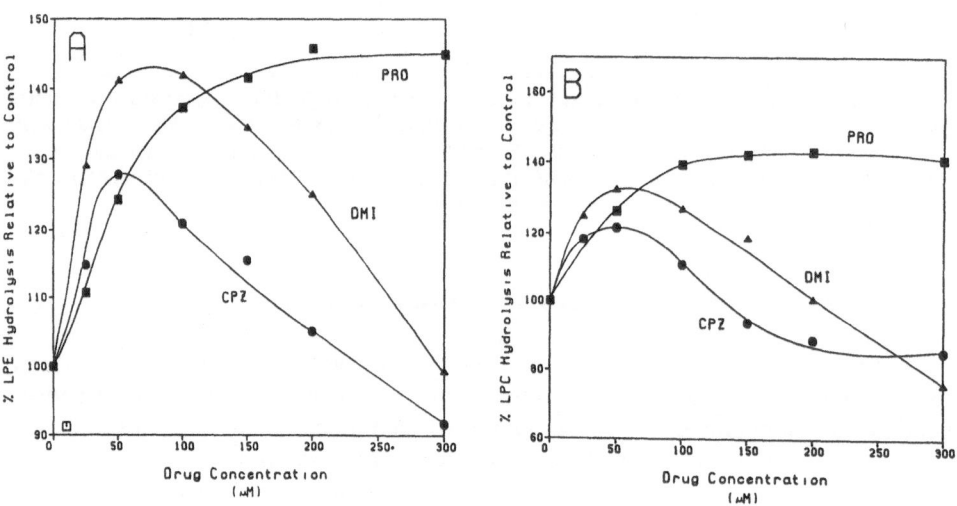

Figure 8. Rat brain microsomal lysophospholipase: Dose-response curves for DMI, CPZ and PRO. [^{32}P]PtdCho and [^{32}P]EPG were prepared by incorporating ^{32}P$_i$ into C6 glioma cell lipids. LysoPtdCho (LPC) was prepared by treatment of PtdCho (PC) with porcine pancreas phospholipase A$_2$. LysoPtdEtn (LPE) was obtained from purified EPG by acid hydrolysis. Enzyme incubations (10 min at 37°C) contained 75 mM Tris·HCl, pH 8.1 and 90 μg of microsomal protein in 200 μl. After partitioning remaining lysophospholipid was counted.

Before attempting purification of rat brain lysophospholipase, we sought information on the mechanism of action of cationic amphiphilic drugs by studying the substrate-velocity dependence in crude microsomal preparations. Increasing concentrations of substrate were inhibitory as also shown in liver (van den Bosch et al., 1968).

Leibowitz-BenGershon et al. (1972) have also reported an "inversion point" in the velocity-substrate curves for the rat brain microsomal enzyme suggesting that micelles of substrate that appear at high concentration are inhibitory and that the monomolecular form of the substrate is preferred. Hill coefficients without added DMI and in the presence of 50 μM (stimulatory) or 500 μM (inhibitory) DMI are respectively 2.02, 1.87, 2.14. This means that the enzyme molecules may have two different binding sites.

If the affinity for the substrate is modified by the binding of the first molecule of the substrate and this were influenced by the drug, one would expect the Hill coefficient to change. Also, kinetics with the same concentrations of DMI as above show an increase in the Vmax for the stimulatory and a decrease for the inhibitory concentrations of the drug (175, 305 and 35 μmol$\cdot$l$^{-1}\cdot$min^{-1} for the control, 50 μM and 500 μM DMI, respectively) and no change in the K_m (3.3 μM). This strongly suggests that DMI does not bind to the active site but might bind to a different, regulatory site of the enzyme molecule. From these findings, a non-competitive mechanism for inhibition and an indirect mechanism for activation of lysophospholipase by cationic amphiphilic drugs can be deduced.

The observed increase in Vmax in presence of stimulatory concentrations of DMI means that more enzyme is made available for the reaction. If the stimulatory capacity of the drugs were due to a detergent-like effect, one would expect a change in the physicochemical state of the substrate (change in the ratio of the monomolecular to the micellar form) leading to a change in the affinity of the enzyme for the substrate. The unchanged apparent K_m in presence of either stimulatory or inhibitory concentrations of DMI excludes this explanation. Cationic amphiphilic drugs might interfere with a "positioning" site of lysophospholipase that could explain the observed kinetic features. Free fatty acids inhibit the enzyme (Leibowitz-BenGershon et al., 1972) and unsaturated fatty acids are more potent than saturated ones in this regard. CPZ, DMI, and PRO are able to counteract this effect in a non-stoichiometric manner (Fig. 9). One might speculate that an apolar part of the enzyme molecule can bind the long chain of both free and esterified fatty acids of the lysocompounds, positioning the substrate for the proper attack on the ester linkage. Cationic amphiphilic drugs might well interfere with such a domain. As yet no support is available for the hypothesis that these cationic amphiphilic drug effects on lysophospholipase have physiological relevance in vivo by changing the amount of lysocompounds in the membranes as part of the mechanism of action of the drugs or as part of their side effects. Nor has it been demonstrated that changes in lysophospholipase activity are responsible in a major way for the redirection of phospholipid metabolism by cationic amphiphilic drugs.

GENERAL CONSIDERATIONS RELATED
TO CATIONIC AMPHIPHILIC DRUG ACTION

A variety of cationic amphiphilic drugs can profoundly alter the normal processes involved in the maintenance of the cell's phospholipid pattern. The question which remains incompletely answered is how the drugs in question, PRO, CPZ, and DMI in the present study, produce the observed phenomena.

It is becoming increasingly recognized that the mechanisms involved must be quite complex and encompass a number of different facets. These must be attributable to

106

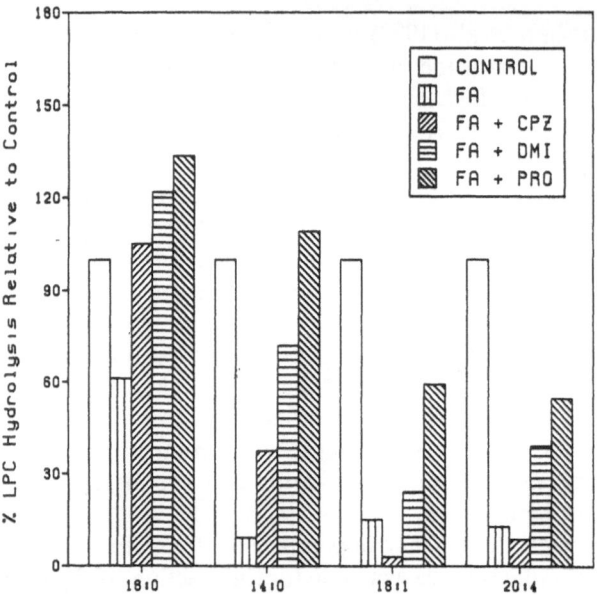

Figure 9. Effects of fatty acids and cationic amphiphilic drugs on lysophospholipase activity. Assays were as in Figure 8. Additions were fatty acids, 100 μM; PRO, 100 μM; CPZ and DMI, 50 μM. LPC = LysoPtdCho.

the physico-chemical nature of the molecules and their ability to alter not only a number of enzymatic reactions involved in phospholipid metabolism but also the arrangement or reactivity of membrane constituents.

Mechanism of Alterations of Phospholipid Metabolism

Hydrophobic drug molecules can bind not only to the lipid components of membranes but also to hydrophobic domains of membrane proteins (Lee, 1977a, b). This can change the fluidity of the lipid (Pang and Miller, 1978; Lee, 1979) and, if charged species are involved, the charge of the membrane. Interactions of this sort can account for the observed influence on lipid metabolism in four ways (a) through complex formation with lipid substrates and lipid molecules capable of modifying the activity of membrane-bound enzymes of lipid metabolism (b) through fluidity changes of the lipids surrounding such enzymes (Rooney and Lee, 1983), (c) through altered local concentrations of charged molecules (such as Ca^{2+} or Mg^{2+} required for enzymatic reactions) as the result of competitive displacement by a positively charged agent, and (d) through direct binding to enzyme proteins. Among the structural characteristics of the drug, critical for the extent of complex formation with phospholipids and resultant interference with their degradation, are hydrophobic forces, provided a positively charged group is present in the side chain (Seydel and Wassermann, 1976; Lüllmann and Wehling, 1979).

It is the conversion of the neutral to the protonated form of the drug in the acidic interior of lysosomes that seems to be responsible for the formation and accumulation of cationic amphiphilic drug-lipid complexes which prevents phospholipid degradation within these organelles and results in the appearance of phospholipidosis (Blohm, 1979).

Cationic amphiphilic drugs that are relatively slowly metabolized and thus not removed before they reach sufficiently high intralysosomal levels are the ones that give rise to this abnormality, but the ones used in this study do not fall into this category.

The aromatic ring protons and the adjacent methylene groups are particularly significant for the binding affinity of the drug to phospholipids. Ring substituents reinforce hydrophobic binding of compounds and may be essential for impairment of phospholipid metabolism. The protonated amino group of the drug and the polarity of the lipid are further factors determining the extent of interaction and hence of inhibition.

Cationic Amphiphilic Drug Interaction with Membranes

The search for mechanisms of action for each drug has been oriented not only towards direct effects on phospholipid metabolism, but also towards the interaction of cationic amphiphilic drugs with membrane components and the influence on their physical properties. Clearly such modifications of the arrangement and characteristics of the lipid bilayer can have indirect effects on the enzymatic reactions involved in the turnover of individual molecules. As stated, preferential electrostatic interaction with polar head groups of acidic phospholipids occurs with cationic drugs which reduces the negative charge on the lipid, reduces the repulsive forces between phospholipids, and causes condensation of the monolayer with uncertain consequences for the functional properties of a biological membrane (Surewicz, 1982). Expansion of the monolayer in the case of drugs which can penetrate and promote separation between phospholipid molecules can also be envisaged.

The neutral form of a drug may actually be more effective in exerting physiological or metabolism-altering effects because of its ability to partition into and freely diffuse within the lipid phase. Lipid solubility of the drugs is therefore of prime importance both in theories of their mechanism of action and in their ability to affect lipid metabolism. The entry of drug molecules into the lipid bilayer has consequences for the physical properties of the membrane which may be responsible in both of these areas. There occurs a lateral expansion of the lipid bilayer (Seeman and Roth, 1972) as well as an increase in its thickness (Ashcroft et al., 1977) and fluidity (Lee, 1976) which would cause indirect changes in the activity of functional membrane proteins. These changes are thought to come about through disruptions of normal lipid-protein interaction or of optimal phase relationships in the lipid which would translate into impairment of function.

The orientation of amphipathic molecules in a membrane bilayer and the extent and firmness of association with its lipid and protein constituents depends on the size, charge, and the relative contribution of the hydrophobic and hydrophilic portions of the compound. Drugs are thus capable of perturbing the lipoprotein bilayer by producing changes in the molecular organization of membrane lipids and a depression of the transition temperature from the gel to the liquid crystalline state (Singer, 1977). A combination of electrostatic and hydrophobic interactions may be responsible for these perturbations, possibly triggered by the increase in molecular order of acidic phospholipids. The degree of order and mobility of the acyl chains of lipids could also be involved.

These general comments are largely relevant under physiological conditions particularly in relation to the accumulation of lipids in tissues when certain drugs are chronically administered. The specific mechanisms involved when individual enzymatic steps are inhibited in assays in vitro presumably include some of the factors mentioned but their complete elucidation must await the results of further experiments. It is obvious that multiple sites along the metabolic pathways are subject to modification by cationic amphiphilic drugs, which are mainly inhibitory in nature but in select instances can also be stimulating (Sturton and Brindley, 1977; Hostetler et al., 1976; Hauser and Pappu, 1982; Bazan et al., 1985).

ACKNOWLEDGMENTS

The studies reported in this paper were supported by research grants NS 06399 and NS 19047 from the U.S. Public Health Service.

REFERENCES

Abdel-Latif AA (1983) Metabolism of phosphoinositides. In: Lajtha A (ed): Handbook of Neurochemistry, Vol. 3. Plenum Publishing Corp, New York, pp. 91-131.

Abdel-Latif AA, Smith JP (1976) Effects of DL-propranolol on the synthesis of glycerolipids by rabbit iris muscle. Biochem Pharmacol 25: 1697-1704.

Ashcroft RG, Coster HGL, Smith JR (1977) The molecular organization of bimolecular lipid membranes. Biochim Biophys Acta 469: 13-22.

Bazan NG, Roccamo de Fernandez AM, Giusto NM, Ilincheta de Boschero MG (1985) Propranolol-induced membrane perturbation and the metabolism of phosphoinositides and arachidonoyl diacylglycerols in the retina. In: Bleasdale J, Eichberg J, Hauser G (eds): Inositol and Phosphoinositides: Metabolism and Regulation. Humana, Clifton, NJ, pp. 67-82.

Blohm TR (1979) Drug-induced lysosomal lipidosis: biochemical interpretations. Pharmacol Rev 30: 593-603.

Bowley M, Cooling J, Burditt SL, Brindley DN (1977) The effects of amphiphilic cationic drugs and inorganic cations on the activity of phosphatidate phosphohydrolase. Biochem J 165: 447-454.

Brindley DN, Bowley M (1975) Drugs affecting the synthesis of glycerides and phospholipids in rat liver. Biochem J 148: 461-469.

Brumley G, van den Bosch H (1977) Lysophospholipase-transacylase from rat lung, isolation and partial purification. J Lipid Res 18: 523-532.

Casals C, Acebal C, Cruz-Alvarez M, Estrada P, Arche R (1982) Lysolecithin: lysolecithin acyltransferase from rabbit lung: Enzymatic properties and kinetic study. Arch Biochem Biophys 217: 422-433.

Cascales C, Mangiapane EH, Brindley DN (1984) Oleic acid promotes the activation and translocation of phosphatidate phosphohydrolase from the cytosol to particulate fractions of isolated rat hepatocytes. Biochem J 219: 911-916.

Choy PC, Farren SB, Vance DE (1979) Lipid requirements for the aggregation of CTP: phosphocholine cytidylyltransferase in rat liver cytosol. Can J Biochem 57: 605-612.

Choy PC, Lim PH, Vance DE (1977) Purification and characterization of CTP: cholinephosphate cytidylyltranferase from rat liver cytosol. J Biol Chem 252: 7673-7677.

Choy PC, Vance DE (1978) Lipid requirements for activation of CTP: phosphocholine cytidylyltransferase from rat liver. J Biol Chem 253: 5163-5167.

Conrad MJ, Singer SJ (1981) The solubility of amphipathic molecules in biological membranes and lipid bilayers and its implications for membrane structure. Biochemistry 20: 808-818.

Dawson RMC (1956) The phospholipase B of liver. Biochem J 64: 192-196.

De Jong JGN, van den Bosch H, Aarsman AJ, van Deenen LLM (1973) Studies on lysophospholipases. II. Substrate specificity of a lysolecithin hydrolyzing carboxylesterase from beef pancreas. Biochim Biophys Acta 296: 105-115.

Drenckhahn D, Lüllmann-Rauch R (1979) Drug-induced experimental lipidosis in the nervous system. Neuroscience 4: 697-712.

Eichberg J, Gates J, Hauser G (1979) The mechanism of modification by propranolol of the metabolism of phosphatidyl-CMP (CDP-diacylglycerol) and other lipids in the rat pineal gland. Biochim Biophys Acta 573: 90-106.

Eichberg J, Shein H, Schwartz M, Hauser G (1973) Stimulation of $^{32}P_i$ incorporation into phosphatidylinositol and phosphatidylglycerol by catecholamines and beta-adrenergic receptor blocking agents in rat pineal organ cultures. J Biol Chem 248: 3615-3622.

Erbland J, Marinetti GV (1962) In vitro metabolism of lysolecithin. Federation Proc 21: 295.

Fiscus WG, Schneider WC (1966) The role of phospholipids in stimulating phosphorylcholine cytidyltransferase activity. J Biol Chem 241: 3324-3330.

Giusto NM, Ilincheta de Boschero MG, Bazan NG (1983) Accumulation of phosphatidic acid from propranolol-treated retinas during short-term incubations. J Neurochem 40: 563-568.

Gross RW, Sobel BE (1982) Lysophosphatidylcholine metabolism in the rabbit heart. J Biol Chem 257: 6702-6708.

Hall M, Taylor SJ, Saggerson ED (1985) Persistent activity modification of phosphatidate phosphohydrolase and fatty acyl-CoA synthetase on incubation of adipocytes with the tumour promoter 12-O-tetradecanoylphorbol 13-acetate. FEBS Lett 179: 351-353.

Hauser G, Eichberg J (1975) Identification of cytidine diphosphatediglyceride in the pineal gland of the rat and its accumulation in the presence of DL-propranol. J Biol Chem 250: 105-112.

Hauser G, Pappu AS (1982) Effects of propranolol and other cationic amphiphilic drugs on phospholipid metabolism In: Horrocks LA, Ansell GB, Porcellati G (eds): Phospholipids in the Nervous System, Vol. 1. Metabolism. Raven Press, New York, pp. 283-300.

Hokin-Neaverson M (1980) Actions of chloropromazine, haloperidol and pimozide on lipid metabolism in guinea pig brain slices. Biochem Pharmacol 29: 2697-2700.

Hostetler KY, Matsuzawa Y (1981) Studies on the mechanism of drug-induced lipidosis. Biochem Pharmacol 30: 1121-1126.

Hostetler KY, Zenner BD, Morris HP (1976) Increased mitochondrial CTP: phosphatidic acid cytidyltransferase in the 777 hepatoma. Biochem Biophys Res Commun 72: 418-425.

Ide H, Nakazawa Y (1985) Phosphatidate phosphatase in rat liver: The relationship between the activities with membrane-bound phosphatidate and aqueous dispersion of phosphatidate. J Biochem (Tokyo) 97: 45-54.

Jarvis AA, Cain C, Dennis EA (1984) Purification and characterization of a lysophospholipase from human amnionic membranes. J Biol Chem 259: 15188-15195.

Kai M, Salway JG, Hawthorne JN (1968) The diphosphoinositide kinase of rat brain. Biochem J 106: 791-801.

Kanoh H, Kondoh H, Ono T (1983) Diacylglycerol kinase from pig brain. J Biol Chem 258: 1767-1774..

Kunze H, Nahas N, Traynor JR, Wurl M (1976) Effects of local anaesthetics on phospholipases. Biochim Biophys Acta 441: 93-102.

Lamb RG, Fallon HJ (1974) Glycerolipid formation from sn-glycerol-3-phosphate by rat liver cell fractions: Role of phosphatidate phosphohydrolase. Biochim Biophys Acta 348: 166-178.

Lands WM (1960) Metabolism of glycerolipids. II. The enzymatic acylation of lysolecithin. J Biol Chem 235: 2233-2237.

Lee AG (1976) Model for action of local anaesthetics. Nature (Lond) 262: 545-548.

Lee AG (1977a) Lipid phase transitions and phase diagrams. I. Biochim Biophys Acta 472: 237-281.

Lee AG (1977b) Lipid phase transitions and phase diagrams. II. Mixtures involving lipids. Biochim Biophys Acta 472: 285-344.

Lee AG (1979) A consumers guide to models of local anesthetic action. Anesthesiology 51: 64-71.

Leibowitz-BenGershon Z, Kobiler I, Gatt S (1972) Lysophospholipases of rat brain. J Biol Chem 247: 6840-6847.

Leibowitz Z, Gatt S (1968) Isolation of lysophospholipase, free of phospholipase activity from rat brain. Biochim Biophys Acta 164: 439-441.

Lüllmann H, Lüllmann-Rauch R, Wassermann O (1975) Drug-induced phospholipidoses. CRC Crit Rev Toxicol 4: 185-218.

Lüllmann H, Lüllman-Rauch R, Wassermann O (1978) Lipidosis induced by amphiphilic cationic drugs. Biochem Pharmacol 27: 1103-1108.

Lüllmann H, Weling M (1979) The binding of drugs to different polar lipids in vitro. Biochem Pharmacol 28: 3409-3415.

Lüllmann-Rauch R (1979) Drug-induced lysosomal storage diseases. In: Dingle JT, Jacques PJ (eds): Lysosomes in Biology and Pathology. Vol. 6, North-Holland, Amsterdam, pp. 49-130.

Martin-Sanz P, Hopewell R, Brindley DN (1985) Spermine promotes the translocation of phosphatidate phosphohydrolase from the cytosol to the microsomal fraction of rat liver and it enhances the effects of oleate in this respect. FEBS Lett 179: 262-266.

Martonosi A, Donley J, Halpin RA (1968) Sarcoplasmic reticulum. III. The role of phospholipids in the adenosine triphosphatase activity and Ca^{++}-transport. J Biol Chem 243: 61-70.

Mitchell MP, Brindley DN, Hübscher G (1971) Properties of phosphatidate phosphohydrolase. Eur J Biochem 18: 214-220.

Pakalapati G, Debuch H (1982) Studies on the liberation of fatty acids from 2-lysophosphatidylcholine by a liver lysosomal enzyme activity from chloroquine-treated rats. Hoppe-Seyler's Z Physiol Chem 363: 573-580.

Pang KY, Miller KW (1978) Cholesterol modulates the effects of membrane perturbers in phospholipid vesicles and biomembranes. Biochim Biophys Acta 511: 1-9.

Pappu AS, Hauser G (1981a) Alterations of phospholipid metabolism in rat cerebral cortex mince induced by cationic amphiphilic drugs. J Neurochem 37: 1006-1014.

Pappu AS, Hauser G (1981b) Changes in brain polyphosphoinositide metabolism induced by cationic amphiphilic drugs in vitro. Biochem Pharmacol 30: 3243-3246.

Pappu AS, Hauser G (1982) Phospholipid metabolism changes in rat tissues in vitro after injections of propranolol. J Pharmacol Exptl Ther 222: 109-115.

Pappu AS, Hauser G (1983) Propranolol-induced inhibition of rat brain cytoplasmic phosphatidate phosphohydrolase. Neurochem Res 8: 1565-1575.

Pappu A, Hostetler KY (1984) Effect of cationic amphiphilic drugs on the hydrolysis of acidic and neutral phospholipids by liver lysosomal phospholipase A. Biochem Pharm 33: 1639-1644.

Pelech SL, Cook HW, Paddon HB, Vance DE (1984a) Membrane-bound CTP: phosphocholine cytidylyltransferase regulates the rate of phosphatidylcholine synthesis in HeLa cells treated with unsaturated fatty acids. Biochim Biophys Acta 795: 433-440.

Pelech SL, Paddon HB, Vance DE (1984b) Phorbol esters stimulate phosphatidylcholine biosynthesis by translocation of CTP: phosphocholine cytidylyltransferase from cytosol to microsomes. Biochim Biophys Acta 795: 447-451.

Pelech SL, Pritchard PH, Brindley DN, Vance DE (1983) Fatty acids promote translocation of CTP: phosphocholine cytidylyltransferase to the endoplasmic reticulum and stimulate rat hepatic phosphatidylcholine synthesis. J Biol Chem 258: 6782-6788.

Pelech SL, Vance DE (1984) Trifluoperazine and chlorpromazine inhibit phosphatidylcholine biosynthesis and CTP: phosphocholine cytidylyltransferase in HeLa cells. Biochim Biophys Acta 795: 441-446.

Reddy PV, Natarajan V, Schmid PC, Schmid HHO (1983) N-Acylation of dog heart ethanolamine phospholipids by transacylase activity. Biochim Biophys Acta 750: 472-480.

Rooney EK, Lee AG (1983) Binding of hydrophobic drugs to lipid bilayers and to the $(Ca^{2+} + Mg^{2+})$-ATPase. Biochim Biophys Acta 732: 428-440.

Seeman P, Roth S (1972) General anesthetics expand cell membranes at surgical concentrations. Biochim Biophys Acta 255: 171-177.

Seydel JK, Wassermann O (1976) NMR studies on the molecular basis of drug-induced phospholipidosis. II. Interaction between several amphiphilic drugs and phospholipids. Biochem Pharmacol 25: 2357-2364.

Shapiro B (1953) Purification and properties of a lysolecithinase from pancreas. Biochem J 53: 663-666.

Singer MA (1977) Interaction of dibucaine and propranolol with phospholipid bilayer membranes — effect of alterations in fatty acyl composition. Biochem Pharmacol 26: 51-57.

Smith ME, Sedgwick B, Brindley DN, Hübscher G (1967) The role of phosphatidate phosphohydrolase in glyceride biosynthesis. Eur J Biochem 3: 70-77.

Sturton RG, Brindley DN (1977) Factors controlling the activities of phosphatidate phosphohydrolase and phosphatidate cytidylyltransferase. Biochem J 162: 25-32.

Surewicz WK (1982) Quinidine is a strong perturber of acidic phospholipid bilayer order and fluidity. Biochim Biophys Acta 692: 315-318.

Tanaka R, Strickland KP (1965) Role of phospholipid in the activation of Na^+, K^+-activated adenosine triphosphatase of beef brain. Arch Biochem Biophys 111: 583-592.

The R, Hasselbach W (1972) Properties of the sarcoplasmic ATPase reconstituted by oleate and lysolecithin after lipid depletion. Eur J Biochem 28: 357-363.

van den Bosch H, Aarsman AJ, De Jong JGN, van Deenen LLM (1973) Studies on lysophospholipases. I. Purificaton and some properties of a lysophospholipase from beef pancreas. Biochim Biophys Acta 296: 94-104.

van den Bosch H, Aarsman AJ, Slotboom AJ, van Deenen LLM (1968) On the specificity of rat liver lysophospholipase. Biochim Biophys Acta 164: 215-225.

van Heusden GPH, Reutelingsperger CPM, van den Bosch H (1981) Substrate specificity of lysophospholipase-transacylase from rat lung and its action on various physical forms of lysophosphatidylcholine. Biochim Biophys Acta 663: 22-33.

Vianen GM, van den Bosch H (1978) Lysophospholipase and lysophosphatidylcholine: lysophosphatidylcholine transacylase from rat lung: Evidence for a single enzyme and some aspects of its specificity. Arch Biochem Biophys 190: 373-384.

Phospholipid research and the nervous system
Biochemical and molecular pharmacology
L.A. Horrocks, L. Freysz, G. Toffano (eds)
Fidia Research Series, vol. 4.
Liviana Press, Padova. © 1986

PHOSPHOLIPID-PROTEIN KINASE C INVOLVEMENT IN THE ADRENERGIC REGULATION OF PINEALOCYTE CYCLIC AMP: A MODEL OF HOW MODULATORS ACT?

David C. Klein, David Sugden, Jiri Vanecek[1], Thomas P. Thomas[2] and Wayne B. Anderson[2]

Section on Neuroendocrinology, Laboratory of Developmental Neurobiology,
National Institute of Child Health and Human Development,
National Institutes of Health, Bethesda, Maryland, USA;
[1] Institute of Physiology, Czechoslovak Academy of Sciences,
Prague, Czechoslovakia;
[2] Laboratory of Tumor Immunology and Biology, National Cancer Institute,
Bethesda, Maryland, 20205.

INTRODUCTION

The mammalian pineal gland is regulated by norepinephrine (NE) released into the pineal perivascular space from sympathetic nerve terminals; there is a 24-hour pattern of release, with high levels occurring at night. This rhythmic release of NE is controlled by a neural circuit which includes a circadian clock and passes through both central and peripheral structures (Klein et al., 1981).

The effects of NE on the pineal gland appear to be mediated to a large degree by the second messenger cyclic AMP. Studies in the rat indicate that cyclic AMP controls the best known function of the pineal gland, the production of melatonin (Klein and Weller, 1973; Klein et al., 1981). Cyclic AMP acts to produce a large nocturnal increase in melatonin production by stimulating the activity of arylalkylamine N-acetyltransferase (NAT, E.C. 2.3.1.85; Voisin et al., 1984). This enzyme converts serotonin to N-acetylserotonin, the precursor of melatonin. The neural system regulating the circadian production of melatonin in the pineal gland is termed the melatonin rhythm generating system (Figure 1).

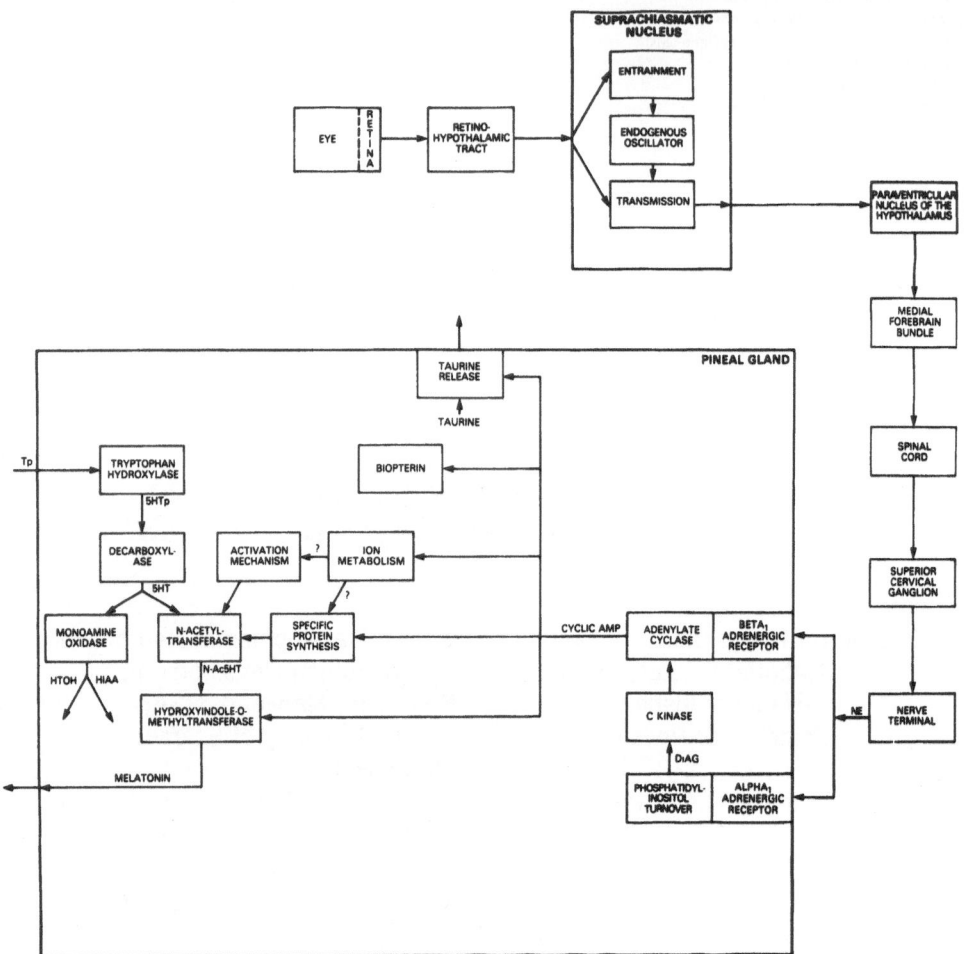

Figure 1. Schematic representation of the melatonin rhythm generating system.

ATYPICAL REGULATION OF PINEAL CYCLIC AMP AND CYCLIC GMP

NE elevates pinealocyte cyclic AMP and cyclic GMP 60- and 400-fold respectively (Figure 2; Strada et al., 1972; Deguchi, 1973; Auerbach et al., 1981; O'Dea and Zatz, 1976; Klein et al., 1979; Vanecek et al., 1985). β-adrenergic stimulation is a prerequisite for the increase of either cyclic nucleotide; α_1-adrenergic stimulation potentiates this effect (Vanecek et al., 1985). The activation of β-adrenoceptors causes a 6-fold increase in cAMP, and a 3-fold increase in cyclic GMP; α_1-adrenergic stimulation potentiates the former about 10-fold and the latter over 100-fold.

This is clearly demonstrated by the pure β-adrenergic agonist isoproterenol (ISO), which has an unusual cyclic AMP dose response curve (Auerbach et al., 1979; Vanecek, et al., 1985, Figure 2); partial stimulation occurs at low concentrations and full stimulation occurs at high concentrations. This reflects the pure β-adrenergic action of ISO

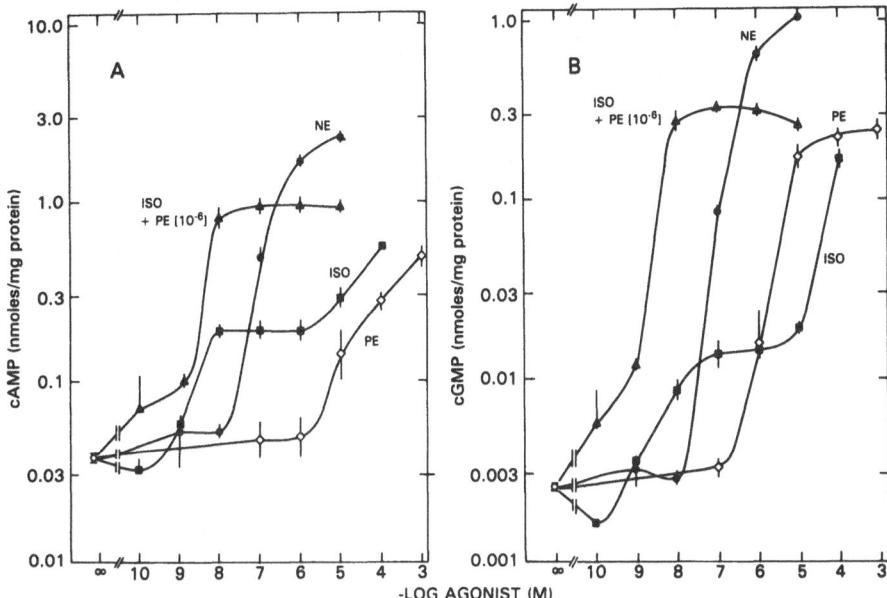

Figure 2. Agonist stimulation of pinealocyte cyclic AMP(A) and cyclic GMP(B). Cells were incubated (100,000 cells; total volume 500 μl) for 15 minutes at 37°C with the indicated concentrations of (-) norepinephrine (NE), phenylephrine (PE), (-)-isoproterenol (ISO) or ISO and PE (1 μM). Each point is the mean of triplicate samples assayed in duplicate. Vertical bars represent the SEM; where absent, the SEM is smaller than the symbol. From Vanecek et al., 1985.

at low concentrations and the non-specific effects of ISO acting through α_1-adrenoceptors at high concentrations. Further, the ISO dose response curve can be made to appear similar to the NE dose response curve by treatment of cells with a low concentration of the α-adrenergic agonist phenylephrine (PE) which has little effect in the absence of ISO.

At high concentrations PE causes a nearly full stimulation of cyclic AMP and cyclic GMP, due to a specific effect mediated by α_1-adrenergic receptors and nonspecific effects mediated by β-adrenergic receptors.

As would be predicted from the studies with adrenergic agonists, it is possible to selectively block the α_1-adrenergic component of the NE stimulation of cyclic AMP with α_1-antagonists; β-adrenergic agonists completely block the NE stimulation of both cyclic nucleotides (Vanecek et al., 1985).

Regulation of Cyclic AMP

It has been well established that β-adrenergic elevation of cyclic AMP reflects the stimulation of adenylate cyclase (Weiss and Costa, 1986; Zatz et al., 1976). A series of observations indicate that α_1-adrenergic potentiation of the stimulation of adenylate cyclase is mediated by a mechanism involving phospholipids and protein kinase C.

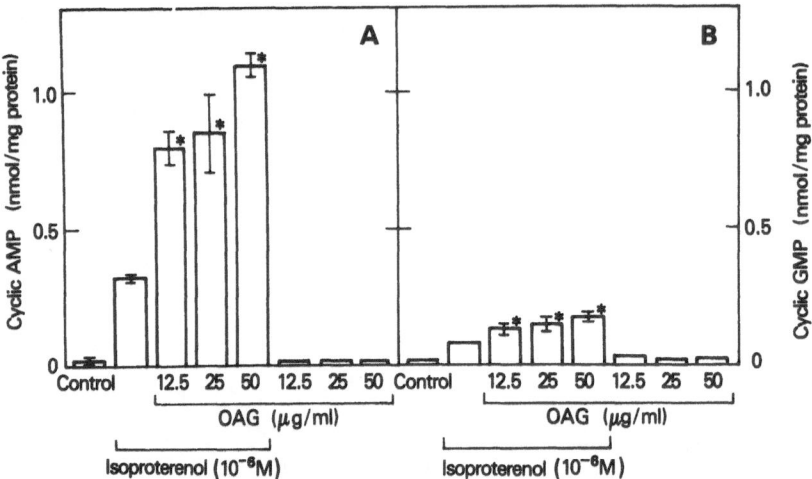

Figure 3. Potentiation of β-adrenergic cyclic nucleotide responses by 1-oleoyl-2-acetyl-*sn*-glycerol (OAG). OAG (10 mg/ml in chloroform) was dried under nitrogen, then suspended in dimethyl sulfoxide (1%) by brief, vigorous sonication on ice immediately before use. OAG (a gift from Dr. Y. Nishizuka) was diluted 25 to 100 fold to give the final concentrations indicated. Pinealocytes (100,000 per tube) were pelleted (1,000xg, 1 min) and resuspended in DMEM without fetal calf serum, then treated for 5 minutes with OAG alone or in combination with ISO. Each bar represents mean ±SEM of three samples. *Significantly different from ISO alone P>0.05 using Duncan's multiple range test. From Sugden et al., 1985.

Role of phospholipids

The presumed role of phospholipids is based on two lines of evidence. First, NE is known to stimulate PtdIns turnover in pinealocytes (Berg and Klein, 1972). This effect was subsequently found to have a pharmacological profile consistent with an α_1-adrenergic mechanism (Smith et al., 1979); also, a high density of α_1-adrenergic receptors was recently described in the pineal gland by direct radioligand binding analysis (Sugden and Klein, 1984).

Second, it has been found that the potentiating effect of PE on ISO-treated pinealocytes is mimicked by the synthetic diacylglycerol, 1-oleoyl-2-acetyl-*sn*-glycerol (Figure 3A; Sugden et al., 1985). Diacylglycerol is generated from PtdIns as a result of increased turnover. Furthermore, the effect of PE on cyclic AMP is also mimicked by several analogs of diacylglycerol, 4-β-phorbol-12-myristate-13-acetate (PMA) and 4-β-phorbol-12, 13-dibutyrate (PDBu; Figure 4A). These compounds act at low concentrations to activate protein kinase C. Their relative potency in amplifying the cAMP response is similar to that seen for activation of protein kinase C in vitro in other systems (Castagna et al., 1982). For example, 4-α-phorbol-12, 13-didecanoate (PDD), an inactive PMA analog in other systems, is also inactive in pinealocytes.

The time course of the effects of ISO in the presence of PMA (Figure 5A) is essentially identical to that of NE (Vanecek et al., 1985), suggesting that PMA and PE act via the same mechanism.

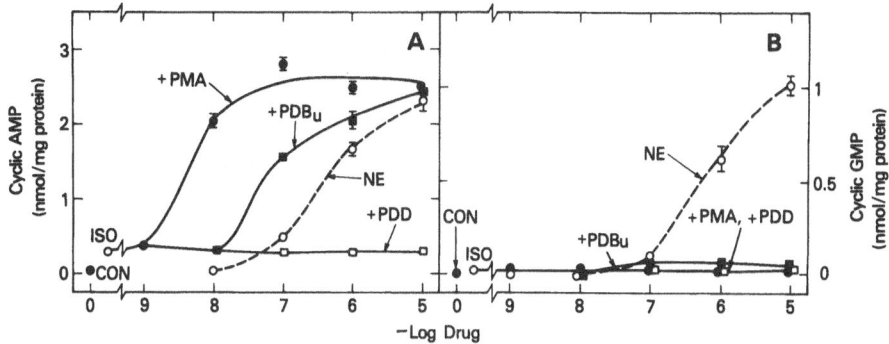

Figure 4. Potentiation of β-adrenergic cyclic nucleotide responses by phorbol esters (A, cyclic AMP; B, cyclic GMP). Pinealocytes (100,000 per tube) were treated for 15 minutes with isoproterenol (ISO, 1 μM) either alone or in combination with 4-β-phorbol-12-myristate-13-acetate (PMA), 4-β-phorbol-12-13-dibutyrate (PDBu) or 4-α-phorbol-12-13-didecanoate (PDD). The dose response curve of NE is shown for comparison. Phorbol ester was added as 100x concentrated solutions in ethanol. All points represent the mean ± SEM of three samples. The absence of error bars indicates that the SEM was less than the area covered by the symbol. From Sugden et al., 1985.

Table 1. *Effect of PE or PMA treatment on pinealocytes on cytosolic and membrane-bound protein kinase C activity*

Expt no	Pretreatment	Protein kinase C activity (pmol ^{32}P incorporated per 3 min per 5×10^6 cells)	
		Cytosol	Membranes
1	Water	13,750	1,000
	Phenylephrine(10^{-6}M)	8,725	5,160
2	Vehicle	13,248	1,395
	PMA(10^{-7}M)	6,286	3,164

Pinealocytes (5×10^6) were treated with phenylephrine (10^{-6}M) for 1 min or with PMA (10^{-7}M) for 15 min. Cells were pelleted (1,000g, 1 min), or washed, then lysed in water (250 μl) containing CaCl$_2$(10^{-5}M) by brief (5 sec) sonication (expt 1) or by repeated aspiration into a syringe through a fine (27 gauge) needle (expt 2). Membrane and cytosol were separated (12,000g, 1 min at 4°C) and 0.5 ml of buffer A(20 mM Tris-HCl, pH 7.5, 2 mM EDTA, 0.5 M EGTA, 2 mM phenylmethylsulfonyl fluoride) was added to each preparation. Cytosol was applied to a DE-52 cellulose column (0.5 × 1 cm) equilibrated with buffer A, and the enzyme eluted with 0.08 M NaCl in buffer A. Membranes were solubilized with Nonidet P-40 (1%) by gentle mixing for 30 min at 4°C. The preparation was centrifuged (Beckman Microfuge, 10,000g, 2 min at 4°C) and the detergent-solubilized preparation was applied to a DE-52 cellulose column and eluted as described for cytosol. Protein kinase C activity was assayed in triplicate as described previously with 0.75 mM Ca^{2+}, in the presence and absence of 24 μg phosphatidylserine, and 1.6 μg diolein. Protein kinase C activity was determined by subtracting the amount of ^{32}P incorporated into histone H1, in the absence of phospholipids. Protein kinase C activity is expressed as pmol ^{32}P incorporated per 3 min per total cytosolic and membrane fraction from 5×10^6 cells.

Role of protein kinase C

The PE → diacylglycerol effect on cyclic AMP appears to be mediated by protein kinase C. This conclusion is based on the results of the direct measurement of protein kinase C in pinealocytes. It was found that translocation of the enzyme from the cytoplasm to the membrane, which is presumed to reflect activation, is caused either by PE or PMA, both of which also potentiate the effects of ISO on cyclic AMP, as discussed above (Sugden et al., 1985).

The mechanism of action of protein kinase C is not clear. One possibility is that it phosphorylates a membrane associated protein involved in the stimulation of cyclic AMP production. This could be either a component of N_S, the GTP binding adenylate cyclase stimulatory protein, or of adenylate cyclase itself. The end result would be an enhanced cyclic AMP response.

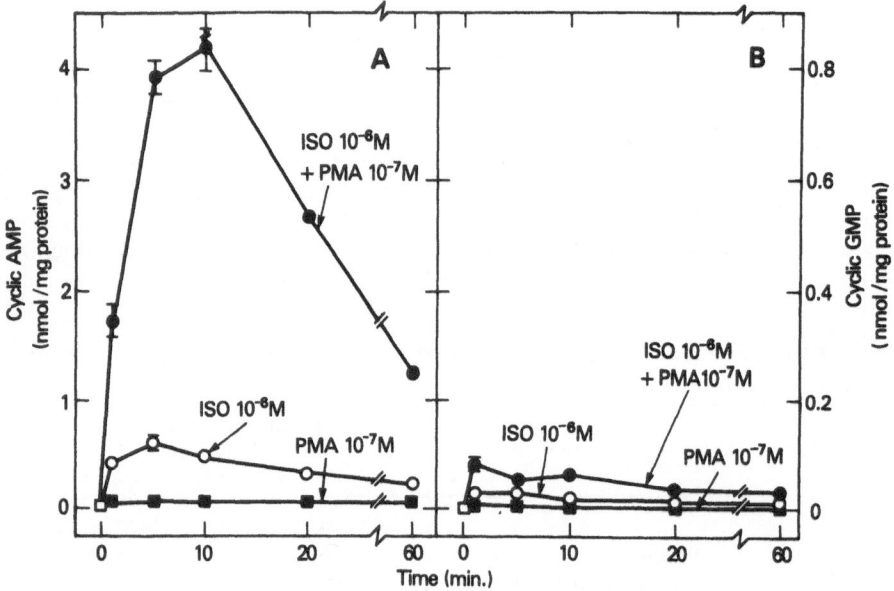

Figure 5. Time course of PMA potentiation of β-adrenergic cyclic nucleotide responses (A, cyclic AMP; B, cyclic GMP). Pinealocytes (100,000/tube) were treated with PMA (10^{-7}M), (—) ISO (10^{-6}M), or PMA (10^{-7}M) plus (—) ISO (10^{-6}M) for the periods indicated. All points represent the mean ±SEM for three samples. From Sugden et al., 1985.

Regulation of Cyclic GMP

The mechanism through which NE stimulates cyclic GMP is not clear. Although it has characteristics similar to the adrenergic stimulation of cyclic AMP, i.e. α_1-adrenergic receptors potentiate the effects of β-adrenoceptors, there are clear differences. For example, there is no indication that NE stimulates guanylate cyclase activity. Similarly, although there is clear evidence to indicate that phospholipids are involved in the α_1-adrenergic potentiation of the β-adrenergic stimulation of cyclic GMP, there is little indication that the effects of PE are mimicked by PMA, or diacylglycerol (Figures 3B, 4B, 5B; Sugden et al., 1985). However, it is generally known that stimulation of cyclic GMP involves a calcium-dependent mechanism, that there is a close association in many systems between cyclic GMP and calcium, and that α_1-adrenoceptors regulate PtdIns turnover and calcium influx in some tissues. This provides reason to suspect that a still to be identified phospholipid mechanism does play a role in the adrenergic regulation of cyclic GMP in the pineal gland.

SEE-SAW SIGNAL PROCESSING

A reciprocal relationship exists between the sensitivity of the cyclic AMP and cyclic GMP responses in the pineal gland (Klein et al., 1979). Stimulus deprivation produced by denervation causes a supersensitive cyclic AMP response and a severely subsensitive cyclic GMP response. This relationship has been termed see-saw signal processing.

It is possible that the supersensitivity of the cyclic AMP response could reflect a tonic elevation of PtdIns turnover, or another metabolic change which would elevate diacylglycerol levels. Perhaps the decrease in cyclic GMP responsiveness is also related to this.

FINAL COMMENT: A MODEL OF HOW MODULATORS ACT?

The studies reviewed here indicate that the stimulation of cyclic AMP in the pineal gland involves α_1-adrenergic stimulated translocation of protein kinase C. This is apparently dependent upon the generation of diacylglycerol which is secondary to increased PtdIns turnover. This appears to potentiate the β-adrenergic stimulation of cyclic AMP which is due to an increase in adenylate cyclase activity. In contrast, this mechanism does not appear to be involved in the α_1-adrenergic potentiation of the β-adrenergic stimulation of cyclic GMP.

It will be of interest to determine if this synergistic cyclic AMP regulatory mechanism occurs in other tissues. Perhaps this type of synergistic mechanism involving protein kinase C potentiation of adenylate cyclase responsiveness is the basis of the interaction of many transmitters or hormones, and bioactive compounds popularly described as modulators, i.e. compounds which have little effect by themselves, but alter the potency of other regulatory agents. This issue is important because if the protein kinase C mechanism were found to be a common means of controlling bioresponses, it would open a new approach to drug therapy.

REFERENCES

Auerbach DA, Klein DC, Woodard C, Auerbach GD (1981) Neonatal rat pinealocytes; Typical and atypical characteristics of [^{125}I]Iodohydroxybenzylpindolol binding and adenosine 3', 5'-monophosphate accumulation. Endocrinol 108: 559-567.

Berg GR, Klein DC (1972) Norepinephrine stimulates ^{32}P incorporation into a specific phospholipid fraction of postsynaptic pineal membranes. J Neurochem 19: 2519-2532.

Castagna M, Takai Y, Kaibuchi K, Sano K, Kikkawa Y, Nishizuka Y (1982) Direct activation of calcium activated phospholipid dependent protein kinase by tumor promoting phorbol esters. J Biol Chem 257: 7847-7851.

Deguchi T (1973) Role of the beta-adrenergic receptor in the elevation of adenosine cyclic 3', 5'-monophosphate and induction of serotonin N-acetyltransferase in rat pineal glands. Mol Pharmacol 9: 184.

Klein DC, Auerbach DA, Namboodiri MAA, Wheler GHT (1981) Indole amine metabolism in the mammalian pineal gland. In: Reiter RJ (ed): The Pineal Gland: Anatomy and Biochemistry. CRC Press, Boca Raton, Florida pp. 199-227.

Klein DC, Sugden D and Weller JL (1983) Postsynaptic α-adrenergic receptors potentiate the β-adrenergic stimulation of pineal serotonin N-acetyltransferase. Proc Natl Acad Sci USA 80: 599-603.

O'Dea RF and Zatz M (1976) Catecholamine-stimulated cyclic GMP accumulation in the rat pineal: Apparent presynaptic site of action. Proc Natl Acad Sci USA 73: 3398-3402.

O'Dea RF, Gagnon C and Zatz M (1978) Regulation of cyclic GMP in the rat pineal and posterior pituitary glands. J Neurochem 31: 733-738.

Smith TL, Eichberg J and Hauser G (1979) Postsynaptic localization of the alpha receptor-mediated stimulation of phosphatidylinositol turnover in pineal gland. Life Sciences 24: 2179-2184.

Strada S, Klein DC, Weller JL and Weiss B (1972) Effect of norepinephrine on the concentration of adenosine 3', 5'-monophosphate in pineal organ culture. Endocrinol 90: 1470-1475.

Sugden D and Klein DC (1984) Rat pineal α_1-adrenoceptors: identification and characterization using [^{125}I]iodo-2-[β-(4-hydroxyphenyl) ethylaminomethyl tetralone. Endocrinol 114: 435-442.

Sugden D, Vanecek J, Klein DC, Thomas TP and Anderson WB (1985) Activation of protein kinase C potentiates isoprenaline-induced cyclic AMP accumulation in rat pinealocytes. Nature 314: 359-361.

Vanecek J, Sugden D, Weller JL and Klein DC (1985) Atypical synergistic α_1- and β-adrenergic regulation of adenosine 3', 5'-monophosphate and guanosine 3', 5'-monophosphate in rat pinealocytes. Endocrinol 116: 2167-2173.

Voisin P, Namboodiri MAA and Klein DC (1984) Arylamine N-acetyltransferase and arylalkylamine N-acetyltransferase in the mammalian pineal gland. J Biol Chem 259: 10913-10918.

Zatz M, Kebabian JW, Romero JA, Lefkowitz RJ and Axelrod J (1976) Pineal beta-adrenergic receptor: correlation of binding of ^{3}H-1-alprenolol with stimulation of adenylate cyclase. J Pharmacol Exp Ther 196: 714-722.

Phospholipid research and the nervous system
Biochemical and molecular pharmacology
L.A. Horrocks, L. Freysz, G. Toffano (eds)
Fidia Research Series, vol. 4.
Liviana Press, Padova. © 1986

THE EFFECTS OF ETHANOL
ON RECEPTOR ACTIVATED PHOSPHOLIPID CASCADES

Fulton T. Crews, Cynthia Theiss,
Robert Raulli and Rueben A. Gonzales

Department of Pharmacology and Therapeutics,
University of Florida School of Medicine,
Box J-267, J. Hillis Miller Health Center, Gainesville, Florida 32610

INTRODUCTION

Ethanol is one of the most widely used drugs in the world. The resulting alcohol abuse and alcoholism is a major health and social problem. An understanding of the actions of ethanol are fundamental to a rational approach to treating alcoholism and to reversing the pathology associated with excessive ethanol consumption.

Although the exact mechanisms of ethanol's actions are not clearly understood, there is considerable evidence that it acts by altering membrane structure. Ethanol is known to partition into the lipid portion of biomembranes (Seeman, 1972). Physiological concentrations of ethanol, e.g. 10-100 mM, are known to disrupt brain, liver, mast cell, and erythrocyte membranes (Goldstein et al., 1982; Rottenberg et al., 1981; Gonzales and Crews, 1985). The ability of ethanol to disrupt mouse brain membranes has been correlated to the genetic sensitivity of mice to ethanol (Goldstein et al., 1982). In addition, the disordering actions of alcohols on membranes have been related to intoxication (Lyon et al., 1981). Furthermore, studies on tolerance and dependence to ethanol have related the changes in the behavioral actions of ethanol to adaptive changes in membrane responses to ethanol (Chin and Goldstein, 1981; Rottenberg et al., 1981; Crews et al., 1983). Thus, the disorganizing and fluidizing actions of ethanol on membranes appear to underlie the behavioral effects of ethanol. Although the link between behavioral and membrane effects is not clear, it is likely to be related to changes in neurotransmitter receptor responses and membrane ion channels.

ETHANOL AND RECEPTOR STIMULATED SECRETION FROM MAST CELLS

The secretion of histamine from mast cells can be stimulated by two or more types of secretagogues. IgE-like secretagogues require extracellular calcium and activate secretion by causing an influx of calcium. Somatostatin and compound 48/80 do not require extracellular calcium and mobilize internal stores of calcium (Crews and Heiman, 1984). Ethanol has been shown to inhibit both IgE-like stimulation of histamine release and somatostatin stimulated release. IgE receptor stimulated secretion with the lectin, concanavalin A, is much more sensitive to inhibition than somatostatin. For example, 100 mM ethanol inhibited concanavalin A stimulated release by 56%, whereas somatostatin stimulated release was reduced only 28% (Crews and Gonzales, 1985). Studies on membrane organization and apparent microviscosity using fluorescent polarization indicated that ethanol alters mast cell membrane structure over the same concentration range that it inhibits release. The changes in membrane structure caused by ethanol correlated ($r^2 = 0.99$) with both the inhibition of IgE-like and somatostatin stimulated release of histamine. These findings suggest that ethanol's effects on membranes can alter receptor function and that some receptors are more sensitive to the membrane actions of ethanol than other receptors. In the case of receptor stimulated histamine release, the IgE receptors which stimulate the influx of calcium are more sensitive to the disruptive actions of ethanol than the receptors which release intracellular stores of calcium. Because the mast cell membranes are altered by ethanol similarly to synaptosomal membranes (Gonzales and Crews, 1985), it is likely that certain neurotransmitter receptors are more sensitive to the membrane actions of ethanol than other receptors.

RECEPTOR ACTIVATION OF PHOSPHOINOSITIDE TURNOVER IN BRAIN

Receptor stimulated changes in lipid metabolism represent a major signal transduction mechanism (Crews, 1982; Berridge, 1984). Recent studies have indicated that the turnover of phosphoinositides initiates a diverse cascade involving both lipids and calcium. Inositol lipid metabolism is coupled to several receptor systems. Three inositol lipids are involved in this receptor activated cascade.

PtdIns is the predominant phosphoinositide in membranes, with PtdIns4P and PtdIns(4,5)P_2 making up less than 10% of the total inositol lipids. Recent studies have suggested that receptor activation leads to the rapid breakdown of PtdIns(4,5)P_2 to form two second messengers, diacylglycerol and Ins(1,4,5)P_3 (Fig. 1; Berridge, 1984). InsP_3 releases internal calcium stores from a variety of cell types (Joseph, 1984). Diacylglycerol can be formed from the hydrolysis of any of the phosphoinositides, including PtdIns(4,5)P_2 (Fig. 1). DAG activates a phosphatidylserine-calcium dependent protein kinase by increasing the affinity of the kinase for its cofactors. This kinase has very high activity in brain tissue and is likely to play a role in neurotransmission. Thus, receptor activated phosphoinositide breakdown is an important signal transduction mechanism involving lipids at several levels.

Cholinergic and Adrenergic Receptors

We have studied receptor stimulated phosphoinositide breakdown in rat brain slices (Gonzales and Crews, 1984, 1985). Both cholinergic and adrenergic receptors activate

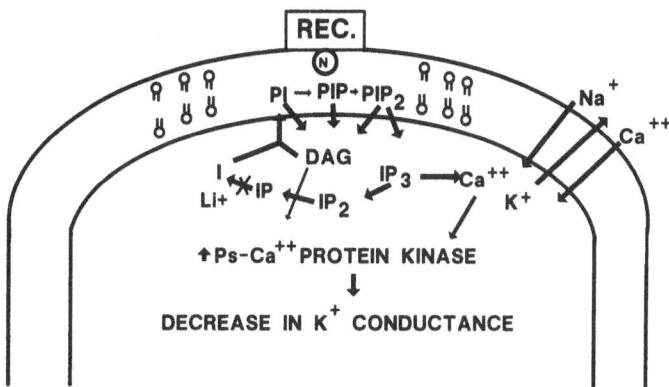

Figure 1. Receptor activated phosphoinositide cascade in neurons. Receptors for a number of agents can activate inositide hydrolysis through a guanine nucleotide binding protein (N). All three phosphoinositides can be hydrolyzed to form diacylglycerol (DAG) and the corresponding phosphorylated inositol sugars. Inositol 1,4,5-trisphosphate (IP$_3$) has mobilized calcium in a number of tissues. Both DAG and IP$_3$ could activate PS/Ca kinase. The resulting phosphorylations could alter membrane conductances. Since the muscarinic cholinergic receptor inhibits potassium conductances in the hippocampus as well as stimulating phosphoinositide hydrolysis, a decrease in potassium conductance may be the change induced by this cascade. See text for more details.

phosphoinositide breakdown in cerebral cortical slices prelabeled with [^{3}H]Ins. Carbachol, a muscarinic agonist, stimulates phosphoinositide breakdown to a greater extent than pilocarpine (Fig. 2). Combinations of carbachol and pilocarpine do not produce any greater stimulation of [^{3}H]Ins phosphate formation than pilocarpine alone. These findings indicate that pilocarpine is a partial agonist at cerebral cortical muscarinic receptors.

Oxotremorine is also a partial agonist for stimulation of phosphoinositide breakdown. Atropine and pirenzepine, muscarinic and selective M$_1$-muscarinic antagonists respectively, both block the response to carbachol suggesting that the receptor is muscarinic and of the M$_1$ muscarinic subtype.

The stimulation of inositol phosphates formation by norepinephrine is blocked by prazosin, but not altered by yohimbine, suggesting that it is mediated through an alpha$_1$-adrenergic receptor (Fig. 3). Clonidine and phenylephrine are both partial agonists. These data indicate that many drugs thought to be full agonists are only partial agonists for cortical phosphoinositide metabolism. Additivity experiments have suggested that in some brain regions these two receptor systems may share the same pool of phosphoinositides (Gonzales and Crews, 1985). Thus, muscarinic and alpha-adrenergic receptors may be on the same neurons in the cerebral cortex and hippocampus, and are likely to activate similar receptor mediated cascades.

Although there are many similarities between muscarinic and alpha stimulated Ins phosphate formation, there are some differences. One difference between muscarinic-cholinergic and alpha-adrenergic stimulated Ins phosphate formation is calcium sensitivity. Whereas carbachol stimulated breakdown of phosphoinositides shows little or no decrease in the absence of extracellular calcium (Gonzales and Crews, 1984; Kendall and Nahorski, 1984), norepinephrine stimulated breakdown is more sensitive to

124

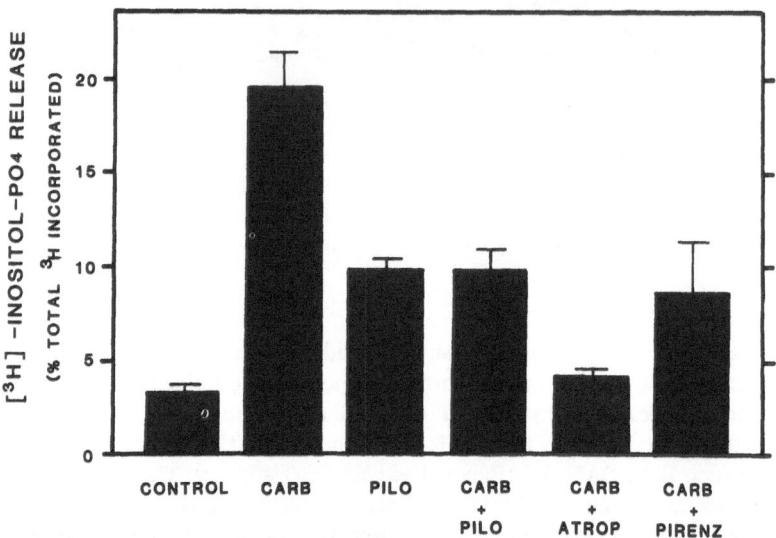

Figure 2. Effects of cholinergic agonists, partial agonists, and antagonists on inositide hydrolysis in rat cortical slices. Rat cerebral cortical slices were labeled with [³H]Ins and receptor stimulated inositide hydrolysis was determined as described in Gonzales and Crews (1984). Antagonists, when present, were added to slices 10 min before agonist stimulation. Each bar represents the mean ± S.E.M. of triplicate values from a single slice preparation. Concentrations of drugs were: carbachol, 1 mM; pilocarpine, 1 mM; atropine, 1 μM; pirenzepine, 1 μM.

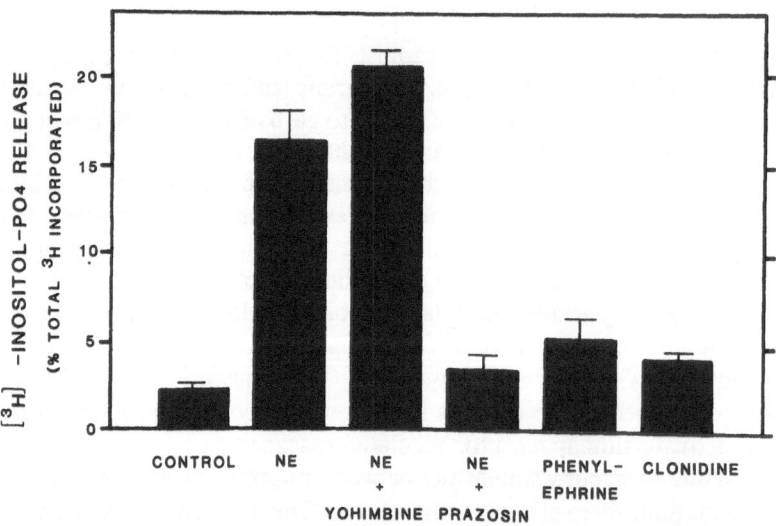

Figure 3. Effects of adrenergic agonists, partial agonists and antagonists on inositide hydrolysis in rat cortical slices. Results were obtained as described in Fig. 2. Drug concentrations were: norepinephrine, 100 μM; yohimbine, 1 μM; prazosin, 1 μM; phenylephrine, 100 μM; clonidine, 100 μM.

extracellular calcium. Kendall and Nahorski (1984) found that removal of extracellular calcium did not alter carbachol stimulated Ins phosphate formation, whereas norepinephrine's response was reduced approximately 20% (p < 0.05). Differences in the time course of alpha-adrenergic stimulated and muscarinic-cholinergic stimulated inositide breakdown have also suggested slightly different coupling mechanisms for these two receptor binding sites to phosphoinositide hydrolysis (Gonzales and Crews, 1985). It is possible that muscarinic receptors are directly coupled to inositide hydrolysis, whereas a portion of alpha-adrenergic receptors are coupled in a calcium-sensitive manner.

Guanine Nucleotide Coupling Proteins and Phosphoinositide Hydrolysis

Recent studies have suggested that muscarinic and alpha-adrenergic receptor binding sites may be modified by guanine nucleotides.

Receptors which increase cyclicAMP formation have been shown to couple to the enzyme adenylate cyclase through a guanine nucleotide binding protein. It is possible that guanine nucleotide binding proteins couple receptor-agonist complexes to phosphoinositide phosphodiesterases, e.g. phosphoinositide phospholipase C(s). To test this hypothesis, we prelabeled rat cortical membranes with [^{3}H]Ins and investigated the actions of Gpp(NH)p on the hydrolysis of phosphoinositides. As shown in Fig. 4, Gpp(NH)p caused a marked hydrolysis of phosphoinositides from isolated membranes in the absence of calcium. These findings suggest that receptors coupled to phosphoinositide hydrolysis are likely to activate phosphoinositide phosphodiesterase(s) through the actions of a guanine nucleotide binding protein. The guanine nucleotide stimulated hydrolysis appears to primarily involve PtdIns(4,5)P. As calcium was added to the buffer, the release of inositol phosphates increased even in the absence of Gpp(NH)p. These findings suggest that there may be two phosphoinositide phosphodiesterases involved in the hydrolysis of phosphoinositides, one which is activated by guanine nucleotides and perhaps by receptor-agonist complexes and a second enzyme which is activated by calcium. Thus, both receptor-agonist complexes and calcium may stimulate phosphoinositide breakdown.

Other Agents Stimulating Phosphoinositide Hydrolysis

Although muscarinic-cholinergic and alpha-adrenergic receptors are the most effective stimulators of phosphoinositide hydrolysis in cerebral cortical slices, several other types of stimulation can also increase Ins phosphate formation. Serotonin (5-HT), potassium (K$^+$), the calcium ionophore - A23187, glutamate, and a variety of peptides have been shown to increase phosphoinositide breakdown in cerebral cortical slices (Table 1). Although the peptides and 5-HT may increase phosphoinositide breakdown through receptor-guanine nucleotide complexes, potassium, A23187, and perhaps glutamate may act by other mechanisms. A23187 is a calcium ionophore which carries calcium into cells. Thus, it is likely that A23187-stimulated release is secondary to increases in intracellular calcium. Glutamate depolarizes cells just as K$^+$ depolarizes cells allowing calcium ion influx. Thus, these three agents may act primarily by causing calcium influx.

Effects of Ethanol on Phosphoinositide Breakdown

To determine the effects of ethanol on the formation of Ins phosphates, cerebral cortical slices were incubated with ethanol and then stimulated with various agents.

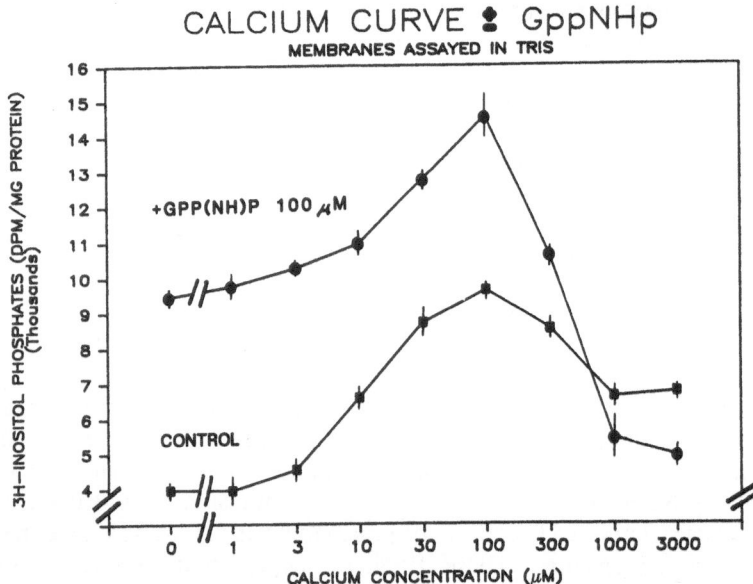

Figure 4. The effect of calcium on Gpp(NH)p stimulated inositide hydrolysis in rat cortical membranes. Cortical slices were labeled with [^{3}H]inositol and subsequently homogenized with a glass-glass homogenizer in Krebs-Ringer buffer containing 1 mM EDTA. The homogenate was centrifuged at $1000 \times g$ for 20 min and the pellet was resuspended in 20 mM Tris to obtain lysed membranes. The membranes were washed twice with Tris before use. [^{3}H]Inositol phosphates released were analyzed as previously described (Gonzales and Crews, 1984). Results represent the means $\pm$ S.E.M. of triplicate incubations of a single membrane preparation. The experiment was repeated with essentially identical results.

Ethanol was found to inhibit norepinephrine, K$^+$, A23187 and glutamate stimulation, but not muscarinic receptors stimulated with carbachol, or 5-HT receptors (Table 1). Norepinephrine, K$^+$, and glutamate stimulation were reduced by 45, 53, and 63%, respectively, by 500 mM ethanol, whereas A23187 stimulation was only reduced 23%. A comparison of the effects of various concentrations of ethanol on membrane microviscosity as determined by membrane polarization and the inhibition of norepinephrine stimulated phosphoinositide hydrolysis suggests that changes in membrane properties parallel the inhibition of norepinephrine stimulated hydrolysis of phosphoinositides (Fig. 5).

There are several possible explanations for the variation in the actions of ethanol on these diverse stimuli. One possible explanation is that the membranes of different neurons have a differential sensitivity to ethanol. Those cells having membranes particularly sensitive to ethanol may be disrupted more than those on cells whose membranes are resistant to ethanol. This suggestion is supported by our finding that alpha-adrenergic stimulated inositide hydrolysis in liver is much more sensitive to inhibition by ethanol than alpha-adrenergic stimulation in brain (Table 1).

Table 1. *Effect of various agonists on inositol lipid hydrolysis and the effect of ethanol in vitro*

	Concentration (µM)	Control			500mM EtOH		
Cortex		*Mean*	±	*S.E.M.*	*Mean*	±	*S.E.M.*
Buffer		2.6	±	0.6	0.6	±	0.4*
NE	100	25.8	±	1.7	14.1	±	1.7**
Carbachol	1000	17.5	±	1.0	15.4	±	1.3
5-HT	100	6.4	±	0.5	5.5	±	0.7
K^+	20mM	12.4	±	0.8	9.8	±	0.8**
A23187	30	12.4	±	0.9	9.6	±	0.8*
Glutamate	1000	8.9	±	0.7	3.3	±	0.7**
Somatostatin	1	3.2	±	0.5		N.D.	
ACTH	1	3.0	±	0.5		N.D.	
Substance P	1	2.9	±	0.6		N.D.	
AVP	1	3.9	±	0.2		N.D.	
Liver					*100 mM EtOH*		
Buffer		1.1	±	0.2	1.0	±	0.4
NE	100	7.5	±	0.5	2.4	±	0.3**
AVP	1	15.0	±	1.4	3.3	±	0.6**

Slices of the indicated tissue were incubated with [^{3}H]Ins to label inositides. The accumulation of [^{3}H]Ins phosphates was measured in the presence of 8 mM lithium by chromatography of aqueous tissue extracts on Dowex-1 columns. Ethanol when present was added to tissue 10 min prior to addition of agonist. Results are presented as percent [^{3}H]Ins phosphates released compared to the total ^{3}H incorporated into lipids (means ± S.E.M. of 2-4 slice preparations each done in triplicate). AVP = arginine vasopressin.
*$p < 0.05$ and **$p < 0.01$ by analysis of variance and Newman-Keuls test for the comparison of control versus ethanol in vitro.

An additional factor may be related to cellular calcium flux. A23187, the calcium ionophore, stimulates phosphoinositide hydrolysis in brain slices as well as a number of other cell types. In addition, K^+ and glutamate are likely to act by depolarizing cells and inducing calcium influx. Studies in GH3 pituitary cells have shown that TRH can directly stimulate the hydrolysis of PtdIns(4,5)P_2 to form InsP_3 and DAG.

However, the calcium ionophore, A23187, stimulated the hydrolysis of PtdIns4P, but not PtdIns(4,5)P_2 (Kolesnick and Gershengorn, 1984). Similar studies in macrophages have shown that A23187-induced calcium influx causes the breakdown of PtdIns4P to form InsP_2, but not breakdown of PtdIns(4,5)P to InsP_3 (Emilsson and Sundler, 1984). These studies suggest that phosphoinositide hydrolysis can occur through two separate mechanisms, e.g. direct receptor coupling and calcium influx. The difference between these two mechanisms appears to be that receptors may preferentially hydrolyze PtdIns(4,5)P, whereas calcium may activate the hydrolysis of primarily PtdIns and PtdIns4P. Our finding that guanine nucleotide stimulated hydrolysis is independent and separate from calcium stimulated hydrolysis is consistent with two mechanisms for phosphoinositide breakdown. This is also consistent with our brain slice experiments, because we have not separated the various Ins phosphates. Thus, it is possible that certain agents may activate inositide hydrolysis by causing calcium influx, whereas other agents, such as carbachol may directly stimulate PtdIns(4,5)P_2 hydrolysis through an agonist-receptor guanine nucleotide coupling protein complex.

128

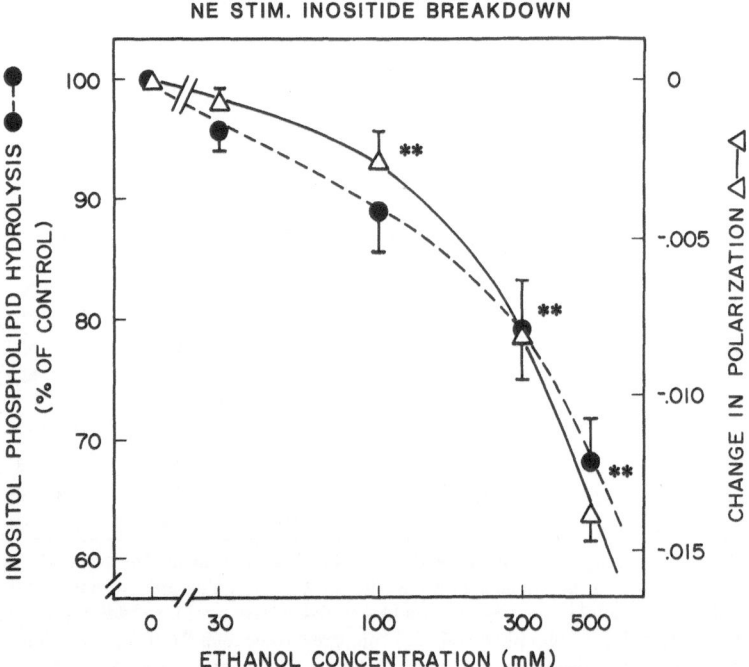

NE STIM. INOSITIDE BREAKDOWN

Figure 5. Comparison of the effects of ethanol in vitro on norepinephrine stimulated inositide hydrolysis and the fluorescence polarization of diphenylhexatriene (DPH). Norepinephrine stimulated inositol phospholipid breakdown was measured in rat cortical slices as described in Fig. 2. Results are presented as means ± S.E.M. of 4 slice preparations, each performed in triplicate. The fluorescence polarization of DPH in synaptosomes was determined as described previously (Gonzales and Crews, 1985). Each point represents the mean ± S.E.M. of 5 separate membrane preparations. **Denotes $p < 0.01$ for the effect of ethanol in vitro at the indicated concentration for both inositide breakdown and change in DPH polarization as determined by analysis of variance and Newman-Keuls test.

Ethanol has been shown to inhibit synaptosomal potassium, glutamate and veratridine-stimulated calcium influx (Stokes and Harris, 1982). Ethanol also inhibits phosphoinositide hydrolysis by K^+, glutamate, and A23187, the three agents which would be most likely to activate phosphoinositide hydrolysis through calcium influx (Table 1). As mentioned above, norepinephrine stimulated hydrolysis is more sensitive to extracellular calcium than carbachol. Similarly, norepinephrine is partially sensitive to ethanol, whereas carbochol is essentially insensitive. Taken together, these findings suggest that part of the norepinephrine response is secondary to calcium influx, that ethanol inhibits calcium influx and thus, phosphoinositide breakdown secondary to calcium influx. Although additional data is needed to support this hypothesis, it is consistent with the differential actions of ethanol on the various agents listed in Table 1 and with the mast cell findings described above.

PHOSPHATIDYLSERINE-CALCIUM DEPENDENT PROTEIN KINASE

Effects of Ethanol on PtdSer/Ca Kinase

The phosphatidylserine-calcium dependent protein kinase has a very high activity in brain and is likely to play a role in neurotransmission (Fig. 1). To determine if ethanol might have an effect on protein kinase activation, we incubated partially purified protein kinase with various concentrations of ethanol. Ethanol inhibits PtdSer/Ca kinase at concentrations greater than 100 mM when saturating concentrations of PtdSer and Ca^{++} are used (Fig. 6). Complete saturation curves for Ca^{++} indicated that increasing concentrations could not overcome the inhibition due to ethanol. Thus, in the presence of 1 M ethanol, the apparent affinity (i.e. K_m, as derived according to Wilkinson, 1961) for calcium in control and ethanol preparations was 5.28 μM and 3.23 μM,

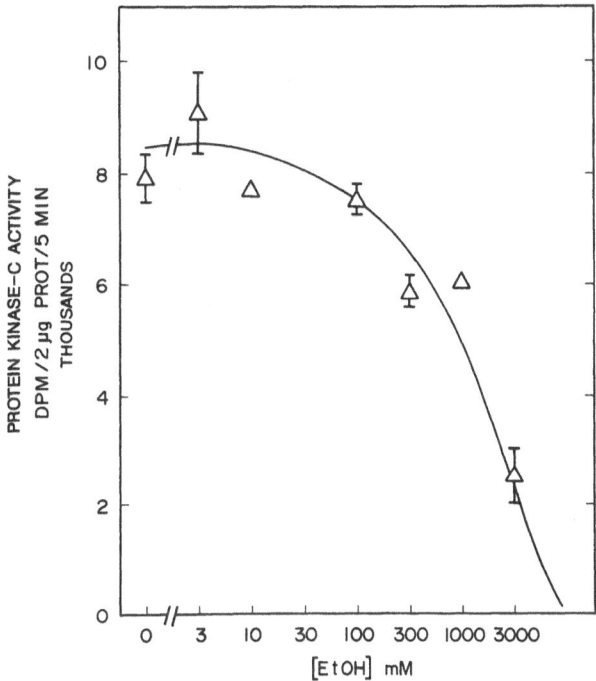

Figure 6. Effects of ethanol on phosphatidylserine-calcium dependent protein kinase activity. A Triton-X-100 extract of rat brain homogenate was purified by ammonium sulfate fractionation, anion exchange chromatography and gel filtration chromatographs as described (Wise et al., 1982).

Protein kinase-C activity was assayed as follows: purified extract is incubated at 37°C for 5 min in 25 mM Pipes buffer pH 6.5, 50 mM mercaptoethanol, 10 mM $MgCl_2$, 250 μM EGTA, 30 mM $CaCl_2$, 30 μg/ml phosphatidylserine, various concentrations of EtOH and 40 μg of histone 1 as the phosphate acceptor. The reaction is started by the addition of 5 μM [γ-^{32}P]ATP. The reaction is stopped by the addition of 10% TCA with 0.25% sodium tungstate. Error bars represent S.E.M., n=3. The IC_{50} for EtOH as determined by probit analysis is 1 M.

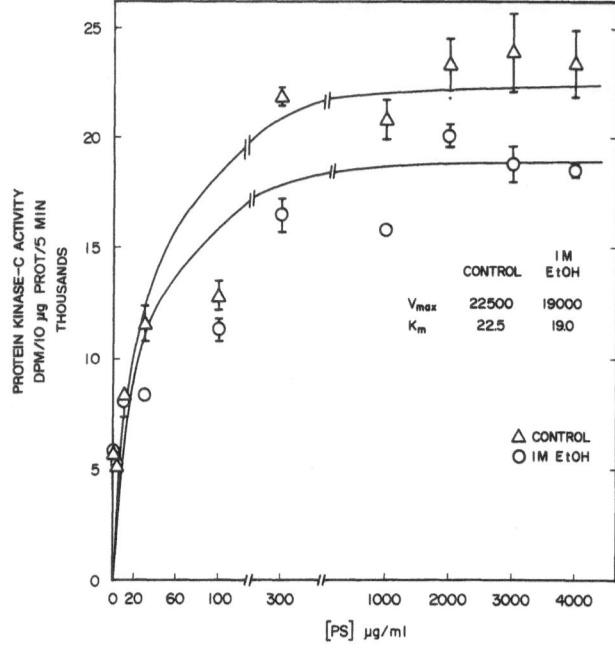

Figure 7. Phosphatidylserine saturation curve in the presence and absence of 1 M EtOH. Rat brain protein kinase was purified by ammonium sulfate fractionation, anion exchange chromatography and gel filtration chromatography as described (Wise et al., 1982). Protein kinase activity was assayed as follows: purified extract is incubated at 37°C for 5 min in 25 mM Pipes buffer pH 6.5, 50 mM mercaptoethanol, 10 mM MgCl$_2$, 250 μM EGTA, 500 μM CaCl$_2$, various concentrations of phosphatidylserine, 1 M EtOH and 40 μg of histone 1 as the phosphate acceptor. The reaction is started by the addition of 5 μM [γ-^{32}P]ATP. The reaction is stopped by the addition of 10% TCA with 0.25% sodium tungstate. Error bars represent S.E.M., n = 3. Kinetic parameters were determined by the nonlinear curve fitting method of Wilkinson (1961).

respectively. The maximum velocity in control and ethanol preparations was 20,600 and 15,300 cpm/10 μg protein/5 min, respectively. Similar data was found with PtdSer (Fig. 7). Ethanol had only a slight effect on apparent affinity. K$_m$ values for control and ethanol treated were 22.5 and 19.0 μg/ml PtdSer, respectively. On the other hand, the maximum velocity was reduced from a control value of 22,500 to 19,000 dpm/10 μg protein/5 min in the presence of ethanol. These data suggest that ethanol may act as a noncompetitive inhibitor of this protein kinase. Athough these concentrations of ethanol are above the physiological levels, it is possible that ethanol might alter calcium activation of the kinase. As mentioned above, PtdSer/Ca kinase can also be activated by DAG formation. To determine the effects of ethanol on DAG activated kinase, experiments were done in the presence of phorbol esters, which are known to mimic the actions of DAG on PtdSer/Ca kinase. Ethanol in concentrations up to 1 M did not alter phorbol ester stimulated PtdSer/Ca kinase in the presence of 30 μg/ml PtdSer and 30 μM Ca. Our finding that high concentrations of ethanol can act as a noncompetitive inhibitor of PtdSer/Ca kinase indicate that the organization of membrane lipids may be important for the interaction of the kinase with its lipid cofactors. However, the high concentrations required to inhibit the kinase and the lack of inhibition in the presence of phorbol esters makes it unlikely that ethanol would acutely alter the activation of PtdSer/Ca kinase in vivo at physiological concentrations.

SUMMARY AND CONCLUSIONS

A number of classical neurotransmitters and peptides stimulate phosphoinositide turnover in brain slices. The pronounced response to cholinergic and adrenergic receptors is probably due to the majority of cells in the slice preparations responding to stimulations. Although both neuronal and glial cells may respond, studies have shown that the neuronal response is several-fold greater than the glial response (Gonzales et al., 1985). The smaller response to peptides is likely to be due to the discrete localization of peptide receptors on certain cells within the slice. Our finding that guanine nucleotides can stimulate PtdIns(4,5)P_2 hydrolysis in isolated brain membranes suggests that receptors may couple to phosphoinositide phosphodiesterase through a nucleotide binding protein. Because this occurs in the absence of calcium, it is possible that there is a calcium independent phosphodiesterase which is coupled to receptors. Calcium also clearly activated phosphoinositide hydrolysis. Thus, it is possible that various receptors can act through two different mechanisms, one that is directly coupled to the phosphodiesterase and one which is indirectly coupled through calcium influx. In any case, a variety of receptors can activate the hydrolysis of phosphoinositides in brain.

The activation of phosphoinositide breakdown by potassium and the calcium ionophore, A23187, is not likely to be due to receptors for these two agents. It is possible that they release an endogenous agent which acts to stimulate a receptor. However, we have not been able to demonstrate inhibition of either of these two agents by receptor antagonists to those transmitters known to stimulate hydrolysis, e.g. cholinergic and adrenergic antagonists. Calcium-free medium and cobalt do block the response to these agents. These data lead us to hypothesize that potassium, A23187, and possibly glutamate stimulate phosphoinositide hydrolysis by inducing calcium influx.

Ethanol inhibits phosphoinositide stimulated hydrolysis by potassium and glutamate and, to a lesser extent, A23187 and norepinephrine. The inhibition of norepinephrine stimulated hydrolysis corresponds with changes in apparent membrane viscosity. Because ethanol inhibits calcium influx in synaptosomes (Stokes and Harris, 1982), it is possible that ethanol inhibits phosphoinositide hydrolysis which is secondary to calcium influx. Additional studies will be required to test these hypothetical mechanisms of phosphoinositide hydrolysis and the actions of ethanol on these responses.

REFERENCES

Berridge MJ (1984) Inositol trisphosphate and diacylglycerol as second messengers. Biochem J 220: 345-360.

Chin JH and Goldstein DB (1981) Membrane-disordering action of ethanol: Variation with membrane cholesterol content and depth of the spin-label probe. Mol Pharmacol 19: 425-431.

132

Crews FT (1982) Rapid changes in phospholipid metabolism during secretion and receptor activation. Int Rev Neurobiol 23: 141-163.

Crews FT and Gonzales R (1985) Correlation of ethanol's membrane action and inhibition of receptor stimulated release from rat mast cells. J Pharmacol Exp Therap, in press.

Crews FT and Heiman AS (1985) Interaction of phospholipid methylation and phosphatidylinositol metabolism in stimulation of secretion. Phospholipids in the Nervous System 2: 87-98.

Crews FT, Majchrowitz E and Meeks R (1983) Changes in cerebral cortical synaptosomal membrane fluidity and composition in ethanol dependent rats. Psychopharmacology 81: 208-213.

Emilsson A and Sundler R (1984) Differential activation of phosphatidylinositol deacylation via diphosphoinositide in macrophages responding to zymosan and ionophore A23187. J Biol Chem 259: 3111-3116.

Goldstein DB, Chin JH and Lyon RC (1982) EtOH disordering of spin-labeled mouse brain membranes: Correlation with genetic determined EtOH sensitivity of mice. Proc Natl Acad Sci USA 79: 4231-4233.

Gonzales RA and Crews FT (1984) Characterization of the cholinergic stimulation of phosphoinositide hydrolysis in rat brain slices. J Neurosci 4: 3120-3127.

Gonzales RA and Crews FT (1985) Cholinergic and adrenergic stimulated inositide hydrolysis in brain: Interaction, regional distribution, and coupling mechanisms. J Neurochem 45: 1076-1084.

Gonzales RA and Crews FT (1985) Effect of ethanol and aging on histamine release and membranes of mast cells. Alcohol 2: 313-316.

Gonzales RA, Feldstein JB, Crews FT and Raizada MK (1985) Receptor mediated inositide hydrolysis is a neuronal response: Comparison of primary neuronal and glial cultures. Brain Res 345: 350-355.

Joseph SK (1984) Inositol trisphosphate: An intracellular messenger produced by Ca^{++} mobilizing hormones. Trends Biochem Sci 10: 420-421.

Kendall DA and Stephan RN (1984) Inositol phospholipid hydrolysis in rat cerebral cortical slices: II. Calcium requirement. J Neurochem 42: 1388-1394.

Kolesnick RN and Gershengorn MC (1984) Ca^{2+} ionophores affect phosphoinositide metabolism differently than thyrotropin-releasing hormone in GH3 pituitary cells. J Biol Chem 259: 9514-9519.

Lyon RC, McComb JA, Scheurs J and Goldstein DB (1981) A relationship between alcohol intoxication and the disordering of brain membranes by a series of short chain alcohols. J Pharmacol Exp Ther 218: 669-675.

Rottenberg H, Waring A and Rubin E (1981) Tolerance and cross-tolerance in chronic alcoholics: Reduced membrane binding of ethanol and other drugs. Science 213: 583-584.

Seeman P (1972) The membrane actions of anesthetics and tranquilizers. Pharmacol Rev 24: 583-655.

Stokes JA and Harris RA (1982) Alcohol and synaptosomal calcium transport. Mol Pharmacol 22: 99-104.

Wilkinson GN (1961) Statistical estimation in enzyme kinetics. Biochem J 80: 324-332.

Wise BC, Raynor RL and Kuo JF (1981) Phospholipid-sensitive Ca-dependent protein kinase from heart. J Biol Chem 257: 8481-8488.

Phospholipid research and the nervous system
Biochemical and molecular pharmacology
L.A. Horrocks, L. Freysz, G. Toffano (eds)
Fidia Research Series, vol. 4.
Liviana Press, Padova. © 1986

EFFECTS OF ETHANOL ON BRAIN PHOSPHOLIPIDS

Grace Y. Sun, Hsueh-Meei Huang and Albert Y. Sun

Sinclair Comparative Medicine Research Farm and Biochemistry Department,
University of Missouri, Columbia, MO 65203 USA

The primary mode of action of ethanol is on the neural membranes. While excessive intake can result in acute intoxication, development of tolerance and physical dependence is frequently associated with long term intake. On withdrawal of the drug, the behavioral manifestation of hyperexcitability is shown in man, as well as in animal models. In spite of the behavioral and physiological effects known to be associated with alcoholism, the molecular site of ethanol action in the central nervous system has not been clearly elucidated.

Meyer-Overton (see Meyerton and Gottlieb, 1926) correlated the potency of anesthetic drugs with their water-oil partition coefficients. Subsequently, it was realized that the potency of these compounds is based on their interaction with biomembranes and their ability to partition between the membrane and the buffer (Seeman, 1972). Due to its high solubility in water, ethanol is regarded as a weak anesthetic agent, and a relatively high concentration is required for nerve-blocking activity. Nevertheless, ethanol can diffuse freely across the membranes and interact with intracellular organelles. During this past decade, intense interest in the physico-chemical interactions between alcohol and biological membranes has developed (Sun, 1979; Michaelis and Michaelis, 1983). Since a comprehensive review on ethanol and membrane lipids has recently been published (Sun and Sun, 1985), this chapter will be devoted mainly to the effects of ethanol on the dynamic aspects of membrane phospholipid metabolism.

BIOPHYSICAL EFFECTS OF ETHANOL ON MEMBRANES

Biophysical techniques such as electron spin resonance have been useful in elucidating the effects of ethanol on membrane physical properties. Using doxylstearic acids as spin probes, Chin and Goldstein (1977a) observed that ethanol and other aliphatic alchohols elicited a decrease in the membrane order parameter. This "fluidizing effect" can be detected at physiologically feasible concentrations of ethanol. The increase in degree of perturbation with aliphatic alcohols of increasing carbon chain lengths clearly indicates that these molecules interact with the membranes by inserting into the hydrophobic region of the lipid bilayer. During chronic exposure, most membranes exhibit adaptation to ethanol by developing resistance to the disordering effects of ethanol and aliphatic alcohols in vitro (Chin and Goldstein, 1977b). Membranes that are adapted to ethanol also may develop cross-tolerance to other drugs.

Measurement with fluorescent probes is another useful technique for assessing the effects of ethanol on membrane physical properties. The compound 1,6-diphenylhexatriene (DPH) is commonly used to probe the hydrophobic region of the membrane core (Shinitzky and Barenholz, 1978). Ethanol at physiological concentrations facilitates the incorporation of DPH into the synaptic plasma membrane (SPM), resulting in an increase in fluorescence intensity (Harris and Schroeder, 1981). On the other hand, membranes obtained from rats after chronic ethanol administration were less responsive to changes in fluorescence depolarization due to in vitro challenge of ethanol (Harris et al., 1984). Using the 2-toluidinonaphthalene-sulfonate (TNS) probe on SPM, Sun and Seaman (1980) found that ethanol produced a biphasic change in the fluorescence intensity of this probe. The biphasic effect disappeared when membranes isolated from ethanol-tolerant animals were used (Sun, 1983).

Although biophysical studies indicate that ethanol alters membrane structure by physical insertion into the lipid bilayer, the possibility remains that the disordering effect may also be associated with changes in protein conformation. Regardless of whether the lipids or proteins are affected, these perturbations are likely to give rise to alteration of membrane functions. The transport enzymes appear to be especially sensitive to the effects of ethanol (Sun and Samorajski, 1970; Sun and Seaman, 1980). Ethanol in vitro exerts a biphasic effect on synaptosomal (Na, K)-ATPase activity, i.e., enhancement of activity at low concentrations (0.1-0.5%) and inhibition at concentrations higher than 1%. The biphasic response seems to correlate well with the behavioral manifestation of low and high alcohol intake in humans. Several investigators have reported an adaptive increase in (Na, K)-ATPase activity after chronic ethanol administration (Israel and Kuriyama, 1971; Sun et al., 1977a; Rangaraj and Kalant, 1978; Beauge et al., 1983). This subacute and chronic response of (Na, K)-ATPase to ethanol is a good example of the kindling hypothesis of alcohol-membrane interaction (Sun, 1983). Because the activity of (Na, K)-ATPase is known to be dependent on the microenvironment of the membrane lipids (Sun and Sun, 1976), adaptation may be brought about by an alteration of the membrane lipids. Besides (Na, K)-ATPase, other membrane-bound enzymes, such as Ca^{2+}-ATPase (Yamamoto and Harris, 1983) and adenylate cyclase (Rabin and Molinoff, 1981; Saito et al., 1985), may also be altered by the chronic effects of ethanol. There are also indications that ethanol alters receptor functions, but the details of these systems will not be described here.

EFFECTS OF ETHANOL ON MEMBRANE CHOLESTEROL CONTENT

Alteration of the physical properties of membranes due to chronic ethanol ingestion suggests that membranes after exposure to ethanol may become more rigid. Cholesterol is an important determinant of the fluid properties of biological membranes

(Demel and DeKruff, 1976). Consequently, alteration of the cholesterol/phospholipid ratio after ethanol ingestion may offer a valid explanation for the changes in physical properties. Such an increase in cholesterol content of erythrocyte membranes and SPM of mice after chronic ethanol feeding has been reported (Chin et al., 1978). However, it was soon realized that this cannot be the sole determinant for the ethanol effects in brain membranes. In fact, sensitivity to ethanol can be demonstrated even when the cholesterol content of membranes from ethanol-treated animals and controls is made equal (Johnson et al., 1979). It is easier to visualize this change in the erythrocytes, because the lipid content in this membrane can easily be affected by the plasma lipids. The cholesterol in brain membranes is exceptionally stable and is not affected easily by dietary factors. This phenomenon was well demonstrated in a recent study by Chin and Goldstein (1985), in which Japanese quail were fed a cholesterogenic diet. The diet resulted in alteration of the cholesterol level of a number of peripheral membranes, but the brain membranes were completely protected from the dietary effect. Therefore, lack of change in the cholesterol/phospholipid ratio in brain membranes after chronic administration of ethanol (Alling et al., 1982; Wing et al., 1982) is not surprising.

EFFECTS OF ETHANOL ON MEMBRANE PHOSPHOLIPID ACYL GROUPS

The importance of phospholipids and their acyl groups in creating a fluid environment for biological membranes has prompted investigations on the effects of chronic ethanol administration on membrane fatty acid composition. If ethanol in vitro enhances the phospholipid acyl group motion in the lipid bilayer, it is reasonable to expect that in the chronic state, fatty acids in membranes are altered towards the more saturated form in response to the disordering effect. This change was first observed by Littleton and John (1977), who administered ethanol to mice by the inhalation technique. However, it was difficult to observe similar changes when rats were given ethanol by other routes (Sun and Sun, 1979; Alling et al., 1982; Smith and Gerhart, 1982; Crews et al., 1983; Sun and Sun, 1983). One reason for the discrepancies may be the differences in the methods for alcohol administration, the type of diets used, and the animal species selected for these studies (Thompson and Reitz, 1978).

As with cholesterol, inability to demonstrate measurable acyl group changes in brain membranes is not surprising, since this organ is well protected from transient disturbances caused by dietary factors. On the other hand, ethanol-induced acyl group changes are commonly found in the peripheral organs. When Sprague-Dawley rats were exposed to ethanol by the inhalation method, La Droitte et al. (1984) observed a decrease in the linoleic acid level of the phosphoglycerides of erythrocyte membranes, and the changes correlated well with the time course for development of tolerance. Alling et al. (1984) also reported similar acyl group changes in erythrocyte membranes of human alcoholics. In fact, the decrease in linoleic acid level with respect to ethanol consumption is more readily shown in the serum lipids (Alling et al., 1980). In this regard, the Sinclair miniature pig has been used as an animal model for alcoholism, because it consumes alcohol voluntarily and exhibits serum lipid changes similar to the human (Foudin et al., 1984).

EFFECTS OF ETHANOL ON PHOSPHOLIPID ACYL GROUP TURNOVER

A number of enzymic routes are known to modulate the turnover of phospholipid acyl groups. In particular, enzymes involved in the deacylation-reacylation mechanism are important in mediating the turnover activity of specific phospholipid pools in the membrane (Lands and Crawford, 1976; Sun et al., 1979; Sun, 1982). In brain, the lysophospholipid acyltransferase(s) preferentially transfers arachidonic acid to lyso PtdCho and lysoPtdIns (Baker and Thompson, 1973; Corbin and Sun, 1978). Recent studies indicate that the Ca^{2+}-dependent phospholipase A_2 in brain membranes preferentially hydrolyzes the same phospholipids, i.e., PtdIns and PtdCho (Kelleher and Sun, 1985). We have reported an increase in acyltransferase activity in rats after chronic ethanol administration (Sun et al., 1977b). The ethanol-induced increase in ATP-dependent transfer of arachidonic acid to PtdCho and PtdIns of rat brain membranes was most evident between 2 and 24 hr after withdrawal but started to return to control level at 72 hr (Sun et al., 1985a). Along this line, John et al. (1985) found an increase in phospholipase A_2 activity in rats made dependent on ethanol by the inhalation procedure. Thus, results from these studies suggest that chronic ethanol results in increased turnover of membrane phospholipid acyl groups through the deacylation-reacylation mechanism. However, one should realize that these changes pertain only to a small yet metabolically active pool of the membrane phospholipids, and an increase in metabolic activity does not necessarily correlate to a detectable change in the bulk acyl composition.

ETHANOL EFFECTS ON MEMBRANE PHOSPHOLIPIDS

The phospholipid head groups are important in providing the membranes with a proper charge distribution and the hydrophilic environment. It is recognized that ethanol and other aliphatic alcohols can alter the metabolism related to turnover of the polar head groups of the phospholipids. This pertains especially to the anionic phospholipids, which are known to provide the proper microenvironment for a number of membrane-bound enzymes (Sun and Sun, 1976). Recently, we found an increase in the acidic phospholipids (PtdH, PtdIns, and PdtSer) in rat brain membranes after administering ethanol in a liquid diet (Sun and Sun, 1983; Sun et al., 1984). Although these changes are not dramatic (e.g., PtdH, 37%; PtdIns, 21%; PtdSer, 13.5%), the changes can be observed in both the SPM and the non-synaptic plasma membrane fractions. Because the activity of (Na, K)-ATPase in brain is sensitive to the membrane lipid microenvironment (Sun and Sun, 1976), the increase in the acidic phospholipids may be one of the underlying mechanisms for the adaptive increase in (Na, K)-ATPase activity after chronic ethanol administration (Sun et al., 1977a; Sun, 1983).

Although the percent increase in PtdSer is very small, this phospholipid constitutes the major acidic phospholipid in brain. Besides its role in (Na, K)-ATPase activity, it is also important in the binding of opiates to their receptors (Abood and Takeda, 1976; Abood et al., 1978). In the liposome form, PtdSer can exert a pharmacological effect on brain (Bruni et al., 1976).

There are indications that PtdSer enhances tyrosine hydroxylase activity (Lloyd and Kaufman, 1974), aminobutyric acid uptake (Chweh and Leslie, 1982), and acetylcholine synthesis (Casamenti et al., 1979) as well. Recently, PtdSer was also shown to be the membrane component responsible for activation of the protein kinase C (Takai et al., 1982). Thus, there is little doubt that if chronic ethanol increases the PtdSer in brain membranes, a number of enzymic processes will be altered. Nevertheless, further experimental evidence would be required to demonstrate a direct relationship.

EFFECTS OF ETHANOL ON POLYPHOSPHOINOSITIDE METABOLISM

Not much is known regarding the effect of chronic ethanol administration on brain phosphoinositide metabolism. A study by Lee et al. (1980) indicated that chronic ethanol increased the turnover of the brain phospholipids, especially the acidic phospholipid species. However, it is not known whether this increase in turnover is related to the receptor-mediated metabolism of the polyphosphoinositides. Recently, several investigators have shown that polyphosphoinositide breakdown by phosphodiesterase is the initial step in the cyclic turnover of PtdIns (Berridge, 1983; Berridge et al., 1983; Brown et al., 1984). This event is mediated by neurotransmitters, receptor agonists, and hormones (see reviews by Downes, 1983; Nishizuka, 1984). Neurotransmitter stimulation of polyphosphoinositide hydrolysis by the phosphodiesterase has been linked to two important physiological events: (1) release of inositol 1,4,5-trisphosphate, which is regarded as a "second messenger" for intracellular Ca^{2+} mobilization (Berridge et al., 1983; Streb et al., 1983; Joseph et al., 1984), and (2) generation of diacylglycerols, which is known to trigger a sequence of intracellular events through protein kinase C (Takai et al., 1982; Nishizuka, 1984). The DAG derived from polyphosphoinositide can either be hydrolyzed by DAG lipase to release free fatty acids (FFA), or they can become phosphorylated to form PtdH (Strosznajder et al., unpublished). The scheme in Fig. 1 depicts the enzymic sequence of the polyphosphoinositide cycle.

It is well-recognized that 3 moles of ATP are needed for each cycle. Thus, ATP availability is important in maintaining the metabolic activity of the cyclic event. During the operation of each cycle, a number of negative charges are created in the membrane through the kinases. Ethanol may alter the dynamic aspects of the membrane polyphosphoinositide metabolism at any of the sites through different mechanisms. Smith et al. (1983) observed an increase in the vasopressin and alpha$_1$ adrenergic stimulation of $^{32}P_i$ incorporation into PtdIns in hepatocytes from ethanol-treated rats. Nevertheless, the muscarinic agonists did not cause a similar increase in $^{32}P_i$ incorporation into synaptosomes of ethanol-tolerant rats (Smith, 1983).

With the indication that acidic phospholipids such as PtdIns and PtdH are increased in brain after chronic ethanol, we started to examine whether other lipids such as polyphosphoinositide and DAG are also altered under the same condition. Results from an earlier study indicate that brain FFA release is suppressed during the hypnotic stage of acute ethanol intoxication (Sun et al., 1983). Although the source of the FFA release is not clear, at least some of the FFA may be derived from the sequential hydrolysis of polyphosphoinositide and DAG. In a subsequent study, the levels of polyphosphoino-

138

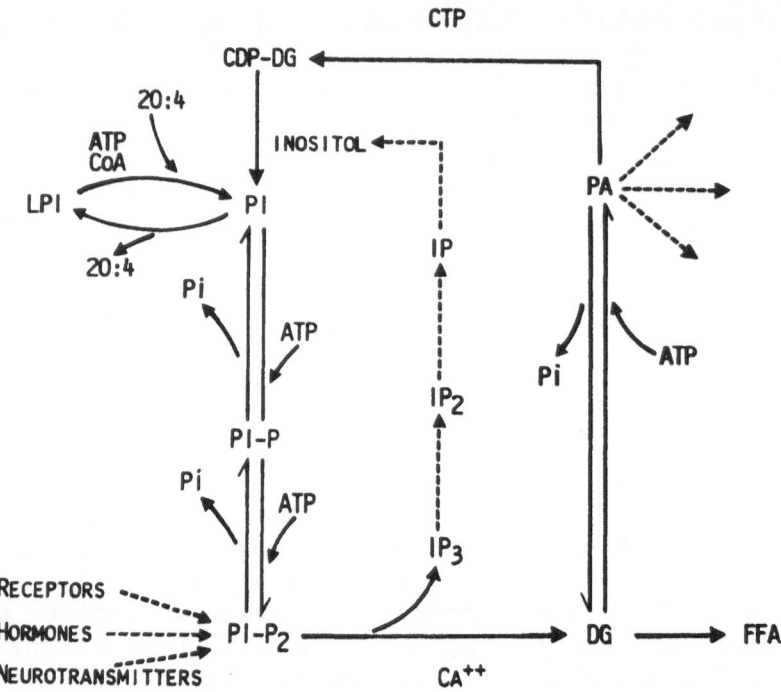

Figure 1. A scheme depicting the current view of the cyclic metabolism of polyphosphoinositides. Abbreviations: DAG (DG); PtdIns (PI); PtdH (PA); Ins$P$ (IP); InsP_2 (IP$_2$); InsP_3 (IP$_3$).

Table 1. *Effect of chronic and acute ethanol administration on degradation of rat brain polyphosphoinositides*

Conditions	Polyphosphoinositide	DAG	FFA
	(μg fatty acid/mg protein)		
Control	1.27 ± 0.09	0.56 ± 0.08	1.17 ± 0.14
	(n = 3)	(n = 5)	(n = 5)
Chronic (16 hr W)	1.25 ± 0.15	0.55 ± 0.08	1.26 ± 0.19
	(n = 5)	(n = 5)	(n = 5)
Acute	1.79 ± 0.15*	0.43 ± 0.01*	0.96 ± 0.08*
	(n = 5)	(n = 5)	(n = 5)

For chronic ethanol administration, Sprague-Dawley rats were given 6 g/kg ethanol via intragastric intubation of a 30% ethanol liquid diet (Similac). Controls received the same diet with an isocaloric amount of sucrose. Animals were on the diets for 3 weeks prior to sacrifice by decapitation. The ethanol group was withdrawn from ethanol 16 hr prior to sacrifice. For acute ethanol administration, animals were given the control diet for three weeks and were given a single dose of ethanol (8 g/kg) by the same route 3 hr prior to sacrifice. Lipids from cerebral cortex homogenates were analyzed by a combination of TLC and GLC procedures. Results are the mean ± SD of the number of animals in parentheses. * Denotes values that are significantly different from controls, $p < 0.01$.

sitide, DAG, and FFA were measured under the same condition of acute intoxication. The results in Table 1 indicate that the decrease in DAG and FFA was concomitant with an increase in polyphosphoinositide level. On the other hand, brain DAG, FFA, and polyphosphoinositide levels were not changed after chronic ethanol administration, suggesting that these animals had developed tolerance to the inhibitory effects of ethanol.

It has long been known that the polyphosphoinositide in brain is highly susceptible to post-mortem degradation (Eichberg and Hauser, 1967). The mechanism of the post-mortem degradation may be similar to the rapid rate of disappearance of polyphosphoinositide detected in brain during ischemia (Sun et al., 1985b). In fact, this event may be considered as an in vivo model for assessing polyphosphoinositide hydrolysis. After decapitation, ATP is rapidly exhausted in brain (Lowry et al., 1964). The rapid depletion of ATP would hamper the ability of PtdIns to replenish the polyphosphoinositide pool, thus constituting a closed system for the phosphodiesterase to act on its endogenous substrate. Along this line, Shah et al. (1984) found that rat pups from mothers that were given ethanol through pregnancy and gestation had less post-mortem hydrolysis of PtdIns(4,5)P_2 than the controls. Their results seem to suggest that these pups have developed resistance to the receptor-mediated event.

In order to accurately assess the effects of ethanol on the responsiveness of the receptor-mediated process in brain membranes, it is necessary to devise a procedure to label the polyphosphoinositide in brain. In one study, we injected $^{32}P_i$ i.p. into young rats (3-4 weeks old). Equilibration for 16 hr resulted in distribution of the label in all major brain phospholipids, including the polyphosphoinositide. To test the effects of ethanol, weanling rats were given ethanol in a liquid diet (Similac) via intragastric intubation (in two daily fractional doses) for 3 weeks. Control rats were given the same diet, except that ethanol was replaced by an isocaloric amount of glucose. During this period, the level of ethanol was increased gradually from 3 g/kg to 6 g/kg. Besides the liquid diet, lab chow was available *ad lib* to both control and alcoholic groups during the third week. After administering the last dose of ethanol or control diet, animals were injected i.p. with 165 μCi $^{32}P_i$ and were sacrificed 16 hr after injection.

After removal of the cerebral cortex, brain tissue was homogenized in sucrose-Tris buffer, and the lipids were extracted by a chloroform-methanol (neutral and acidified) procedure. When the radioactivity of the neutral and acidic organic phases was measured, an increase was found in label of the acidic chloroform-methanol extract at the expense of the neutral chloroform-methanol extract (Fig. 2). The increase was the same in total homogenates, as well as in synaptosomes, plasma membranes, and myelin isolated from cerebral cortex. Because most of the polyphosphoinositides were extracted into the acidic chloroform-methanol fraction, the increase in label may represent an increase in labeling of these phospholipids. This was further substantiated by separating the lipids by two-dimensional TLC. As shown in Fig. 3, ethanol induced an increase in labeling of polyphosphoinositide (31%) and PtdH (22%), whereas the labeling pattern of other phospholipids was not changed. A similar increase in labeling of polyphosphoinositide was also found in the non-synaptic plasma membrane fraction (Fig. 4). From this study, we can conclude that chronic ethanol treatment gave rise to an increase in labeling of the polyphosphoinositide in brain.

A subsequent study was designed to determine whether the increase in polyphosphoinositide labeling is due to an adaptive mechanism or is a manifestation of the hypersensitivity during withdrawal. The experimental protocol was similar to that described above, except that after chronic ethanol treatment, animals in the ethanol

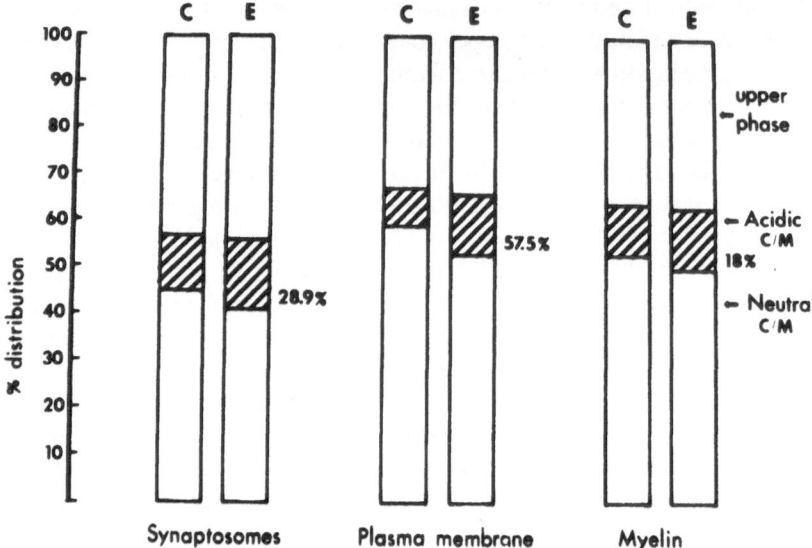

Figure 2. Distribution of $^{32}P_i$ radioactivity recovered in brain subcellular fractions of control (C) and chronic ethanol (E) rats. The label was injected i.p. into rats 16 hr prior to sacrifice. Experimental details are the same as in Table 1. Results are expressed as the percent of radioactivity recovered in the aqueous upper phase, neutral and acidic chloroform-methanol extracts. Percentages indicated are the increase in radioactivity of the acidic chloroform-methanol fraction.

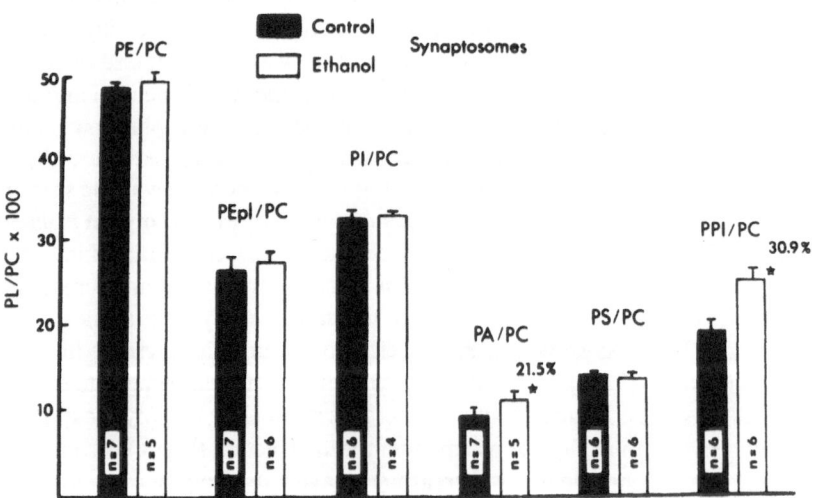

Figure 3. Ratio of radioactivity of individual phospholipids relative to PtdCho (PC) in synaptosomes of cerebral cortex from control and ethanol-treated rats. Results are the mean ± S.D. of 6 animals in each group. Values that are significantly different ($p < 0.05$, one-way analysis of variance) are denoted by an asterisk. Abbreviations: PtdEtn (PE); PtdIns (PI); PtdH (PA); PtdSer (PS); polyphosphoinositide (PPI); ethanolamine plasmalogen (PEpl).

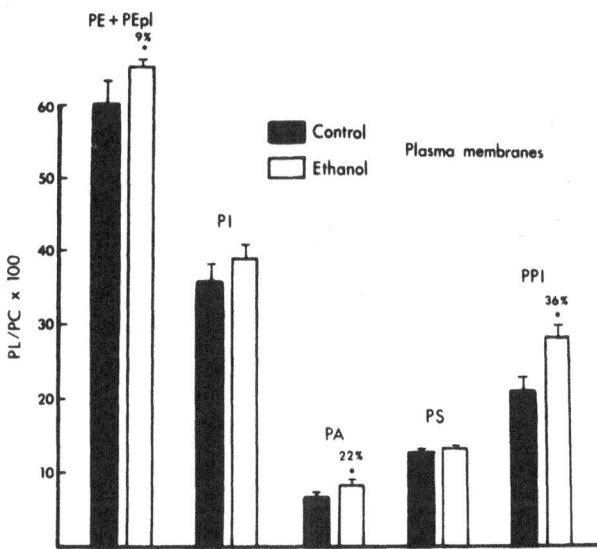

Figure 4. Ratio of radioactivity of individual phospholipids (relative to PtdCho) in plasma membranes of cerebral cortex from control and ethanol-treated rats. Details are the same as in Fig. 3.

group were sacrificed at 2 and 24 hr after the last feeding. All animals were injected with ^{32}Pi 16 hr prior to sacrifice. As shown in Fig. 5, ^{32}P-labeling of polyphosphoinositide in synaptosomes of the ethanol-treated group was increased only at 24 hr after ethanol withdrawal. Synaptosomes from the 2 hr withdrawal group showed an increase in incorporation of label into PtdIns (28%) and PtdH (20%), but this effect disappeared after 24 hr. In the non-synaptic plasma membrane fraction, an increase in labeling of polyphosphoinositide was observed in both the 2 hr and 24 hr withdrawal groups (41% at 2 hr and 86% at 24 hr), but the labeling of PtdIns and PtdH was not altered. Therefore, the increase in polyphosphoinositide labeling may represent metabolic changes with respect to withdrawal, whereas the increase in labeling of PtdIns and PtdH at 2 hr may represent adaptation towards more active turnover of the cyclic event. Since the labeling pattern of PtdSer was not altered, the increase in level of this phospholipid after chronic ethanol probably is not due to altered de novo biosynthesis.

The results of the above studies indicate that both acute and chronic ethanol administration alter the metabolism of polyphosphoinositide in brain. Further identification of the molecular site of ethanol action on this cyclic event would be the logical approach for future research. Like the adenylate cyclase system (Saito et al., 1985), the mechanism of polyphosphoinositide metabolism is likely to be very complex, and ethanol may act on multiple enzymic sites. Nevertheless, an understanding of the mechanism would greatly enhance our knowledge of the dynamic aspects of membrane phospholipids and how they are affected by drugs.

SUMMARY

Ethanol is known to exert its primary mode of action on the neural membranes. Although the molecular mechanisms of ethanol-membrane interactions have not been

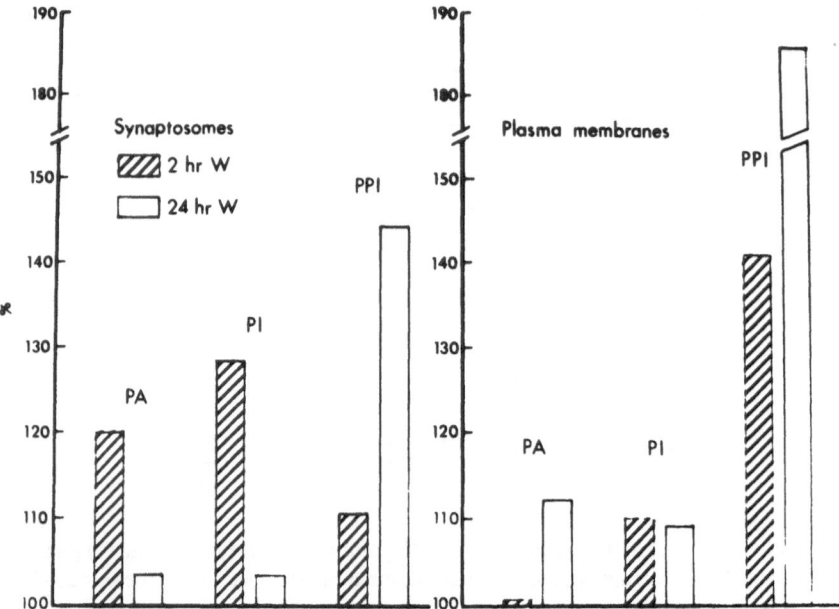

Figure 5. Percent increase in ³²P-labeling of PtdH (PA), PtdIns (PI) and polyphosphoinositides (PPI) in rat brain subcellular fractions after chronic ethanol treatment. Animals were sacrificed 2 and 24 hr after the last feeding of ethanol. Values are the percent increase relative to controls (100%).

clearly elucidated, there are indications that membranes from animals after chronic ethanol administration become more resistant to the fluidizing effect of ethanol in vitro. These adaptive changes may be the biochemical basis for development of tolerance and physical dependence.

There is much evidence indicating that changes in membrane biochemical and biophysical properties are due largely to alterations of the membrane lipids. In this regard, alteration of the membrane cholesterol/phospholipid ratio and phospholipid acyl group unsaturation have been investigated extensively. We have reported an increase in acidic phospholipids (PtdSer, PtdIns, PtdH) in brain membranes after chronic ethanol treatment. The increase in acidic phospholipids may help to explain the adaptive increase in (Na, K)-ATPase activity, as well as changes in other membrane activities. Acute and chronic ethanol also may affect the cyclic mechanism of the receptor-mediated turnover of polyphosphoinositide. This interaction may involve a number of enzymic sites. In the acute intoxicated stage, inhibition of the phosphodiesterase activity in brain is shown by a decrease in DAG and FFA and an increase in polyphosphoinositide. On the other hand, the levels of these lipids are not greatly altered after chronic ethanol administration. A series of experiments were carried out to examine the effect of chronic ethanol on brain polyphosphoinositide metabolism. These studies in vivo were made possible by prelabeling the brain tissue with ³²Pᵢ. A specific increase of labeled PtdIns and PtdH was observed in the prodromal stage after chronic ethanol administration, whereas an increase in ³²P-labeling of the polyphosphoinositide was found after 16-24 hr (withdrawal stage). It is envisaged that alteration of the receptor-mediated poly-phosphoinositide turnover by ethanol will lead to changes in other intracellular events such as calcium metabolism and protein phosphorylation. These possibilities should be explored in detail.

ACKNOWLEDGMENTS

The research reported from our laboratory was supported in part by grants AA02054 and AA06661 from NIAAA. We thank Dr. Ding Lee for technical assistance and Diane Torres and Dr. Laurie Foudin for preparing the manuscript.

REFERENCES

Abood LG, Takeda F (1976) Enhancement of stereospecific opiate binding to neural membranes by phosphatidyl serine. Eur J Pharmacol 39: 71-77.

Abood LG, Salem N, MacNeil M, Butler M (1978) Phospholipid changes in synaptic membranes by lipolytic enzymes and subsequent restoration of opiate binding with phosphatidyl serine. Biochim Biophys Acta 530: 35-46.

Alling C, Balldin J, Kahlson K, Olsson R (1980) Decreased linoleic acid in serum lecithin after ethanol abuse. Subst Alcohol Act/Mis 1: 557-563.

Alling C, Liljequist S, Engel J (1982) The effect of chronic ethanol administration on lipids and fatty acids in subcellular fractions of rat brain. Med Biol 60: 149-154.

Alling C, Gustavsson C, Kristensson A, Wallerstadt S (1984) Changes in fatty acid composition of major glycerophospholipids in erythrocyte membranes from chronic alcoholics during withdrawal. Scand J Clin Lab Invest 44: 283-289.

Baker RR, Thompson W (1973) Selective acylation of 1-acylglycerophosphorylinositol by rat brain microsomes. Comparison with 1-acylglycerophosphorylcholine. J Biol Chem 248: 7060-7065

Beauge F, Stibler H, Kalant H (1983) Brain synaptosomal (Na$^+$and K$^+$)-ATPase activity as an index of tolerance to ethanol. Pharmacol Biochem Behav 18: 519-524.

Berridge MJ (1983) Rapid accumulation of inositol trisphosphate reveals that agonists hydrolyze polyphosphoinositides instead of phosphatidylinositol. Biochem J 212: 849-858.

Berridge MJ, Dawson RMC, Downes CP, Heslop JP, Irvine RF (1983) Changes in the levels of inositol phosphates after agonist-dependent hydrolysis of membrane phosphoinositides. Biochem J 212: 473-482.

Brown E, Kendall DA, Nahorski SR (1984) Inositol phospholipid hydrolysis in rat cerebral cortical slices: I. Receptor characterization. J Neurochem 42: 1379-1387.

Bruni A, Toffano G, Leon A, Boarato E (1976) Pharmacological effects of phosphatidylserine liposomes. Nature 260: 331-333.

Casamenti F, Mantovani P, Amaducci L, Pepeu G (1979) Effect of phosphatidylserine on acetylcholine output from the cerebral cortex of the rat. J Neurochem 32: 529-533.

Chin JH, Goldstein DB (1977a) Effects of low concentrations of ethanol on the fluidity of spin-labeled erythrocyte and brain membranes. Mol Pharmacol 13: 434-441.

Chin JH, Goldstein DB (1977b) Drug tolerance in biomembranes: A spin label study of the effects of ethanol. Science 196: 684-685.

Chin JH, Parsons LM, Goldstein DB (1978) Increased cholesterol content of erythrocyte and brain membranes in ethanol-tolerant mice. Biochim Biophys Acta 513: 358-363.

Chin JH, Goldstein DB (1985) Cholesterol blocks the disordering effects of ethanol in biomembranes. Lipids 19: 929-935.

144

Chweh AY, Leslie SW (1982) Phosphatidylserine enhancement of [^{3}H]γ-aminobutyric acid uptake by rat whole brain synaptosomes. J Neurochem 38: 691-695.

Corbin DR, Sun GY (1978) Characterization of the enzymic transfer of arachidonoyl groups to 1-acyl-phosphoglycerides in mouse synaptosome fraction. J Neurochem 30: 77-82.

Crews FT, Majchrowicz E, Meeks R (1983) Changes in cortical synaptosomal plasma membrane fluidity and composition in ethanol-dependent rats. Psychopharmacol 81: 208-213.

Demel RA, Dekruiff B (1976) The function of sterols in membranes. Biochim Biophys Acta 457: 109-132.

Downes CP (1983) Inositol phospholipids and neurotransmitter receptor signalling mechanisms. Trends Neurol Sci 9: 313-316.

Eichberg J, Hauser G (1967) Concentrations and disappearance post mortem of polyphosphoinositides in developing rat brain. Biochim Biophys Acta 144: 415-422.

Foudin L, Tumbleson ME, Sun AY, Geisler RW, Sun GY (1984) Ethanol consumption and serum lipid profiles in Sinclair(S-1) miniature swine. Life Sci 34: 819-826.

Harris RA, Schroeder F (1981) Ethanol and the physical properties of brain membranes. Fluorescence studies. Mol pharmacol 20: 128-137.

Harris RA, Mitchell MA, Hitzemann RJ (1984) Physical properties of brain membrane from ethanol tolerant-dependent mice. Mol Pharmacol 15: 401-409.

Israel Y, Kuriyama K (1971) Effects of in vivo ethanol administration on adenosine triphosphatase activity of subcellular fractions of mouse brain and liver. Life Sci 10: 591-599.

John GR, Littleton JM, Nhamburo PT (1985) Increased activity of Ca^{2+}-dependent enzymes of membrane lipid metabolism in synaptosomal preparations from ethanol-dependent rats. J Neurochem 44: 1235-1241.

Johnson DA, Lee NM, Cooke R, Loh HH (1979) Ethanol-induced fluidization of brain lipid bilayers: required presence of cholesterol in membranes for the expression of tolerance. Mol Pharmacol 15: 739-746.

Joseph SK, Thomas AP, Williams RJ, Irvine RF, Williamson JR (1984) Myo-inositol 1, 4, 5-trisphosphate. A second messenger for the hormonal mobilization of intracellular Ca^{2+} in liver. J Biol Chem 259: 3077-3081.

Kelleher JA, Sun GY (1985) Enzymic hydrolysis of arachidonoylphospholipids by rat brain synaptosomes. Neurochem Int 7: 825-831.

La Droitte PH, Lamboeuf Y, de Saint Blanquat G (1984) Membrane fatty acid changes and ethanol tolerance in rat and mouse. Life Sci 35: 1221-1229.

Lands WEM, Crawford CG (1976) Enzymes of membrane phospholipid metabolism in animals. In: Martonosi A (ed): The enzymes of biological membranes, Vol. 2. Plenum Press, New York, pp. 3-85.

Lee NM, Friedman HJ, Loh HH (1980) Effect of acute and chronic ethanol treatment on rat brain phospholipid turnover. Biochem Pharmacol 29: 2815-2818.

Littleton JM, John G (1977) Synaptosomal membrane lipids of mice during continuous exposure to ethanol. J Pharm Pharmacol 29: 579-580.

Lloyd T, Kaufman S (1974) The stimulation of partially purified bovine caudate tyrosine hydroxylase by phosphatidyl-L-serine. Biochem Biophys Res Commun 59: 1262-1269.

Lowry OH, Passonneau JV, Hasselberger FX, Schulz DW (1964) Effect of ischemia on known substrates and cofactors of the glycolytic pathway in brain. J Biol Chem 239: 18-25.

Meyerton HH, Gottlieb R (1926) Experimental Pharmacology as a Basis for Therapeutics, Ed 2 (Henderson VE, trans), JB Lippincott Co., Philadelphia p. 121.

Michaelis EK, Michaelis ML (1983) Physico-chemical interactions between alcohol and biological membranes. In: Smart et al. (eds): Research Advances in Alcohol and Drug Problems, vol 7. Plenum Press, New York, pp. 127-173.

Nishizuka Y (1984) Protein kinases in signal transduction. Trends Biochem Sci 9: 163-166.

Rabin RA, Molinoff PB (1981) Activation of adenylate cyclase by ethanol in mouse striatal tissue. J Pharmacol Exp Ther 216: 129-134.

Rangaraj N, Kalant H (1978) Effects of ethanol withdrawal stress and amphetamine on rat brain (Na$^+$ + K$^+$)-ATPase. Biochem Pharmacol 27: 1139-1144.

Saito T, Lee JM, Tabakoff B (1985) Ethanol's effects on cortical adenylate cyclase activity. J Neurochem 44: 1037-1044.

Seeman P (1972) The membrane actions of anesthetics and tranquilizers. Pharmacol Rev 24: 583-655.

Shah IR, Uma S, Ramakrishnan CV, Hauser G (1984) Effect of ethanol on rat brain polyphosphoinositides. J Neurochem 42: 873-874.

Shinitzky M, Barenholz Y (1978) Fluidity parameters of lipid regions determined by fluorescence polarization. Biochim Biophys Acta 515: 367-394.

Smith TL (1983) Influence of chronic ethanol consumption on muscarinic cholinergic receptors and their linkage to phospholipid metabolism in mouse synaptosomes. Neuropharmacology 22: 661-663.

Smith TL, Gerhart MJ (1982) Alterations in brain lipid composition of mice made physically dependent to ethanol. Life Sci 31: 1419-1425.

Smith TL, Vickers AE, Brendel K, Yamamura HI (1983) Influence of chronic ethanol treatment on alpha$_1$-adrenergic and vasopressin receptor-stimulated phosphatidylinositol synthesis in isolated rat hepatocytes. Biochem Pharmacol 32: 3059-3063.

Streb H, Irvine RF, Berridge MJ, Schulz I (1983) Release of Ca^{2+} from a nonmitochondrial intracellular store in pancreatic acinar cells by inositol-1, 4, 5-trisphosphate. Nature 306: 67-69.

Sun AY (1979) Biochemical and biophysical approaches in the study of ethanol-membrane interaction. In: Majchrowicz E, Noble EP (eds): Biochemistry and pharmacology of ethanol, vol 2. Plenum Publishing Corp., New York, pp. 81-100.

Sun AY (1983) The kindling effect of ethanol on neuronal membranes. In: Sun GY, Bazan N, Wu JY, Porcellati G, Sun AY (eds): Neural Membranes. Humana Press, Clifton, New Jersey, pp. 317-340.

Sun AY, Samorajski T (1970) Effects of ethanol on the activity of adenosine triphosphatase and acetylcholinesterase in synaptosomes isolated from guinea-pig brain. J Neurochem 17: 1365-1372.

Sun AY, Seaman RN (1980) Physicochemical approaches to the alcohol-membrane interaction in brain. Neurochem Res 5: 537-545.

Sun AY, Sun GY (1976) Functional roles of phospholipids of synaptosomal membrane. In: Porcellati G, Amaducci L, Galli C (eds): Functional and metabolism of phospholipids in CNS and PNS. Plenum Press, New York, pp. 169-197.

Sun AY, Seaman RN, Middleton CC (1977) Effects of acute and chronic alcohol administration on brain membrane transport systems. In: Gross MM (ed): Alcohol intoxication and withdrawal, Vol. 3A. Pleum Publishing Corp., New York, pp. 123-138.

Sun GY (1982) Metabolic turnover of arachidonoyl groups in brain membrane phosphoglycerides. In: Horrocks LA, Ansell GB, Porcellati G (eds): Phospholipids in the nervous system: Metabolism, Vol. 1. Raven Press, New York, pp. 75-89.

Sun GY, Sun AY (1979) Effect of chronic ethanol administration on phospholipid acyl groups of synaptic plasma membrane fraction isolated from guinea pig brain. Res Commun Chem Pathol Pharmacol 24: 405-408.

Sun GY, Sun AY (1983) Chronic ethanol administration induced an increase in phosphatidylserine in guinea pig synaptic plasma membranes. Biochem Biophys Res Commun 113: 262-268.

Sun GY, Sun AY (1985) Ethanol and membrane lipids. Alcohol: Clin Exp Res 9: 164-180.

Sun GY, Creech DM, Corbin DR, Sun AY (1977) The effect of chronic ethanol administration on arachidonoyl transfer to 1-acyl-glycerophosphorylcholine in rat brain synaptosomal fraction. Res Commun Chem Pathol Pharmacol 16: 753-756.

Sun GY, Su KL, Der OM, Tang W (1979) Enzymic regulation of arachidonate metabolism in brain membrane phosphoglycerides. Lipids 14: 229-235.

Sun GY, Tang W, Lee D-Z, Sun AY (1983) Effects of acute and chronic ethanol administration on the free fatty acids of rat cerebral cortex. Neurochem Pathol 1: 137-146.

Sun GY, Huang H-M, Lee D-Z, Sun AY (1984) Increased acidic phospholipids in rat brain membranes after chronic ethanol administration. Life Sci 35: 2127-2133.

Sun GY, Kelleher JA, Sun AY (1985a) Effects of chronic ethanol administration and withdrawal on incorporation of arachidonate into membrane phospholipids. Neurochem Int 7: 491-495.

Sun GY, Tang W, Huang SF-L, Foudin L (1985b) Is phosphatidylinositol involved in the release of free fatty acids in cerebral ischemia? In: Bleasdale JE, Eichberg J, Hauser G (eds): Inositols and phosphoinositides: metabolism and biological regulation. Humana Press, Clifton, New Jersey, pp. 511-527.

Takai Y, Minakuchi R, Kikkawa U, Sano K, Kaibuchi K, Yu B, Matsubara T, Nishizuka Y (1982) Membrane phospholipid turnover, receptor function and protein phosphorylation. Prog Brain Res 56: 287-301.

Thompson JA, Reitz RC (1978) Effects of ethanol ingestion and dietary fat levels on mitochondrial lipids in male and female rats. Lipids 13: 540-550.

Wing DR, Harvey DJ, Hughes J, Dunbar PG, McPherson KA, Paton WDM (1982) Effects of chronic ethanol administration on the composition of membrane lipids in the mouse. Biochem Pharmacol 31: 3431-3439.

Yamamoto HA, Harris RA (1983) Effects of ethanol and barbiturates on Ca^{2+}-ATPase activity of erythrocyte and brain membranes. Biochem Pharmacol 32: 2787-2791.

Phospholipid research and the nervous system
Biochemical and molecular pharmacology
L.A. Horrocks, L. Freysz, G. Toffano (eds)
Fidia Research Series, vol. 4.
Liviana Press, Padova. © 1986

EFFECTS OF ETHANOL ON CONCENTRATION AND ACYL GROUP COMPOSITION OF ACIDIC PHOSPHOLIPIDS

C. Alling and L. Gustavsson

Dept of Psychiatry and Neurochemistry, University of Lund, Sweden

INTRODUCTION

Ethanol exposure causes changes in the lipid composition of biological membranes. Such findings are, however, inconsistent among species, among subcellular fractions, and in relation to duration and dose of ethanol exposure. Changes in lipid concentrations as well as disturbed acyl group patterns have been attributed to an adaptation of the membranes to the fluidizing effect of ethanol. It was recently shown that the concentration of acidic phospholipids was increased in brains from ethanol exposed animals (Alling et al.; 1983, Sun and Sun, 1983, Alling et al., 1984a; Sun et al., 1984). Previously, Lee et al. (1980) had found that the turnover of PtdSer and PtdIns was enhanced in ethanol tolerant animals. Another study has also indicated that the fatty acid pattern was more disturbed in PtdSer than in other phospholipids (Harris et al., 1984) following ethanol exposure. Taken together, these discoveries open up new vistas for research on ethanol effects on brain membrane because acidic lipids seem to be especially important for the function of membrane proteins. Best known is the strong relationship between Na^+, K^+-ATPase activity and sulphatide concentration (Karlsson, 1982). Evidence has been presented favouring a role for PtdSer for the opiate receptor function (Abood and Hoss, 1975; Abood and Takeda, 1976; Abood et al., 1978). The molecular species of PtdSer containing docosahexaenoic acid [22:6(n-3)] were especially important since removal of that fatty acid caused a fall in opiate binding. In addition, it has been claimed that muscarinic receptors depend on acidic phospholipids like PtdSer, PtdH, and PtdIns but not on neutral phospholipids (Aronstam et al., 1977). GABA uptake is stimulated by PtdSer both when the endogenous PtdSer was increased via the base exchange reaction (De Medio et al., 1977, 1980) and when it was added to synaptosome preparations (Chweh and Leslie, 1982). There is also reasonable evidence for a role for PtdSer in the glutamate receptor function (Foster et al., 1982). An ethanol modulating action on these receptors (Hoffman et al., 1982; Charness et al., 1983) may therefore be due to the environmental acidic phospholipids. The finding that the acidic lipid phosphatidylethanol is pathologically formed in rat organs during ethanol exposure (Alling et al., 1984a; Benthin et al., 1985) stimulated a more extensive exploration of the acidic phospholipids. This report focuses on the acidic phospholipids in rat organs and in human platelets after ethanol exposure.

MATERIALS AND METHODS

Subjects

In experiment I, male Sprague-Dawley rats (n = 9) received ethanol in their drinking water for five months (Alling et al., 1982). Between 16 and 60 days of age the rats were administered 8% (w/v) ethanol and between 60-150 days of age 16% (w/v) ethanol. This procedure resulted in a daily intake of ethanol of about 8g/kg body weight/24 hours. Control rats (n = 9) received water without ethanol. In experiment II female Sprague-Dawley rats were fed on either of two semisynthetic diets for three generations (Alling et al., 1984b). The diets were identical except for the content of essential fatty acids (EFA). Diet A was deficient in EFA (0.3 energy-%), and diet B contained a normal amount of EFA (3.0 energy-%). Half of each EFA-group (n = 6 in each final subgroup) received daily intraperitoneal injections of ethanol (3 g/kg body weight) for 23 days. Control rats were injected with isocaloric amounts of glucose. Two hours after the last injection the rats were killed by decapitation, and brain and other organs were dissected. All organs were stored at -20° until analysis.

Human male alcoholics, aged 34-62 years, who filled criteria of being addicted, having a present history of more than 7 days of heavy drinking and admitted within 24 hours after last alcohol intake were selected from an intake ward. Male hospital staff with average alcohol habits were selected as a control group (age 21-56 years) (Alling et al., 1985).

Preparation of Subcellular Fractions from Rat Brain

In experiment I, myelin, synaptosomes, and mitochondria were isolated for analysis (Alling et al., 1982). A 10% (w/v) brain homogenate was prepared in 0.32 M sucrose with an all-glass homogenizer. After differential centrifugation the resulting P_2-pellet was separated into fractions containing myelin, synaptosomes, and mitochondria on a discontinuous density sucrose gradient (Whittaker and Barker, 1972). The fractions were purified three times by suspension in 0.156 M KCl and centrifugation at 17500 × g for 30 minutes.

Dissection of Different Regions from Rat Brain

Rats were killed by decapitation under ether anaesthesia. The cerebellum was removed and the forebrain separated from brainstem by an incision through the cerebral peduncles. Brainstem was separated from spinal cord 5 mm further down. Forebrain and brainstem were used for analyses.

Isolation of Platelets

Venous blood was collected in EDTA-containing tubes from chronic alcoholics (n = 17), both at the day of admittance and after one week of detoxification, and control persons (n = 12). The samples were centrifuged at $2000 \times g$ for 10 minutes. The platelets were separated from other blood cells on a Ficoll-Paque[R] (Pharmacia, Sweden) density gradient (Boyum, 1968). The cell-rich interface under the plasma layer was purified from lymphocytes by centrifugation at $150 \times g$ for 10 minutes. The platelets in this supernatant and from the plasma were finally isolated by centrifugation at $2000 \times g$ for 15 minutes (Alling et al., 1985).

Assays

Preparation of Lipid Extracts

Brain tissue and platelets were homogenized with distilled water in an all-glass homogenizer and extracted with chloroform, methanol, and water in a final volume ratio of 4:8:3 (Svennerholm and Fredman, 1980). The sample was centrifuged at $1000 \times g$ for 10 minutes and re-extracted once. The extract was purified from compounds with low molecular weight by gel filtration on a Sephadex G-25 column (Wells and Dittmer, 1963). The lipids were separated into a neutral and an acidic fraction by ion-exchange chromatography on a DEAE-Sepharose column in acetate-form (Nilsson et al., 1981). The acidic eluate was separated from potassium acetate (included in the elution medium) by a further Sephadex G-25 step.

Quantitative Analysis

Lipid phosphorus (Bartlett, 1959) and cholesterol (Franey and Amador, 1968) concentrations were determined with colorimetric methods. Individual phospholipid classes were determined by reflectance densitometry on HPTLC plates after visualization with a molybdenum blue reagent. The acidic phospholipids were separated in the solvent system chloroform: acetone: methanol: acetic acid: water 50:15:10:10:5 (volume ratios).

Qualitative Analysis

Phospholipids were isolated by preparative TLC for determination of the fatty acid composition. The solvent system used for the acidic phospholipids was chloroform: acetone: methanol: acetic acid: water 50:20:15:10:5 and the detection was performed with bromphenolblue reagent. Fatty acid methyl esters were prepared and analysed by gas-liquid chromatography (Svennerholm, 1968).

RESULTS AND DISCUSSION

Concentration and Fatty Acid Composition of Acidic Phospholipids in Rat Organs

Low doses of ethanol in drinking water over an extended period of time (8 months) did not influence the concentration of total phospholipids in brain (Alling et al., 1982). Higher doses (3 g/kg intraperitoneally, daily, 3 weeks) induced increased levels of phospholipids in brain and in kidney (Alling et al., 1983; Alling et al., 1984a). A separation of the phospholipids into a neutral and an acidic fraction revealed that the increase was confined to the acidic fraction (Alling et al., 1983). Similar findings have been reported by Sun and Sun (1983) who found PtdSer increased in relation to neutral phospholipids in guinea pig synaptic membranes and PtdSer, PtdIns, and PtdH increased in relation to the protein concentration of rat brain plasma membranes (Sun et al., 1984). In mice, on the other hand, no increase of PtdSer was found after 7 days administration of an ethanol containing liquid diet (Harris et al., 1984). Obviously factors such as species, mode of ethanol administration, and membrane type must be considered when the results are interpreted and a generalized increase of acidic phospholipids can therefore not be assumed.

The fatty acid compositions of PtdSer and PtdIns are characterized by large proportions of stearic (18:0) and arachidonic [20:4(n-6)] acids. Neither PtdSer nor PtdIns revealed any differences in fatty acid composition between the ethanol-treated rats and the control rats among different organs (Tables 1 and 2). In brain, the concentration

Table 1. *Proportions of major fatty acids in PtdIns from different organs of rats, treated i.p. with ethanol for 3 weeks*

	Brain				Kidney			
	Control		Ethanol		Control		Ethanol	
	M	SD	M	SD	M	SD	M	SD
16:0	8.9	1.20	8.8	0.70	11.7	1.49	11.7	0.75
18:0	41.1	2.21	40.5	1.58	40.9	1.48	40.9	0.67
18:1	8.3	1.78	9.0	0.45	6.7	0.45	7.0	1.25
20:4(n-6)	35.4	4.41	36.7	1.75	30.9	1.14	30.3	1.52
22:6(n-3)	3.2	0.31	2.9	0.35	1.2	0.16	1.3	0.10

	Skeletal Muscle				Liver			
	Control		Ethanol		Control		Ethanol	
	M	SD	M	SD	M	SD	M	SD
16:0	4.0	1.69	4.4	1.10	4.0	0.42	4.2	0.52
18:0	47.4	2.64	48.1	1.29	49.7	1.56	51.2	1.71
18:1	3.8	0.48	4.6	0.64	2.2	0.43	2.9	0.33
20:4(n-6)	28.5	1.36	27.4	1.04	36.8	1.51	30.8	1.64
22:6(n-3)	3.3	0.44	3.0	0.26	1.0	0.18	1.7	0.42

Table 2. *Proportions of major fatty acids in PtdSer from brain and kidney of rats, treated i.p. with ethanol for 3 weeks*

	Brain				Kidney			
	Control		Ethanol		Control		Ethanol	
	M	SD	M	SD	M	SD	M	SD
16:0	2.2	0.25	1.8	0.35	5.0	0.59	4.9	0.43
18:0	44.8	0.51	45.2	1.01	44.6	1.60	45.3	1.08
18:1	19.9	0.78	20.2	0.56	8.6	1.10	6.8	0.67
20:4(n-6)	2.8	0.12	3.4	0.22	30.9	1.87	32.4	1.36
22:6(n-3)	22.1	2.14	22.7	1.53	1.6	0.17	1.8	0.14

of 22:6(n-3) is high in PtdSer. Other laboratories have reported both increased (Sun and Sun, 1983) and decreased proportions (Harris et al., 1984) of this fatty acid in PtdSer of brain membranes after ethanol exposure. Further studies are required before definite conclusions can be drawn concerning the inertness of the fatty acid pattern of acidic phospholipids, however one aspect of importance in this context is the dietary regime. A low dietary level of linoleic acid induces formation of the fatty acid 20:3 (n-9) which is incorporated in membrane lipids. This effect is more pronounced following ethanol; in our animals, it was most obvious in PtdIns and PtdEtn (Figure 1). These findings indicate that ethanol may decrease the ability of individual phospholipids to maintain a constant fatty acid composition.

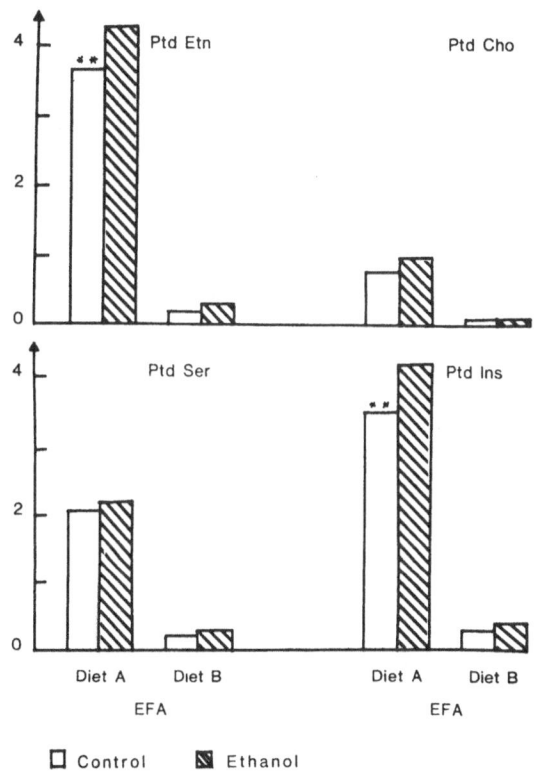

Figure 1. Proportions of the fatty acid 20:3(n-9) in PtdEtn, PtdCho, PtdSer and PtdIns from brain of rats treated i.p. with ethanol (hatched bars) or glucose (open bars) and fed either an EFA-deficient diet (Diet A) or a control diet (Diet B).

Different Regions of Rat Brain

To determine if the increase in concentration of acidic phospholipids in whole rat brain was located to a specific brain region, rats (n = 6) were treated with daily intraperitoneal injections of ethanol for 21 days. Brainstem and forebrain were analysed. The concentration of lipid phosphorus in the acidic fraction was significantly increased in forebrain after ethanol exposure (15.7 ± 0.74 μmol/g wet weight compared to 11.6 ± 1.16 μmol/g wet weight in the control group). A smaller but still statistically significant increase among the acidic phospholipids was found in brainstem (22.1 ± 0.93 μmol lipid phosphorus/g wet weight in ethanol group compared to 19.8 ± 1.43 in the control group). The concentrations of the individual subclasses of acidic phospholipids in forebrain were determined by densitometry. The increase of acidic phospholipids in forebrain after ethanol exposure was partly due to the formation of PtdEt (3.0 μmol/g wet weight). The concentrations of PtdH and cardiolipin were enhanced, but those lipids could not be quantified by densitometry due to insufficient separation on the HPTLC-plates. The ratio PtdSer/PtdIns was decreased (1.8 ± 0.42 in the ethanol group compared to 2.4 ± 0.29 in the control group). PtdSer was decreased whereas PtdIns was unaffected.

The larger increase of the amount of acidic phospholipids in forebrain than in brainstem indicates that there are differences in the effect of ethanol on different membrane structures. Brainstem consists to a great extent of myelin and forebrain is enriched in neurons. This is also illustrated by the higher lipid content in brainstem than in forebrain. It is possible that the dynamic structures of gray matter are more readily affected by ethanol intoxication than the structures in white matter with a lower phospholipid turnover rate.

Formation of Phosphatidylethanol (PtdEt)

HPTLC analysis of lipid extracts from brains of ethanol treated rats resulted in the discovery of an abnormal phospholipid (Alling et al., 1983). The concentration of the compound was highest in membrane-dense organs such as kidney and brain. It was also present in liver and skeletal muscle but not in serum. High speed centrifugation of a kidney homogenate from ethanol treated rats revealed that the abnormal lipid was localized in membrane structures.

Structural analyses were performed on the abnormal lipid to determine its chemical identity (Alling et al., 1984a). The lipid bound to an anion exchange resin and therefore had acidic properties. The Rf-value of the abnormal phospholipid differed from those of other acidic phospholipids on HPTLC-plates (Table 3). Quantitative analysis of lipid

Table 3. *Rf-values for different acidic phospholipids compared to phosphatidyl-ethanol in a HPTLC system. Solvent used was chloroform: acetone: methanol: acetic acid: water 50:15:15:10:5*

Phospholipid	Rf-Value
phosphatidylethanol	0.72
phosphatidylinositol	0.22
phosphatidylserine	0.27
phosphatidylglycerol	0.42
cardiolipin	0.53
phosphatidic acid	0.55
bis(monoacylglycero)phosphate	0.57

phosphorus and fatty acids revealed a molar ratio of 1:2. The fatty acids in the lipid were both saturated and unsaturated which indicated that it is a glycerophospholipid. The fact that phospholipase D could catalyse incorporation of primary alcohols (Yang et al., 1967) led us to propose that the abnormal phospholipid was phosphatidylethanol. This hypothesis was strengthened in the results of several experiments. PtdEt was synthesized on one hand in a phospholipase D-catalysed incubation system and on the other by a reaction between PtdH and diazoethane. Both of the prepared PtdEt had the same chromatographic mobility as the abnormal lipid in several HPTLC solvent systems (Fig. 2) (Alling et al., 1984a). Further evidence for the phosphatidylethanol structure was provided by GC-MS-analysis (Alling et al., 1984a; Benthin et al., 1985).

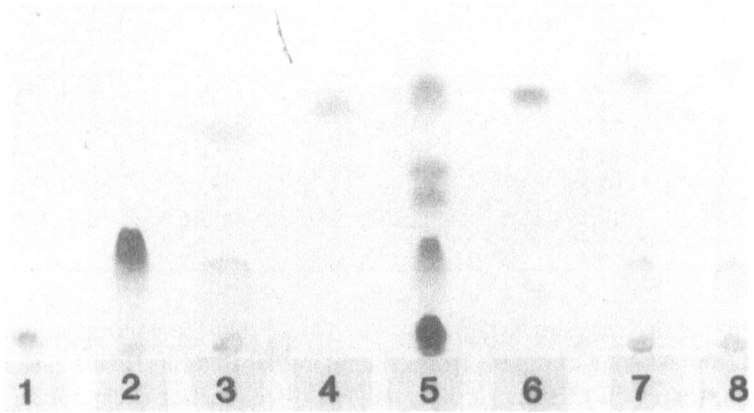

Figure 2. Chromatographic mobilities in a HPTLC system (chloroform: methanol: acetic acid 80:25:1) for (1). PtdCho reference, (2). PtdH reference, (3). Phosphatidylmethanol, (4). Phosphatidylethanol formed in a phospholipase D incubation system (see text), (5). Lipid extract from brain of ethanol treated rat, (6). Phosphatidylethanol formed in a diazoethane system (see text), (7). Phosphatidylpropanol, (8). Phosphatidylbutanol.
Nos. 3, 7 and 8 were also prepared in a phospholipase D system.

To examine some of the characteristics of PtdEt, its fatty acid composition in different organs from ethanol-treated rats (Table 4) was analysed. Palmitic and oleic acids dominated in brain; palmitic, oleic, and arachidonic acids in kidney; whereas palmitic and stearic acids had the highest proportions in liver and heart muscle. The low proportion of linoleic acid which is characteristic for brain phospholipids was also seen in PtdEt from brain. The organ-specific fatty acid composition and the fact that PtdEt was not present in rat serum indicates that this lipid is formed within the respective organ and not transported from a common pool. To determine if PtdEt was formed from a specific phospholipid the fatty acid compositions of PtdEt and other major phospholipids from brain were compared (Figure 3). The fatty acid composition of PtdEt was not identical to any of the other lipids. Because of the high content of oleic acid in PtdEt it also could not be a simple mixture from the whole phospholipid pool. It is possible that the abnormal lipid is formed from some of the minor phospholipids not analysed or from specific molecular species present at particular localisations in the lipid membrane.

Table 4. *Proportions of major fatty acids in phosphatidylethanol from different organs of rats, treated i.p. with ethanol for 3 weeks*

	Brain		Kidney	
	M	SD	M	SD
16:0	27.8	2.22	23.5	1.03
18:0	14.1	1.33	14.8	0.51
18:1	33.7	0.90	22.3	0.79
18:2(n-6)	1.2	0.12	8.0	0.52
20:4(n-6)	9.7	1.02	22.6	0.73
22:6(n-3)	7.4	0.99	2.3	0.18

	Heart Muscle		Liver	
	M	SD	M	SD
16:0	27.7	1.85	20.8	1.01
18:0	28.7	3.33	29.9	1.83
18:1	14.0	0.92	11.7	0.89
18:2(n-6)	7.8	1.89	9.2	0.66
20:4(n-6)	14.7	1.27	15.6	0.70
22:6(n-3)	2.4	0.43	5.5	0.37

After the above series of experiments, we found that the concentration of PtdEt increases during freezing of organs from ethanol intoxicated rats (Benthin et al., 1985). The concentration of PtdEt in kidney analysed directly after decapitation was 0.02 μmol/g wet weight (the animals were given a single dose of 3 g ethanol/kg body weight and killed after 2 hours). Freezing at -20°C for seven days increased the amount to 0.8 μmol/g. There was no further significant increase beyond this time. The optimum temperature for the formation of PtdEt was -10°C and at -80°C no formation occurred at all. It is probable that the freezing effect is a result of locally increased concentrations of ethanol due to microcrystallization of water (Benthin et al., 1985). Consequently, the concentration in vivo of PtdEt is lower than we have earlier reported (Alling et al., 1984a). It is, however, present in a significant amount and probably affects the properties of the lipid membrane. Further investigations are necessary to examine if PtdEt is confined to specific membrane structures. Another remarkable finding was that the concentration of PtdEt increased only when the organs were frozen intact and not as a homogenate. This indicates a requirement of an undisturbed membrane structure for the reaction to occur.

There are three main mechanisms possible for the formation of PtdEt in vivo (Fig. 4). Either of two enzymes could catalyse the reaction: phospholipase D or base exchange enzymes. Both these enzymes are reported to exist in mammal brain (Buchanan and Kanfer, 1980a,b; Chalifour and Kanfer, 1982; De Vries et al., 1983). The base exchange enzyme consists of three different enzymatic activities which selectively incorporate ethanolamine, serine, and choline into phospholipids (Buchanan and Kanfer, 1980b). On the other hand phospholipase D has been reported to catalyse formation of PtdEt in vitro (Yang et al., 1967). We therefore conclude that a phospholipase D-catalysed reaction is the most probable mechanism for the formation of PtdEt in vivo.

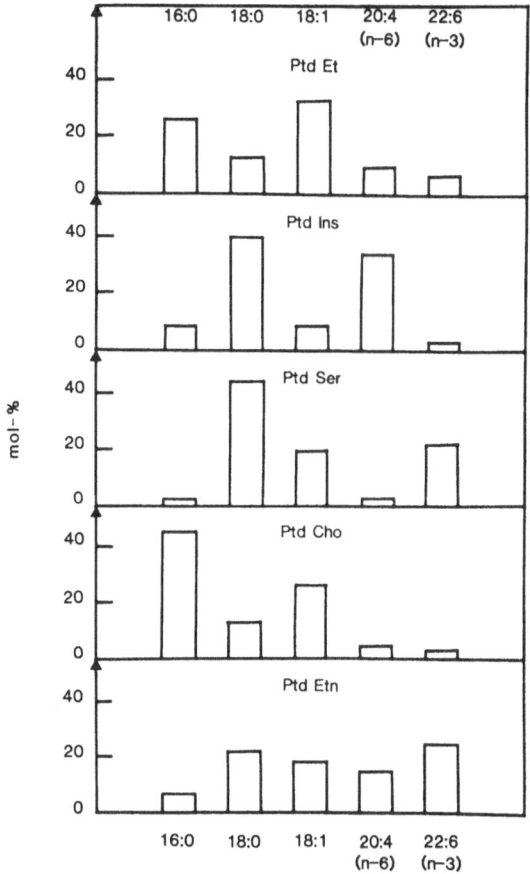

Figure 3. Proportions of major fatty acids in phosphatidylethanol and four other phospholipids from brain of rats treated i.p. with ethanol for 3 weeks.

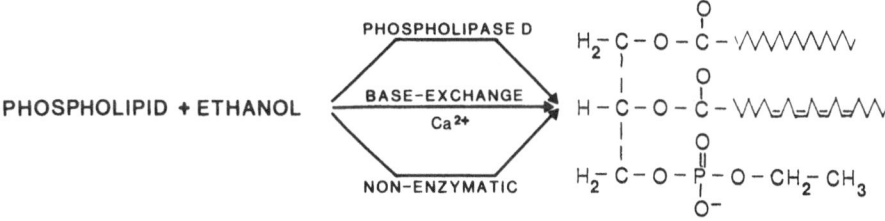

Figure 4. Alternative mechanisms for the formation of phosphatidylethanol.

Human Platelets

The platelet, being the best available model for synaptosomes when it comes to human studies, is known to contain neurotransmitters and to have membrane receptors (Pletscher and Laubscher, 1980; Fenn et al., 1983). A major problem when platelets are used for such studies is the preparation steps where high purity must be traded off

156

against low yield. By a modification of the Boyum method (1968) we have reduced the lymphocyte admixture by 50-75% (Alling et al., 1985).

The acidic phospholipids constituted about 1/6 of total phospholipids in platelets. The ratio between anionic and neutral phospholipids in platelets from alcoholics at admittance and after one week of detoxification did not differ from the control group (Fig. 5). The ratios between total phospholipids and cholesterol concentrations and total phospholipids and protein concentrations were also the same among the groups (Fig. 5) indicating a constant concentration of acidic phospholipids in platelets after an abuse period. Fatty acid compositions of PtdSer and PtdIns were also the same among the groups and revealed no significant differences. As in organs, the two fatty acids stearic (18:0) and arachidonic [20:4(n-6)] dominated. Docosahexaenoic acid [22:6(n-3)] constituted less than 1% in PtdIns and about 2% in PtdSer (Table 5).

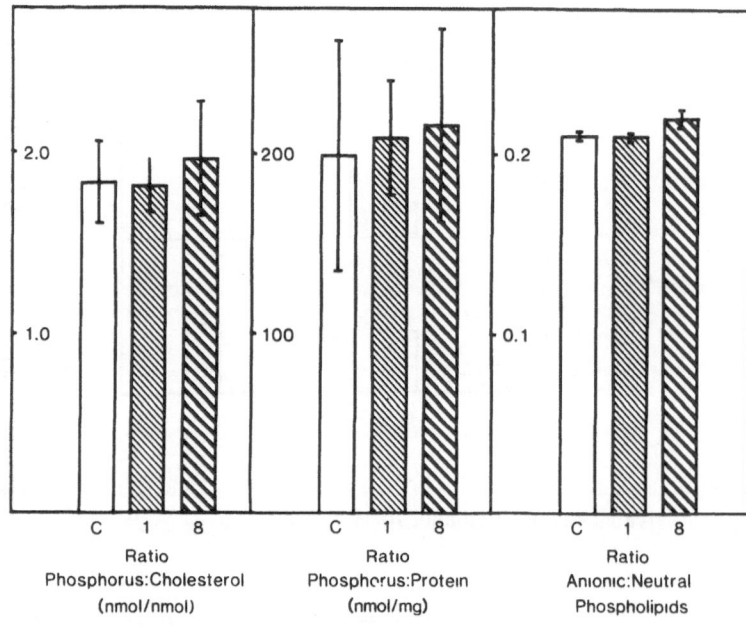

Figure 5. Concentrations of neutral and anionic phospholipids and cholesterol in platelets of alcoholics at admittance (1) and after one week of detoxification (8) compared to controls (C).

Table 5. *Proportions of major fatty acids in PtdSer and PtdIns from platelets of alcoholics at admittance (Alc 1) and after one week of detoxification (Alc 8)*

	16:0		18:0		18:1		20:4(n-6)	
	M	SD	M	SD	M	SD	M	SD
PtdSer								
Controls(n = 12)	4.5	3.6	40.6	4.0	23.4	3.3	16.6	3.3
Alc 1(n = 17)	2.7	1.8	41.1	3.6	24.9	3.3	16.7	3.8
Alc 8(n = 17)	2.5	2.0	40.0	2.8	27.1	2.7	15.8	2.9
PtdIns								
Controls(n = 12)	3.9	1.0	43.2	2.7	12.5	5.0	31.9	4.5
Alc 1(n = 17)	4.7	2.0	42.3	2.6	14.0	4.8	31.2	5.5
Alc 8(n = 17)	5.4	1.6	39.5	3.5	14.7	5.8	29.9	5.0

The lipid compositions of platelets and of synaptosomes have similarities such as the percentage of acidic phospholipids of total phospholipids and their fatty acid composition. One notable exception is the low proportion of 22:6(n-3) in platelet PtdSer. In spite of the similarities the present study has not revealed pathological findings in human platelets from alcoholics. Platelets may therefore be less suitable for assaying abnormalities among acidic phospholipids after ethanol abuse and other cell types could prove to be more useful.

CONCLUSIONS

Previous reports and results from the present study clearly indicate that acidic phospholipids are involved in the ethanol-induced disordering of brain lipids. Such changes are most probably confined to membranes and are more pronounced in neuron rich regions of the brain. Whether the acidic phospholipids are primarily affected by the perturbing action of ethanol or secondarily disturbed by abnormal function of membrane proteins has not been elucidated. Some brain membrane receptors are obviously interacting with acidic phospholipids and the findings are therefore of great concern with regard to membrane function.

Only a limited number of different cell membranes are available in human studies. Because of many similarities between synaptosomes and platelets, the latter has often been used as a model for nervous tissue. Among such similarities are storage of amines in vesicles, active uptake and transport of some neurotransmitters, and presence of receptors. The present study has not, however, demonstrated that platelets could be used as a peripheral marker for changes among brain acidic phospholipids. Factors such as ethanol toxicity on bone marrow and differences in platelet age and size between patients and controls complicate the interpretation of results. Blood cells with longer life time may turn out to be more suitable for studies in alcoholics.

ACKNOWLEDGMENTS

This study was supported by grants from The Swedish Medical Research Council (project no. 05249), The Swedish Life Insurance Company, Committee for Medical Research and The Albert Pålsson Foundation.

REFERENCES

Abood LG and Hoss W (1975) Stereospecific morphine adsorption to phosphatidyl serine and other membranous components of brain. Eur J Pharmacol 32: 66-75.

Abood LG and Takeda F (1976) Enhancement of stereospecific opiate binding to neural membranes by phosphatidylserine. Eur J Pharmacol 39: 71-77.

Abood LG, Salem N, MacNeil M and Butler M (1978) Phospholipid changes in synaptic membranes by lipolytic enzymes and subsequent restoration of opiate binding with phosphatidylserine. Biochim Biophys Acta 530: 35-46.

Alling C, Liljequist S and Engel J (1982) The effect of chronic ethanol administration on lipids and fatty acids in subcellular fractions of rat brain. Med Biol 60: 149-154.

Alling C, Gustavsson L and Änggård E (1983). An abnormal phospholipid in rat organs after ethanol treatment. FEBS Letters 152: 24-28.

Alling C, Gustavsson L, Månsson JE, Benthin G and Änggård E (1984a) Phosphatidylethanol formation in rat organs after ethanol treatment. Biochim Biophys Acta 793: 119-122.

Alling C, Becker W, Jones AW and Änggård E (1984b) Effects of chronic ethanol treatment on lipid composition and prostaglandins in rat fed essential fatty acid deficient diets. Alcoholism. Clin Exp Res 8: 238-242.

Alling C, Jönsson G, Gustavsson L, Jensen L and Simonsson P (1986) Anionic glycerophospholipids in platelets from alcoholics. Drug Alcohol Dependence, 16: 309-320.

Aronstam RS, Abood LG and Baumgold J (1977) Role of phospholipids in muscarinic binding by neural membranes. Biochem Pharmacol 26: 1689-1695.

Bartlett GR (1959) Phosphorus assay in column chromatography. J Biol Chem 234: 466-468.

Benthin G, Änggård E, Gustavsson L and Alling C (1985) Formation of phosphatidylethanol in frozen kidneys from ethanol treated rats. Biochim Biophys Acta. 835: 385-389.

Boyum A (1968) Isolation of mononuclear cells and granulocytes from human blood. Scand J Clin Lab Invest suppl 97: 77-89.

Buchanan AG and Kanfer JN (1980a) Topographical distribution of base exchange activities in rat brain subcellular fractions. J Neurochem 34: 720-725.

Buchanan AG and Kanfer JN (1980b) The effects of various incubation temperatures, particulate isolation and possible role of calmodulin on the activity of the base exchange enzymes of rat brain. J Neurochem 35: 814-822.

Chalifour R and Kanfer JN (1982) Fatty acid activation and temperature perturbation of rat brain microsomal phospholipase D. J Neurochem 39:299-305.

Charness ME, Gordon AS and Diamond I (1983) Ethanol modulation of opiate receptors in cultured neural cells. Science 222:1246-1248.

Chweh AY and Leslie SW (1982) Phosphatidylserine enhancement of [^{3}H] γ-aminobutyric acid uptake by rat whole brain synaptosomes. J Neurochem 38:691-695.

De Medio GE, Hamberger A, Sellström A and Porcellati G (1977) The phospholipid base-exchange system as a possible modulator of γ-aminobutyric acid transport in brain cells. Neurochem Res 2:469-484.

De Medio GE, Trovarelli G, Hamberger A and Porcellati G (1980) Synaptosomal phospholipid pool in rabbit brain and its effect on GABA-uptake. Neurochem Res 5:171-179.

DeVries GH, Chalifour RJ and Kanfer JN (1983) The presence of phospholipase D in rat central nervous axolemma. J Neurochem 40:1189-1191.

Fenn GC, Lynch MA, Nhamburo PT, Caberos L and Littleton JM (1983) Comparison of effects of ethanol on platelet function and synaptic transmission. Pharmacol Biochem Behav 18:37-43.

Foster AC, Fagg GE, Harris EW and Cotman CW (1982) Regulation of glutamate receptors: possible role of phosphatidylserine. Brain Res 242:374-377.

Franey RJ and Amador E (1968) Serum cholesterol measurement based on ethanol extraction and ferric chloride — sulfuric acid. Clin Chim Acta 21:255-263.

Harris RA, Baxter DM, Mitchell MA and Hitzemann RJ (1984) Physical properties and lipid composition of brain membranes from ethanol tolerant-dependent mice. Molecular Pharmacol 25:401-409.

Hoffman PL, Urwyler S and Tabakoff B (1982) Alterations in opiate receptor function after chronic ethanol exposure. J Pharm Exp Ther 222:182-189.

Karlsson KA (1982) Glycosphingolipids and surface membranes. In: Chapman D (ed): Biological Membranes, Academic Press, London, pp 1-74.

Lee NM, Friedman HJ and Loh HH (1980) Effects of acute and chronic ethanol treatment on rat brain phospholipid turnover. Biochem Pharmacol 29:2815-2818.

Nilsson O, Fredman P, Klinghardt GW, Dreyfus H and Svennerholm L (1981) Chloroquine-induced accumulation of gangliosides and phospholipids in skeletal muscles. Eur J Biochem 116:565-571.

Pletscher A and Laubscher A (1980) Use and limitations of platelets as models for neurones. In: Rotman A, Meyer FA, Gitler C and Silberberg A (eds), Platelets: Cellular Response Mechanisms and their Biological Significance. John Wiley and Sons, New York, pp 267-276.

Sun GY and Sun AY (1983) Chronic ethanol administration induced an increase in phosphatidylserine in guinea pig synaptic plasma membranes. Biochem Biophys Res Commun 113:262-268.

Sun GY, Huang HM, Lee DZ and Sun AY (1984) Increased acidic phospholipids in rat brain membranes after chronic ethanol administration. Life Sciences 35:2127-2133.

Svennerholm L (1968) Distribution and fatty acid composition of phosphoglycerides in normal human brain. J Lipid Res 9:570-579.

Svennerholm L and Fredman P (1980) A procedure for the quantitative isolation of brain gangliosides. Biochim Biophys Acta 617:97-109.

Wells MA and Dittmer JC (1963) The use of Sephadex for the removal of nonlipid contaminants from lipid extracts. Biochemistry 2:1259-1263.

Whittaker VP and Barker LA (1972) The subcellular fractionation of brain tissue with special reference to the preparation of synaptosomes and their component organelles. In: Fried R (ed), Methods in Neurochemistry, vol. 2, Marcel Dekker, New York, pp 1-52.

Yang SF, Freer S and Benson AA (1967) Transphosphatidylation by phospholipase D. J Biol Chem 242:477-484.

Phospholipid research and the nervous system
Biochemical and molecular pharmacology
L.A. Horrocks, L. Freysz, G. Toffano (eds)
Fidia Research Series, vol. 4.
Liviana Press, Padova. © 1986

THE CONTROL AND FUNCTION
OF INOSITIDE-METABOLIZING ENZYMES

R.F. Irvine, A.J. Letcher, D.J. Lander and R.M.C. Dawson

Department of Biochemistry, AFRC Institute of Animal Physiology,
Babraham, Cambridge CB2 4AT, UK

INTRODUCTION

Recent advances in our understanding of the role of inositides in intracellular signalling (Michell et al., 1981; Berridge, 1984a; Berridge and Irvine, 1984) have highlighted the need to understand the mechanisms by which the cellular levels of inositides, and hence indirectly of the second messengers generated from them, are controlled. A stimulation of PtdInsP_2 phosphodiesterase is the likely initial event occurring after cell activation by a large number of agonists (Fig. 1). As a result of phosphodiesteratic cleavage of PtdInsP_2 the two second messengers, diacylglycerol (Nishizuka, 1984) and InsP_3 (Berridge, 1984a; Berridge and Irvine, 1984), are formed. The former activates protein kinase C (Nishizuka, 1984) and the latter primarily mobilizes intracellular calcium (Berridge and Irvine, 1984), and these two pathways frequently, but not always (Labarca et al., 1984; Danthurluri and Deth, 1984), act synergistically to produce a final cell activation (Kaibuchi et al., 1984; Rink et al., 1983; Putney et al., 1984). Although the acute control of the levels of these intracellular messengers lies in the regulation of PtdInsP_2 phosphodiesterase activity, the regulation of diacylglycerol and InsP_3 over the longer term (after the first few seconds of stimulation) lies in the hands of the enzymes which generate PtdInsP_2, and those which degrade diacylglycerol and InsP_3. It is the possible mechanism for regulation of these enzymes which are discussed here.

Inositide Kinases

These enzymes maintain levels of PtdIns$P$ and PtdInsP_2 in resting and stimulated cells (Fig. 2). Their possible control by feedback of products (Michell et al., 1973; Ir-

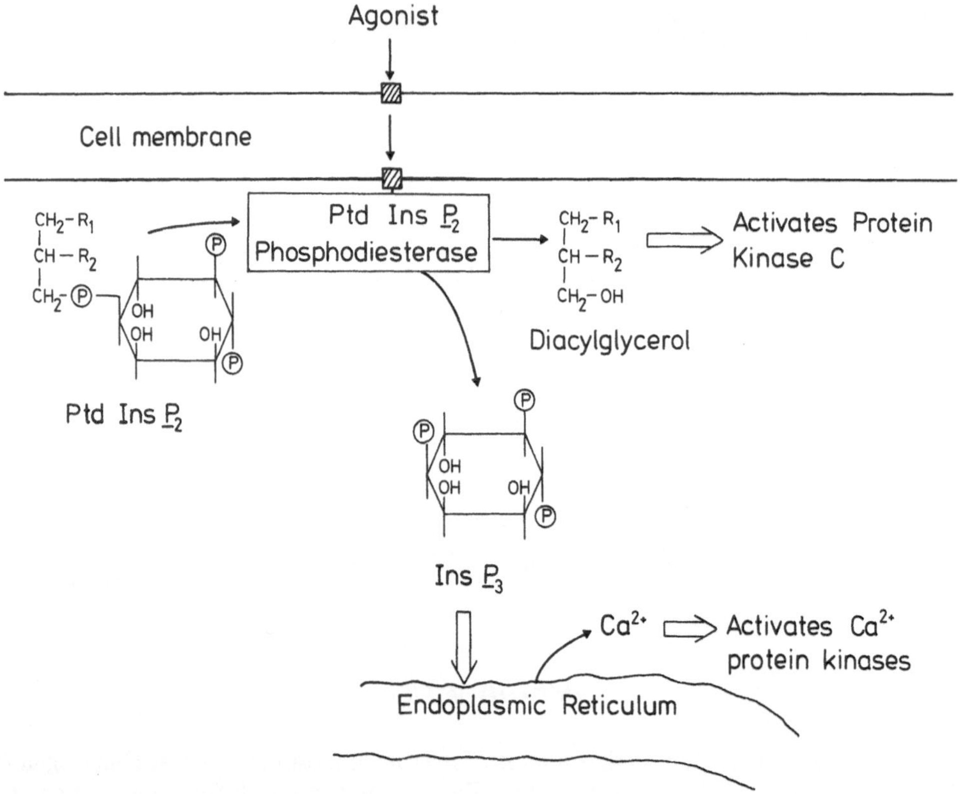

162

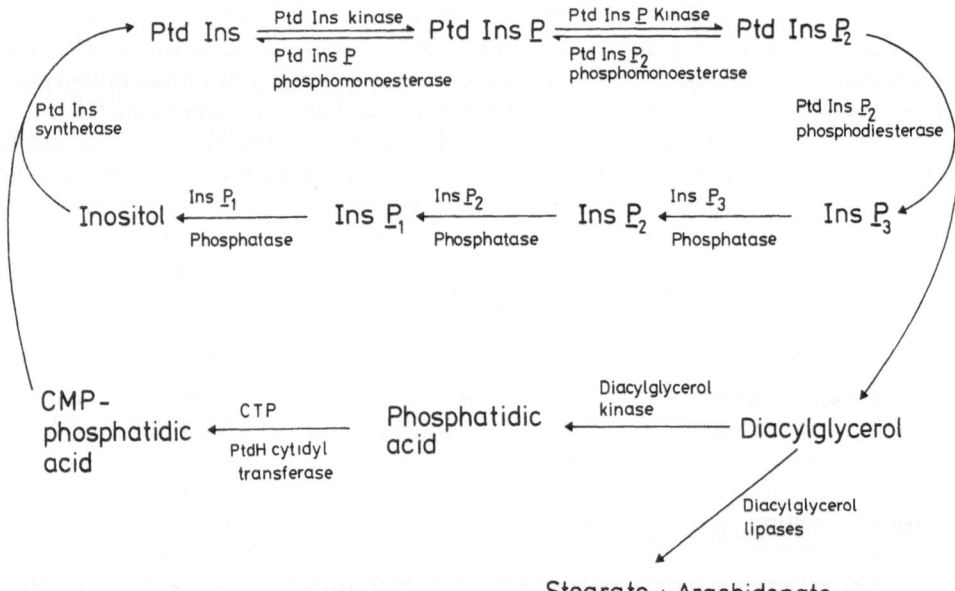

Agonist

Cell membrane

Ptd Ins P_2
Phosphodiesterase

CH_2-R_1
$CH-R_2$
CH_2-OH

Diacylglycerol

Activates Protein
Kinase C

CH_2-R_1
$CH-R_2$
CH_2- ⓅOH
OH
OH

Ptd Ins P_2

Ins P_3

Ca^{2+} Activates Ca^{2+}
protein kinases

Endoplasmic Reticulum

Figure 1. Generation of second messengers from inositides.

Ptd Ins $\xrightleftharpoons[\text{Ptd Ins } P \text{ phosphomonoesterase}]{\text{Ptd Ins kinase}}$ Ptd Ins P $\xrightleftharpoons[\text{Ptd Ins } P_2 \text{ phosphomonoesterase}]{\text{Ptd Ins } P \text{ Kinase}}$ Ptd Ins P_2

Ptd Ins
synthetase

Ptd Ins P_2
phosphodiesterase

Inositol $\xleftarrow[\text{Phosphatase}]{\text{Ins } P_1}$ Ins P_1 $\xleftarrow[\text{Phosphatase}]{\text{Ins } P_2}$ Ins P_2 $\xleftarrow[\text{Phosphatase}]{\text{Ins } P_3}$ Ins P_3

CMP-
phosphatidic
acid $\xleftarrow[\text{PtdH cytidyl transferase}]{\text{CTP}}$ Phosphatidic
acid $\xleftarrow[]{\text{Diacylglycerol kinase}}$ Diacylglycerol

Diacylglycerol
lipases

Stearate + Arachidonate

Figure 2. Pathways of Inositide Metabolism.

vine, 1982) or availability of substrate are reviewed elsewhere (Irvine, 1982; Irvine et al., 1984a). It is important to note that the subcellular distribution of PtdIns kinase, which has been reported in several locations other than the plasma membrane (Collins and Wells, 1983; Smith and Wells, 1983; Jergil and Sundler, 1983), could also determine PtdInsP_2 distribution; PtdIns$P$ is, on present evidence, predominantly soluble (Irvine, 1982), and therefore one might expect it to act only where its substrate is formed. Thus it is of interest to ask whether PdtIns$P$ is ever formed to the exclusion of PtdInsP_2. Several authors have reported the presence of the enzymes that make and dephosphorylate PtdIns$P$ in, for example nuclear membranes (Smith and Wells, 1983; 1984) or sarcoplasmic reticulum (Varsanyi et al., 1982) and have therefore proposed a function for this lipid in those membranes. That may indeed be so, but if PtdIns$P$ kinase is soluble and able to act on its substrate whatever the lipid structure (Irvine et al., 1984a,b), then PtdInsP_2 will inevitably be formed to some extent.

Thus, the multifunctional role of inositides (Irvine and Dawson, 1980; Berridge 1981; Hawthorne, 1983) may extend to other intracellular locations and may involve both PtdIns$P$ and PtdInsP_2. One or both of PtdIns kinase and PtdIns$P$ kinase are activated on cell stimulation (Taylor et al., 1984: de Courcelles et al., 1984; Halenda and Feinstein, 1984; see also Aloyo et al., 1983) and the effect of cyclic nucleotides (Torda, 1972; Enyedi et al., 1984) on their activity is an interesting area for exploration. It may be that both kinase C and cyclic nucleotide-dependent protein kinase can phosphorylate the inositide kinases and hence control their activity. Clearly, we have much more to learn about these enzymes in resting and stimulated tissues.

PtdIns$P$ and PtdInsP_2 Phosphomonoesterases

The regulation of these enzymes is as yet unclear (Irvine, 1982; Irvine et al., 1984a) despite the obvious point that wherever alterations of inositide kinases have been reported (see previous paragraph, above), it is equally likely that it may be the phosphomonoesterases rather than the kinases which are regulated. PtdIns$P$ phosphomonoesterase and PtdInsP_2 phosphomonoesterase are both classically soluble and Mg^{2+} - dependent (Dawson and Thompson, 1964; Irvine, 1982), but recent reports of distinct membrane-bound and Mg^{2+}-independent activities (for example, Mack and Palmer, 1984; Seyfred et al., 1984) may throw open an entirely new view of the regulation of inositide levels by these enzymes. They have mostly been studied in brain, and this tissue may very well be abnormal in the levels and distributions of inositide-metabolizing enzymes.

PtdIns Synthetase

Several recent reports have documented increases in PtdIns synthesis de novo following cell stimulation (Farese et al., 1983). Whether this is due to a direct stimulation of the enzyme, or of greater availability of its indirect substrate, phosphatidic acid, by increased acylation of glycerophosphate is not entirely clear. Some agonists which induce increases in inositide synthesis are not "classical" agonists in that they do not stimulate a phosphodiesterasic hydrolysis of PtdInsP_2 (but see Chapman et al., 1983) and it is an open question as to whether the two groups of agonists exert their effects on inositide synthesis by entirely different mechanisms.

In summarizing the enzymes discussed so far, it is apparent that there are huge gaps in our knowledge of their control, their sub-cellular location and their interactions. The recent renewal of emphasis on the polyphosphoinositides (Michell et al., 1981) and on PtdInsP_2 in particular in view of its dual second-messenger producing function (Nishizuka, 1984; Berridge, 1984a; Berridge and Irvine, 1984), will in turn bring about an increased amount of study on these enzymes. But again it should be emphasized that this second messenger generation may not be the only function of PtdInsP_2, and that polyphosphoinositides may have other intracellular functions as yet only hinted at.

PtdInsP_2 Phosphodiesterase

Recently this enzyme has become to be considered as the primary receptor-stimulated enzyme, and the one responsible for the acute control of the production of InsP_3 and diacylglycerol. The possible ways in which it may be regulated by substrate structure and availability (Irvine et al., 1984a,b) and/or G-proteins (Haslam and Davidson, 1984a,b; Berridge and Irvine, 1984) are discussed elsewhere. Recently the suggestion (Low and Weglicki, 1983; Irvine et al., 1984b) that one enzyme, or group of enzymes (Hirasawa et al., 1982; Low and Weglicki, 1983; Low et al., 1984) can hydrolyse all three inositides has received direct confirmation with a purified enzyme (Wilson et al., 1984). The reason why the enzyme is heterogenous remains unknown and further explorations of different tissues (Low et al., 1984) may help to resolve this. Clearly, for example, there is a difference between the soluble rat brain enzyme and membrane-bound human red cell enzyme (Downes and Michell, 1982; Irvine et al., 1984b), as compared with the membrane-bound enzymes of sea-urchin eggs (Whittaker and Irvine, 1984) or rabbit red cells (Quist, 1985) in that the former pair will not hydrolyse a membrane-bound substrate under physiological Mg^{2+} and K^+ concentrations until the Ca^{2+} is at a non-physiological concentration of 0.1 mM or more, whereas the latter two are active at 5-10 μM, a Ca^{2+} concentration not unlikely in a stimulated cell. Whether these differences are due to enzyme heterogeneity, or to an absence or presence of nucleotide-binding proteins, is an interesting area of exploration.

Inositol 1,4,5-trisphosphate

Following the original proposal that Ins(1,4,5)P_3 is the second messenger for Ca^{2+} mobilization (Berridge, 1983), considerable experimental evidence in its favour has been accumulated (Berridge and Irvine, 1984). The acute control of its production lies in the activity of PtdInsP_2 phosphodiesterase, but the initial pool of PtdInsP_2 will quickly be used up following stimulation, and it is after the first 15-30 seconds or so that the regulation of the other enzymes of inositide metabolism, discussed above, come into play. Recent work from Drummond's laboratory (Drummond and Raeburn, 1984) has shown just how important to a cell's metabolism is the maintenance of adequate PtdInsP_2 levels, as even under conditions of PtdIns depletion, the resting levels of polyphosphoinositides are maintained.

Ins(1,4,5)P_3 is catabolised by a plasma-membrane bound enzyme (Downes et al., 1982; Storey et al., 1984; Seyfred et al., 1984), though other subcellular locations for this enzyme cannot be ruled out. The enzyme is clearly very specific in that only the 5 phosphate is removed (Downes et al., 1982) and Ins(1,3,4)P_3 is not hydrolysed (Irvine et al., 1984c), and its potency is illustrated by the remarkably tight control kept of Ins(1,4,5)P_3 levels in stimulated cells (see, for example, Berridge, 1983; Irvine et al., 1985). It is not known whether this enzyme is regulated, or whether it is simply catabolic and limited by substrate only. Clearly a defect in its activity could have a profound effect on cell function (Berridge, 1984b).

Diacylglycerol

The production of this second messenger (Nishizuka, 1984) is subject to the same constraints as Ins(1,4,5)P_3, though the production of Ins(1,3,4)P_3 may represent a means by which the diacylglycerol/Ca^{2+} stoichiometry could be increased (Irvine et al., 1985). The removal of diacylglycerol is controlled by a number of factors. Probably the major inactivation pathway is by phosphorylation to phosphatidate, but the regulation of diacylglycerol kinase is an unstudied and unknown area; it is an enzyme that is, like so many discussed here, part soluble and part membrane bound (Kanoh and Åkesson, 1978) and an association with microtubules has never been disproved (Daleo et al., 1976). A long term regulation of diacylglycerol by limiting the further metabolism of phosphatidate has been documented by Drummond and Raeburn (1984). Chronic depletion of tissue inositol levels (by lithium) resulted in a raising of control and stimulated diacylglycerol levels. Similarly, the whole story of phosphatidate phosphohydrolase regulation (see Butterworth et al., 1984) could be brought to bear on the control of diacylglycerol levels after stimulation by agonists which modify inositide metabolism, and could partly explain some of the apparent cyclic-nucleotide or ethanol interactions with these agonists because the diacylglycerol levels following stimulation may be modified by cellular phosphatidate phosphohydrolase activity.

The degree to which diacylglycerol lipase regulates diacylglycerol levels, and in turn the extent to which inositide contributes to arachidonate release (Irvine, 1982b), is another problem area. In some tissues rich in diacylglycerol lipase the effect of this enzyme may be considerable (see, for example, Mauco et al., 1984), but in others this enzyme may be of little significance either in generating arachidonate or in controlling diacylglycerol levels.

A final point concerning diacylglycerol as a second messenger concerns the location of its production. In the "classic" inositide response it will be confined entirely to the plasma membrane. If we accept that kinase C is in part a soluble enzyme caused to be membrane-bound by the appearance of diacylglycerol (Nishizuka, 1984) then its activation will be localised to that membrane. Treatment of cells with phorbol esters, which will go into all membranes however, may activate kinase C in membranes not normally active in this respect.

This may explain some of the more bizarre effects of phorbol esters, and calls for some caution in the interpretation of such experiments. We know nothing about the possible generation of diacylglycerol in intracellular membranes (see above) and here again may be another entirely new field to be explored.

CONCLUSIONS

With the remarkable increase in interest in the polyphosphoinositides as generators of second messengers comes the realisation that our knowledge of the enzymology of the synthesis and removal of these second messengers is scanty at present. Entirely new questions are posed, and exciting areas for exploration in the field of inositide enzymology present themselves.

REFERENCES

Aloyo VJ, Zwiers H and Gispen WH (1983) J Neurochem 41: 649-653.

Berridge MJ (1981) Mol Cell Endocrinol 24: 115-140.

Berridge MJ (1983) Biochem J 212: 849-858.

Berridge MJ (1984a) Biochem J 220: 345-360.

Berridge MJ (1984b) Biotechnology 2: 541-546.

Berridge MJ and Irvine RF (1984) Nature 312: 315-321.

Best L and Malaisse WJ (1984) Arch Biochem Biophys 234: 253-257.

Butterworth SC, Martin A and Brindley BN (1984) Biochem J 222: 487-493.

Chapman BA, Wilson JS, Colley PW, Picola RC and Somes JB (1983) Biochem Biophys Res Commun 115: 771-776.

Collins CA and Wells WW (1983) J Biol Chem 258: 2130-2134.

Daleo GR, Piras MM and Piras MR (1970) Eur J Biochem 68: 339-346.

Danthurluri NR and Deth RC (1984) Biochem Biophys Res Commun 125: 1103-1109.

Dawson RMC and Thompson W (1964) Biochem J 91: 244-250.

De Chaffoy de Courcelles D, Roevens P and Van Belle H (1984) FEBS Lett 173: 389-393.

Downes CP, Mussat MC and Michell RH (1982) Biochem J 203: 169-177.

Drummond AH and Raeburn CA (1984) Biochem J 224: 129-136.

Enyedi A, Fargo A, Sarkdi B and Gardos G (1984) FEBS Lett 176: 235-238.

Farese RV, Barnes DE, Davis JS, Standaert ML and Pollet RJ (1984) J Biol Chem 259: 7094-7100.

Halenda SP and Feinstein MB (1984) Biochem Biophys Res Commun 124: 507-513.

Haslam RJ and Davidson MML (1984a) FEBS Lett 174: 90-95.

Haslam RJ and Davidson MML (1984b) J Receptor Res 4: 605-629.

Hawthorne JN (1983) Bioscience Rep 3: 887-904.

Hirasawa K, Irvine RF and Dawson RMC (1982) Biochem J 205: 437-442.

Irvine RF (1982a) Cell Calcium 3: 295-309.

Irvine RF (1982b) Biochem J 204: 3-16.

Irvine RF and Dawson RMC (1980) Biochem Soc Trans 8: 376-377.

Irvine RF, Letcher AJ, Lander DJ and Dawson RMC (1985) in: Bleasdale J E, Eichberg J and Hauser G (eds): Inositol and Inositides, Humana Press, Clifton, NY USA, pp. 123-135.

Irvine RF, Letcher AJ and Dawson RMC (1984b) Biochem J 218: 177-185.

Irvine RF, Letcher AJ, Lander DJ and Downes CP (1984c) Biochem J 223: 237-243.

Irvine RF, Ånggård EE, Letcher AJ and Downes CP (1985) Biochem J 229: 505-511

Jergil B and Sundler R (1983) J Biol Chem 258: 7968-7973.

Kaibuchi K, Takai Y, Sawamura M, Hoshijima M, Fujikara M and Nishizuka Y (1983) J Biol Chem 258: 6701-6704.

Kanoh H and Åkesson B (1978) Eur J Biochem 85: 225-232.

Labarca R, Janowsky A, Patel J and Paul SM (1984) Biochem Biophys Res Commun 123: 703-709.

Low MG and Weglicki W (1983) Biochem J 215: 325-334.

Low MG, Carrol RC and Weglicki WB (1984) Biochem J 221: 813-820.

Mack SE and Palmer FBStC (1984) J Lipid Res 25: 75-85.

Mauco G, Dangelmaier C and Smith JB (1984) Biochem J 224: 933-940.

Michell RH, Harwood JL, Coleman R and Hawthorne JN (1967) Biochim Biophys Acta 144: 649-658.

Michell RH, Kirk CJ, Jones LM, Downes CP and Creba JA (1981) Phil Trans R Soc B 296: 123-137.

Nishizuka Y (1984) Nature 308: 693-698.

Putney JW, McKinney JS, Aub DL and Leslie BA (1984) Molec Pharmacol 26: 261-266.

Quist EE (1985) Arch Biochem Biophys 236: 140-149.

Rink TJ, Sanchez A and Hallam TJ (1983) Nature 305: 317-319.

Seyfred MA, Farrell LE and Wells WW (1984) J Biol Chem 259: 13204-13208.

Smith CD and Wells WW (1983) J Biol Chem 258: 9368-9373.

Smith CD and Wells WW (1984) Arch Biochem Biophys 235: 529-537.

Storey DJ, Shears SB, Kirk CJ and Michell RH (1984) Nature 312: 374-376.

Taylor MV, Metcalfe JC, Hesketh TR, Smith GA and Moore SP (1984) Nature 312: 462-465.

Torda C (1972) Biochim Biophys Acta 286: 389-395.

Varsanyi M, Tolle H, Heilmeyer MG, Dawson RMC and Irvine RF (1983) EMBO J 2: 1543-1548.

Whitman MR, Epstein J and Cantley L (1984) J Biol Chem 259: 13652-13655.

Whittaker M and Irvine RF (1984) Nature 312: 636-639.

Wilson DB, Bross TE, Hofmann SL and Majerus PW (1984) J Biol Chem 259: 11718-11724.

Phospholipid research and the nervous system
Biochemical and molecular pharmacology
L.A. Horrocks, I. Freysz, G. Toffano (eds)
Fidia Research Series, vol. 4.
Liviana Press, Padova. © 1986

DIACYLGLYCEROLS AND ARACHIDONIC ACID IN THE MOLECULAR PATHOGENESIS OF BRAIN INJURY

Nicolas G. Bazan, Dale L. Birkle, T. Sanjeeva Reddy and Robert E. Vadnal

Louisiana State University School of Medicine, LSU Eye Center
and Department of Psychiatry, New Orleans, LA

INTRODUCTION

Stimulation or injury promotes an accumulation of free fatty acids (FFA) and diacylglycerols in the central nervous system (Bazan, 1970, 1971; Bazan and Rakowski, 1970; Bazan et al., 1971; Aveldano and Bazan 1975a, b, 1979; Bazan, 1976). The accumulated lipids are potentially harmful to excitable membranes by mechanisms such as FFA inhibition of membrane-bound enzymes (Rhoads et al., 1983) and lipid peroxidation (Yoshida et al., 1982). The accumulation of FFA following a single electroconvulsive shock (ECS) is transient, returning to normal levels within 5 min after the shock (Bazan and Rakowski, 1970). Similarly, the FFA increase in gerbil brain due to ischemia induced by bilateral carotid occlusion can be reversed to normal levels after reperfusion (Yoshida et al., 1980, Bhakoo et al., 1984). In the case of prolonged ischemia or during bicuculline-induced status epilepticus, the accumulation of FFA and DAG continues to rise and may lead to irreversible brain damage (Aveldano and Bazan, 1975a, b; DeMedio et al., 1980; Shiu et al., 1983; Tang and Sun, 1982, 1985; Rodriguez de Turco et al., 1983). Irreversible damage may occur because of selective degradation of functionally critical lipid classes or molecular species of lipids (e.g. polyphosphoinositides) in synaptic membranes and/or other cell membranes, such as mitochondria (Bazan et al., 1971; Bazan, 1976). In brain, accumulation of DAG and FFA (especially polyunsaturates) as a result of catabolism of polyphosphoinositides, may disturb interneuronal communication by altering levels of secondary messengers

(DAG itself and water-soluble inositol polyphosphates) and the activity of protein kinase C. Protein kinase C in brain, which can be activated by polyunsaturated fatty acids (Murakami and Routtenberg, 1985) and unsaturated DAG (Nishizuka et al., 1984a), appears to play a critical role in stimulus-secretion coupling (Kawahara et al., 1980; Wooten and Wrenn, 1984; Putney et al., 1984).

METHODOLOGICAL CONSIDERATIONS

Brain Sampling and Basal FFA and DAG Levels in Brain

The method of sampling the brain is critical to assessment of the endogenous content of DAG and FFA. In addition to ischemia and stimulation, trauma, such as scraping the skull periosteum, will cause changes in brain FFA (Politi et al., 1985). Several methods have been used to facilitate the rapid inactivation of brain enzymes: a) rapid removal and homogenization of the brain in chloroform: methanol, which usually takes about 30 sec for the rat (Aveldano and Bazan, 1975a,b); b) fixation of decapitated heads in liquid N_2 (Aveldano and Bazan, 1979; DeMedio et al., 1980; Tang and Sun, 1982); c) liquid N_2 fixation of the brain in situ (Ponten et al., 1973); and d) fixation of the brain by microwave irradiation (Cenedella et al., 1975; Galli and Spagnuolo, 1976; Soukup et al., 1978). In addition to rapid fixation, ideally the method of sampling brain should minimize trauma, such as that caused by opening the cranial bones to subsequently apply liquid N_2 (e.g. Ponten et al., 1973), inasmuch as the surgery itself may affect lipid levels (Politi et al., in press). Microwave irradiation may be the method of choice because it allows rapid fixation with minimal confounding effects of stress and trauma. Galli and his associates (1976) used low power microwave irradiation to fix the brain in 3 sec. More recently, high power microwave irradiation has been shown to inactivate brain enzymes involved in neurotransmitter and cyclic nucleotide metabolism within one sec (Stavinova et al., 1977; Schneider et al., 1981; Fujiwara et al., 1978; Schneider, 1984). Thus, through the use of high power microwave irradiation techniques, one may be able to measure levels of FFA and DAG closer to the condition in vivo.

Another important consideration in the evaluation of the basal levels of FFA and DAG is the extraction, isolation and quantitation of lipids in the tissue (Bazan and Bazan, 1975). For example, acidic extraction solvents or exposure to silicic acid columns can cause an artifactual increase in the level of FFA. If the contact with the silicic acid column is prolonged, phospholipid hydrolysis occurs. This degradation is undetectable as phospholipid loss because of the relatively large mass, but very significant in terms of FFA generated (Bazan and Joel, 1970). Purified FFA and DAG should be separated carefully to prevent cochromatography with other lipids. Triacylglycerol can contaminate the FFA fraction and cholesterol can contaminate the DAG fraction (Shiu and Nemoto, 1981; Shiu et al., 1983). Also, because DAG isomerizes in the presence of acid or silica gel, it is important to recover both the naturally-occuring 1,2-DAG and the 1,3-DAG artifact (Aveldano and Bazan, 1974).

Rapid Fixation of the Brain and Basal Levels of FFA and DAG

Recently, we have used high power microwave irradiation (6.0 kW, 2450 MHz) for the rapid fixation of rat and mouse brain. The results show that the levels of FFA and DAG vary with time of microwave irradiation, which is largely dependent on the size of the animals. For example, rats weighing 275-320 g required about 2 sec irradiation to achieve a stable level. However, for rats weighing 100-200 g and mice weighing 25-30 g, fixation of the brain was possibile in 1.5 and 0.8 sec, respectively. The levels of FFA and DAG were low (25-30 nmol/g wet weight) and were much lower than values previously reported in the literature. The major fatty acids were palmitic (16:0), stearic (18:0) and oleic (18:1) acids. Free arachidonic (20:4, n-6) and docosahexaenoic (22:6, n-3) acids were undetectable, but very low levels (2-4 nmol/g brain) of arachidonate were recovered in the DAG fraction. The differences in FFA and DAG between 2-3 sec of irradiation and less than 1.5 sec irradiation reflect release of these labile lipids during the irradiation period. This clearly demonstrates the importance of fixing the brain as rapidly as possible. It is advisable either to anesthetize the animals prior to irradiation or to acclimate them to the microwave holder to avoid stress-induced accumulation of these lipids. Unanesthetized animals showed significantly higher values for arachidonic acid, showing that there is a stress-induced release of this fatty acid. Thus, high power microwave irradiation seems to serve as an excellent technique for the rapid fixation of brain. However, one should keep in mind the importance of animal size and microwave irradiation time in achieving a complete and rapid fixation of the brain.

EFFECTS OF SHORT-TERM STIMULATION ON BRAIN LIPIDS

Effect of a ECS on Rat Brain FFA and DAG

Previous studies have shown a transient increase in rat brain FFA (Bazan and Rakowski, 1970; Bazan, 1970) and in DAG (Aveldano and Bazan, 1979) after ECS. In the present studies, we have evaluated this phenomenon at very short time points. Rats (100-120 g) were subjected to a single ECS using a Grass S48 stimulator (Grass Medical Instruments, Quincy, MA) with a stimulation rate of 150 pulses/sec at 130 V with a 750 msec pulse train duration. The animals were fixed rapidly using high power microwave irradiation (6.0 kW at 2450 MHz for 1.5 sec). There was a significant increase in all FFA after 30 sec of ECS. Levels of FFA began to return to normal 60 sec after ECS and reached normal levels by 5 min. A significant increase was seen in arachidonic and docosahexaenoic acids (10-15 fold accumulation over control values), followed by stearic (3-5 fold), oleic (2-3 fold) and palmitic (1.5-2 fold) acids. As compared to FFA, the ECS-induced accumulation of DAG was small but significantly higher than controls for the stearoyl-arachidonoyl species. The levels of these lipids also returned to control values 5 min after ECS.

Effect of ECS on Content and Composition of Phosphoinositides

Because there was a significant ECS-induced increase in stearoyl and arachidonoyl species of both FFA and DAG, we analyzed the acyl group composition of inositol phospholipids to determine whether these lipids contributed to the accumulation. While extracting these lipids by the method of Hauser and Eichberg (1973), we noticed a sizable loss of polyphosphoinositides into the neutral chloroform: methanol extract (Reddy, Horrocks and Bazan, manuscript in preparation). Therefore, the neutral and acidified chloroform: methanol extracts were pooled to obtain a quantitative measure of polyphosphoinositides. The recovery of polyphosphoinositides obtained by this method was *higher* than that reported previously (Hauser and Eichberg, 1973).

172

There was a significant decrease in the content of stearate and arachidonate in phosphatidylinositol 4,5-bisphosphate (PtdIns(4,5)P_2) after a single ECS. No significant changes were seen in the content of palmitate and docosahexaenoate. The content of fatty acids in phosphatidylinositol (PtdIns) and phosphatidylinositol 4-phosphate (PtdIns4P) was not affected by a single ECS. Thus, ECS appears to cause the specific degradation of PtdIns(4,5)P_2. Catabolism of PtdIns(4,5)P_2 could account for the accumulation of the stearoyl-arachidonoyl species of FFA and DAG. These are the first results showing that ECS causes the loss of the stearoyl-arachidonoyl species of PtdIns(4,5)P_2 and to a minor extent PtdIns4P, but not PtdIns in brain. This suggests that there may be a transient massive degradation of the receptor-linked PtdIns(4,5)P_2 that could, in turn, affect cell signaling. The increase in FFA and DAG enriched in polyunsaturated fatty acids and the loss of PtdIns(4,5)P_2 correlate with the seizure event, and reflect a stimulation of the phosphoinositide cycle (Fig. 1).

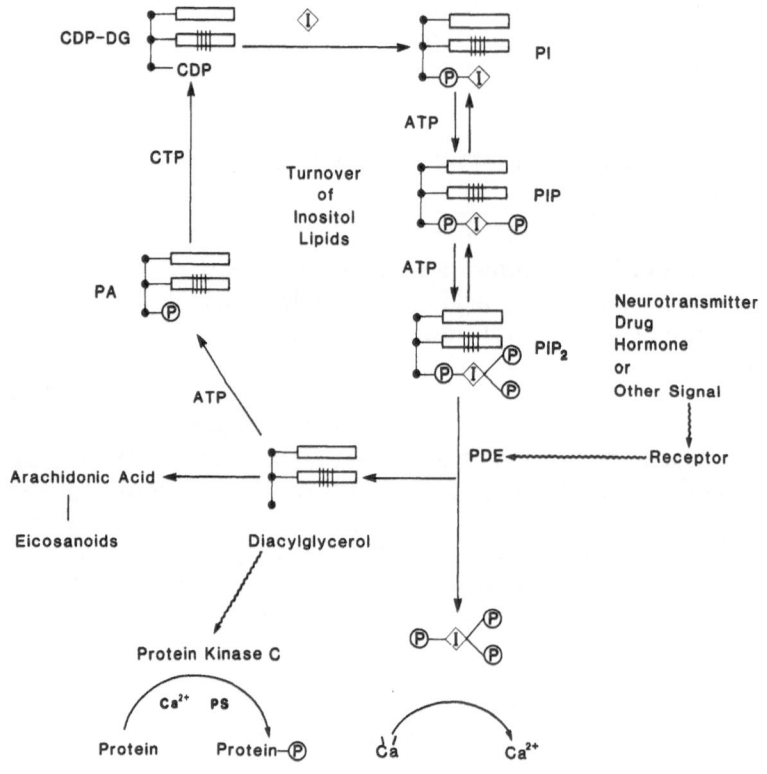

PHYSIOLOGICAL RESPONSES

Figure 1. The phosphoinositide cycle. Receptor stimulation causes the degradation of PtdIns(4,5)P_2 (PIP$_2$) to DAG and InsP_3 mediated by a phosphodiesterase (PDE, phospholipase C). DAG is reconverted to phosphoinositides via phosphatidic acid (PA), or further degraded by DAG lipase to yield free arachidonic acid. The second messengers, DAG and InsP_3, stimulate the activity of protein kinase C and increase levels of intracellular calcium.

Effects of Lithium on Phosphoinositide Metabolism during Seizures

Allison and Stewart (1971) first noted an effect of lithium on the inositol cycle. They found that rats injected subcutaneously with 10 mEq/kg of lithium showed a marked reduction in myo-inositol levels in cerebral cortex, and later discovered that levels of myo-inositol-1-phosphate (Ins$P$) increased. These effects could be reversed with atropine (Allison et al., 1976). The effects of lithium are the result of inhibition of the conversion of Ins$P$ to inositol by myo-inositol-1-phosphatase (Hallcher et al., 1980). This classical effect has been proposed by Berridge et al. (1982) as a mechanism which "amplifies" the production of inositol phosphates from precursor phosphoinositides. Myo-inositol-1,4,5-trisphosphate (InsP_3) and DAG have been proposed as second messengers (Nishizuka, 1984; Berridge, 1984). The effects of various inositol phosphates continue to be explored and there has been a recent report of two isomers of InsP_3: inositol-1,4,5-trisphosphate and inositol-1,3,4-trisphosphate (Irvine et al., 1984). The complete picture is not clear at this time, but the effects of lithium on this system are very powerful, and occur at levels which are in the same therapeutic range used to treat manic-depressive disorders.

Lithium causes alterations in the level of inositol phosphates derived from the phosphoinositide cycle. However, no effects were found previously on the lipid components of the phosphoinositide cycle (PtdIns, PtdIns4P, PtdIns(4,5)P_2) (Sherman et al., 1985). Because of the importance of the phosphoinositide cycle as a second messenger system and the possible contributions of this cycle in the pathogenesis of manic-depressive disorders and in seizure phenomena, we have explored the effects of lithium on this cycle, both the lipid and water-soluble components, using ECS as the stimulus producing the desired model of the hyper-functioning neuron.

Rat brain was prelabeled with [^{3}H]inositol by intraventricular injection 18-22 hr prior to sacrifice. Two groups of rats also were given LiCl (10 mEq/kg) subcutaneously with the radiolabel. Two groups of rats received saline. One group of rats, which was pretreated with lithium, received ECS, as did one group pretreated with saline. Animals receiving electroshock (122 V, 155 pulses/sec, scalp electrodes) were sacrificed 25 sec after onset of seizure by high power head-focused microwave irradiation (6.5 kW, 1.5 sec). Lipid extracts obtained from the cerebrum were analyzed by thin layer chromatography, the radiolabel quantified by scintillation counting and the mass by phosphorus assay. In the ECS-saline group, [^{3}H]inositol labeling of PtdIns(4,5)P_2 was reduced 53% and mass was reduced 27%. Labeling of PtdIns4P was reduced 39% and mass reduced 27%. In the ECS-lithium group of rats, there were no significant changes in radiolabel or mass compared to sham controls. Lithium in the absence of ECS had no effect on radiolabel or mass of phosphoinositides. Thus, lithium prevented the ECS-induced phosphoinositide breakdown, demonstrating a powerful effect on the lipid phase of the phosphoinositide cycle.

EFFECTS OF IRREVERSIBLE TRAUMA ON BRAIN LIPIDS

Effect of Status Epilepticus on Phosphoinositide Turnover

While ECS represents a short-term, reversible alteration in brain lipid metabolism, status epilepticus causes irreversible effects resulting in tissue damage. Previous studies have shown a large and long-lasting accumulation of polyunsaturated FFA and DAG in rat brain during bicuculline-induced status epilepticus (Siesjo et al., 1982; Rodriguez et al., 1983). We have investigated the turnover of phosphoinositides in this preparation. Rat brain was prelabeled in vivo by intraventricular injection of 10-20 μCi ^{32}P. The radiolabel diffused rapidly throughout the brain and steady-state levels of labeling of phospholipids were obtained in approximately 30 min. Rats were mechanically ventilated, given 10 mg/kg (i.p.) bicuculline and sacrificed by microwave irradiation during the status epilepticus phase (4-5 min after drug administration). Bicuculline treatment resulted in an enhanced turnover of phosphoinositides. There was a 25% to 55% increase in labeling of PtdIns(4,5)P_2, PtdIns4P, PtdIns and phosphatidic acid, as compared to vehicle controls. Bicuculline also increased levels of inositol-1,4-bisphosphate (InsP_2) and InsP_3. Because enhanced phosphoinositide turnover has been linked to activation of muscarinic cholinergic receptors, we also studied the interaction of atropine, a muscarinic antagonist, on the inositol lipid effect. Rats were pretreated with atropine (150 mg/kg, i.p.) prior to injection of bicuculline. Atropine antagonized the bicuculline-induced stimulation of phosphoinositide turnover, but did not affect the accumulation of InsP_2 and InsP_3. In control animals, atropine diminished the basal incorporation of ^{32}P into phosphoinositides and increased the labeling of phosphatidylcholine/ phosphatidylethanolamine. These data indicate several interesting points. The basal turnover of phosphoinositides appears to be at least partially under muscarinic control. Status epilepticus activates the turnover of phosphoinositides by increasing catabolism, as reflected in increased levels of InsP_2 and InsP_3, and by increasing synthesis. This increased synthesis is mainly under muscarinic control, at least indirectly, as atropine pretreatment attenuated the effect of bicuculline. Increased catabolism, resulting in accumulation of InsP_2 and InsP_3, may not be muscarinically mediated because atropine had no effect.

Effect of Post-Decapitation Ischemia on Phosphoinositide Turnover

Prolonged cerebral ischemia also results in irreversible damage to brain tissue. We have investigated the effect of various periods of post-decapitation ischemia on the activity of PtdIns kinase and PtdIns4P kinase in synaptic plasma membranes. These kinases catalyze the conversion of PtdIns to PtdIns4P to PtdIns(4,5)P_2. They utilize ATP, are Mg^{2+}-dependent, and are stimulated by Ca^{2+}. We found that increasing periods of ischemia resulted in increased activity of PtdIns and PtdIns4P kinases, as measured by incubation of synaptic plasma membranes with [^{32}P]ATP. Because these assays were done under optimal in vitro conditions in terms of substrate and cofactor availability, the results indicate that ischemia alters the activity of the kinase (e.g. the affinity of the enzymes for Ca^{2+}, Mg^{2+}, ATP or phospholipid may be increased).

ARACHIDONOYL COENZYME A (CoA) SYNTHETASE AND DOCOSAHEXAENOYL CoA SYNTHETASE IN STIMULATED NEURAL MEMBRANES

One of the mechanisms proposed for the accumulation of FFA in brain after stimulation is a failure in the activation and reacylation of released FFA (Bazan, 1976; Bazan et al., 1984a). We recently have analyzed the activities and kinetic parameters of long-chain acyl CoA synthetase in neural membranes isolated from control rats and rats undergoing bicuculline-induced status epilepticus. Long-chain acyl CoA synthetase converts fatty acids to thiol esters, which are in turn acylated to lysophospholipids by acyl transferases. This reaction is an obligatory step in the utilization of FFA for various reactions such as acylation, elongation and desaturation.

Control rats treated with drug vehicle or rats treated with bicuculline (10 mg/kg body wt) were killed by decapitation 7 min after treatment. Brains were removed and dissected into cerebrum, cerebellum and brainstem, which were homogenized for the isolation of synaptic plasma membranes and microsomes (Reddy and Bazan, 1985). The apparent K_m and Vmax of arachidonoyl CoA synthetase and docosahexaenoyl CoA synthetase were assayed. The synthesis of arachidonoyl CoA was 3-4 fold higher than the production of docosahexaenoyl CoA in microsomes from the various brain regions. In contrast, the rate of synthesis of both acyl CoA derivatives was similar in synaptic plasma membranes. Bicuculline-induced seizures did not significantly alter the apparent K_m and Vmax for the synthesis of arachidonoyl CoA or docosahexaenoyl CoA either in microsomes or in synaptic plasma membranes. These results suggest that there were no alterations in vivo in the tertiary structure of the long-chain acyl CoA synthetase due to seizures, which could result in changes in the apparent kinetics. However, these assays were done in the presence of saturating concentrations of all cofactors. In vivo, the activity of long-chain acyl CoA synthetase may be different due to the lack of certain cofactors such as ATP, which decreases immediately after seizures or ischemia (Chapman et al., 1977; Yoshida et al., 1982). Thus, the possible contribution of decreased fatty acid activation in vivo (due to the lack of cofactors) to accumulation of FFA cannot be disregarded.

SYNTHESIS OF OXYGENATED METABOLITES OF ARACHIDONIC AND DOCOSAHEXAENOIC ACIDS IN BRAIN

Aside from the putative role of free polyunsaturated fatty acids as modulators of enzymes such as protein kinase C, these fatty acids are precursors to a variety of oxygenated metabolites with biological functions. There are many reports of the conversion of arachidonic acid to prostaglandins (PGs) by brain cyclooxygenase (for review see Wolfe and Coceani, 1979; Wolfe, 1982). More recently, the synthesis in brain of lipoxygenase reaction products, i.e. hydroxyeicosatetraenoic acids (HETEs) and leukotrienes (LTs), has been reported (Sautebin et al., 1978; Spagnuolo et al., 1979; Susuki et al., 1983; Moskowitz et al., 1984; Adesuyi et al., 1985). The production of prostaglandins in brain is increased by neuronal stimulation, trauma, seizures and hypoxia. This effect appears to be a result of increased substrate availability (release of free arachidonic acid) and in the case of pentylenetetrazol-induced seizures, an alteration in cyclooxygenase resulting in decreased K_m for the fatty acid substrate (Lysz et al., 1985). The physiological and pathological modulation of lipoxygenase activity in brain has been investigated only recently. Suzuki et al. (1983) reported increased levels of 5-HETE in CSF from human patients with subarachnoid hemorrhage. Moskowitz et al. (1984) reported increased levels of LTs (immunoreactive slow-reacting substance) in gerbil brain following carotid ligation and reperfusion. Adesuyi et al. (1985) have shown 12-HETE levels in mouse brain are increased 300-fold by homogenization of the tissue. Their data indicate that about half of the 12-HETE in brain is derived from platelets in the cerebral vasculature.

Recently, we examined the oxygenation of arachidonic and docosahexaenoic acids in the rat brain. We also studied the effects of bicuculline-induced status epilepticus on the production of these metabolites. In these experiments, rat brain was prelabeled in vivo by the intraventricular injection of radiolabeled arachidonic or docosahexaenoic acid. Thirty minutes later, rats were mechanically ventilated, then treated with bicuculline or drug vehicle. After 4 min, rats were sacrificed by microwave irradiation, or by decapitation followed by subcellular fractionation. We found that the synaptosomal fraction was most active in producing lipoxygenase and cyclooxygenase reaction products of arachidonic acid. All major prostaglandins were detected. LTB$_4$, 5-HETE and 12-HETE also were produced. Bicuculline treatment caused an increase in lipoxygenase activity in both microsomes and synaptosomes. There also was a seizure-induced stimulation of PG synthesis, particularly in the microsomal fraction. Due to the design of these experiments, the quantitation of eicosanoids was done about two hours postmortem. This means that status epilepticus caused some long lasting alteration in the control of eicosanoid biosynthesis. Furthermore, the effect on lipoxygenase activity was somewhat specific for synaptosomes and more pronounced than the effect on PG synthesis. HETEs and LTs may play an important role as endogenous pro- or anticonvulsants, as has been proposed for PGs (Seregi et al., 1984; Forstermann et al., 1984).

Recently, there have been reports of the lipoxygenation of docosahexaeonic acid in platelets (Aveldano and Sprecher, 1983), neutrophils (Fischer et al., 1984) and retina (Bazan et al., 1984b). The production of hydroxydocosahexaenoic acids (HDHEs) in retina is enhanced in canine ceroid lipofuscinosis; a hereditary neuronal degeneration (Reddy et al., 1985). We have investigated the capacity of the brain to produce these metabolites, because the synaptic membrane is highly enriched in docosahexaenoic acid. We found that rat brain produces a variety of trihydroxy-, dihydroxy-, monohydroxy- and hydroperoxy-derivatives of docosahexaenoic acid. Moreover, the synthesis of these metabolites is doubled during bicuculline-induced status epilepticus. A similar profile of products was observed when monkey cerebral synaptosomes were incubated with radiolabeled docosahexaenoic acid. In this preparation in vitro, the contribution of free radical peroxidation of the fatty acid could be assessed. Therefore, it appears that brain, like retina, has the capability for enzymatic lipoxygenation of docosahexaenoic acid. Although the physiological role of HDHEs has not been discerned, the apparent stimulation of their synthesis in ceroid lipofuscinosis and status epilepticus indicates a relationship to neuronal function and neuropathological responses.

CONCLUSIONS

It is apparent from numerous studies and the data described above that brain stimulation, either short- or long-term, causes major alterations in membrane lipids. Short- and long-term stimulation appear to affect the same types of lipids, i.e. stimulation of the phosphoinositide cycle, increased FFA and DAG, and increased fatty acid oxygenation. Although the mechanisms of these effects are not clear, the use of pharmacological tools, such as atropine and lithium, will allow progress in this important area. Of particular interest is the question of recovery versus irreversible changes in brain lipids, because this issue will be a key point in the design of effective therapeutic regimens for the treatment of epilepsy, head injury and stroke.

ACKNOWLEDGMENTS

This work was supported in part by NIH grants EY05121, EY04428 and EY07073 and by the Epilepsy Foundation of America and the Esther A. and Joseph Klingenstein Fund, Inc., New York.

REFERENCES

Adesuyi SA, Cockrell CS, Gamache DA, Ellis EF (1985) Lipoxygenase metabolism of arachidonic acid in brain. J Neurochem 45: 770-776.

Allison JH, Blisner NW, Holland WH, Hipps PP, Sherman WR (1976) Increased brain myo-inositol 1-phosphate in lithium-treated rats. Biochem Biophys Res Comm 71: 664-670.

Allison JH, Stewart MA (1971) Reduced brain inositol in lithium-treated rats. Nature 233: 267-268.

Aveldano MI, Bazan NG (1974) Free fatty acids, diacyl- and triacylglycerols and total phospholipids in vertebrate retina: Comparison with brain, choroid and plasma. J Neurochem 23: 1127-1135.

Aveldano MI, Bazan, NG (1975a) Differential lipid deacylation during brain ischemia in a homeotherm and a poikilotherm. Content and composition of free fatty acids and triacylglycerols. Brain Res 100: 99-110.

Aveldano MI, Bazan NG (1975b) Rapid production of diacylglycerols enriched in arachidonate and stearate during early brain ischemia. J Neurochem 25: 919-920.

Aveldano MI, Sprecher H (1983) Synthesis of hydroxy fatty acids from 4,7,10,13,16,19-[1-^{14}C] docosahexaenoic acid by human platelets. J Biol Chem 258: 9339-9343.

Aveldano de Caldironi MI, Bazan NG (1979) α-Methyl-p-tyrosine inhibits the production of free arachidonic acid and diacylglycerols in brain after a single electroconvulsive shock. Neurochem Res 4: 213-221.

Bazan NG (1970) Effects of ischemia and electroconvulsive shock on free fatty acid pool in the brain. Biochim Biophys Acta 218: 1-10.

Bazan NG (1971) Changes in free fatty acids of brain by drug-induced convulsions, electroshock and anesthesia. J Neurochem 18: 1379-1385.

Bazan NG (1976) Free arachidonic acid and other lipids in the nervous system during early ischemia and after electroshock. Adv Exp Med Biol 72: 317-335.

Bazan NG, Bazan HEP (1975) Analysis of free and esterified fatty acids in neural tissues using gradient-thickness thin-layer chromatography. In: Marks N and Rodnight R (eds): Research Methods in Neurochemistry, Vol III, Plenum Press, New York, pp. 309-324.

Bazan NG, Giusto NM (1983) Anoxia-induced production of methylated and free fatty acids in retina, cerebral cortex, and white matter. Comparison with triglycerides and with other tissues. Neurochem Pathol 1: 17-41.

Bazan NG, Joel CD (1970) Gradient-thickness thin-layer chromatography for the isolation and analysis of trace amounts of free fatty acids in large lipid samples. J Lipid Res 11: 42-47.

Bazan NG, Rakowski H (1970) Increased levels of brain free fatty acids after electroconvulsive shock. Life Sci 9: 501-507.

Bazan NG, Bazan HEP, Kennedy WG, Joel CD (1971) Regional distribution and rate of production of free fatty acids in rat brain. J Neurochem 18: 1387-1393.

178

Bazan NG, Politi E, Rodriguez de Turco EB (1984a) Endogenous pools of arachidonic acid-enriched membrane lipids in cryogenic brain edema. In: Go KG, Baethmann A (eds): Recent Progress in the Study of Brain Edema, Plenum Press, New York, pp. 203-212.

Bazan NG, Birkle DL, Reddy TS (1984b) Docosahexaenoic acid (22:6, n-3) is metabolized to lipoxygenase reaction products in the retina. Biochem Biophys Res Commun 125: 741-747.

Berridge MJ (1984) Inositol trisphosphate and diacylglycerol as second messengers. Biochem J 220: 345-360.

Berridge MJ, Downes CP, Hanley MR (1982) Lithium amplifies agonist-dependent phosphatidylinositol responses in brain and salivary glands. Biochem J 206: 587-595.

Bhakoo KK, Crockard HA, Lascelles PT (1984) Regional studies of changes in brain fatty acids following experimental ischemia and reperfusion in the gerbil. J Neurochem 43: 1025-1031.

Cenedella RJ, Galli C, Paoletti R (1975) Brain free fatty acid levels in rats sacrificed by decapitation versus focused microwave irradiation. Lipids 10: 290-293.

Chapman AG, Meldrum BS, Siesjo B (1977) Cerebral metabolic changes during prolonged epileptic seizures in rats. J Neurochem 28: 1025-1035.

DeMedio GE, Goracci G, Horrocks LA, Lazarewicz JW, Mazzari S, Porcellati G, Strosznajder J, Trovarelli G (1980) The effect of transient ischemia on fatty acid and lipid metabolism in the gerbil brain. Ital J Biochem 29: 412-432.

Fischer S, Schacky CV, Siess W, Shasser T, Weber PC (1984) Uptake, release and metabolism of docosahexaenoic acid in human platelets and neutrophils. Biochem Biophys Res Comm 120: 907-918.

Forstermann U, Seregi A, Hertting G (1984) Anticonvulsive effects of endogenous prostaglandins formed in brain of spontaneously convulsing gerbils. Prostaglandins 27: 913-923.

Fujiwara M, Watanabe Y, Katayama Y, Shirakabe Y (1978) Application of high-powered microwave irradiation for acetylcholine analysis in mouse brain. Eur J Pharmacol 51: 299-301.

Galli C, Spagnuolo C (1976) The release of brain free fatty acids during ischemia in essential fatty acid-deficient rats. J Neurochem 26: 401-404.

Hallcher LM, Sherman WR (1980) The effects of lithium ion and other agents on the activity of myo-inositol-1-phosphatase from bovine brain. J Biol Chem 225: 10896-10901.

Hauser G, Eichberg J (1973) Improved conditions for the preservation and extraction of polyphosphoinositides. Biochim Biophys Acta 326:201-209.

Irvine RF, Letcher AJ, Lander DJ, Downes CP (1984) Inositol trisphosphates in carbachol-stimulated rat parotid glands. Biochem J 223: 237-243.

Kawahara Y, Takai Y, Minakuchi R, Sano K, Nishizuka Y (1980) Phospholipid turnover as a possible transmembrane signal for protein phosphorylation during human platelet activation by thrombin. Biochem Biophys Res Comm 97: 309-317.

Lysz TW, Centra M, Keeting P (1985) Increased brain fatty acid cyclooxygenase activity is associated with drug induced clonic convulsions. Soc Neurosci Abstr, Vol II, part 1, pp 748.

Moskowitz MA, Kiwak KJ, Hekimian K, Levine L (1984) Synthesis of compounds with properties of leukotrienes C_4 and D_4 in gerbil brains after ischemia and reperfusion. Science 224: 886-889.

Murakami K, Routtenberg A (1985) Direct activation of purified protein kinase C by unsaturated fatty acids (oleate and arachidonate) in the absence of phospholipids and Ca^{2+}. FEBS Lett 192: 189-193.

Nishizuka Y (1984) Turnover of inositol phospholipids and signal transduction. Science 225: 1365-1370.

Politi LE, Rodriguez de Turco EB, Bazan NG (1985) Dexamethasone effect on free fatty acid and diacylglycerol accumulation during experimentally-induced vasogenic brain edema. Neurochem Pathol 3: 253-274.

Ponten U, Ratcheson RA, Salford LG, Siesjo BK (1973) Optimal freezing conditions for cerebral metabolites in rats. J Neurochem 21: 1127-1138.

Putney JW Jr, McKinney JS, Aub DL, Leslie BA (1984) Phorbol ester-induced protein secretion in rat parotid gland: Relationship to the role of inositol lipid breakdown and protein kinase C activation in stimulus-secretion coupling. Molec Pharmacol 26: 261-266.

Reddy TS, Bazan NG (1985) Synthesis of arachidonoyl coenzyme A and docosahexaenoyl coenzyme A in synaptic plasma membranes of cerebrum and microsomes of cerebrum, cerebellum, and brain stem of rat brain. J Neurosci Res 13: 381-390.

Reddy TS, Birkle DL, Armstrong D, Bazan NG (1985) Change in content, incorporation and lipoxygenation of docosahexaenoic acid in retina and retinal pigment epithelium in canine ceroid lipofuscinosis. Neurosci Lett 59: 67-72.

Rhoads DE, Osburn LD, Peterson NA, Raghupathy E (1983) Release of neurotransmitter amino acids from synaptosomes: Enhancement of calcium-independent efflux by oleic and arachidonic acids. J Neurochem 41: 531-537.

Rodriguez de Turco EB, Morelli de Liberti S, Bazan NG (1983) Stimulation of free fatty acid and diacylglycerol accumulation in cerebrum and cerebellum during bicuculline-induced status epilepticus. Effect of pretreatment with α-methyl-p-tyrosine and p-chlorophenylalanine. J Neurochem 40: 252-259.

Sautebin L, Spagnuolo C, Galli G (1978) A mass fragmentographic procedure for the simultaneous determination of HETE and $PGF_{2\alpha}$ in the central nervous system. Prostaglandins 16: 985-988.

Schneider DR, Felt BT, Goldman H (1981) Microwave radiation energy: A probe for the neurobiologist. Life Sci 29: 643-653.

Schneider HH (1984) Brain cAMP response to phosphodiesterase inhibitors in rats killed by microwave irradiation or decapitation. Biochem Pharmacol 33: 1690-1693.

Seregi A, Forstermann U, Hertting G (1984) Decreased levels of brain cyclo-oxygenase products as a possible cause of increased seizure susceptibility in convulsion-prone gerbils. Brain Res 305: 393-395.

Siesjo BK, Ingvar M, Westerberg E (1982) The influence of bicuculline-induced seizures on free fatty acid concentrations in cerebral cortex, hippocampus and cerebellum. J Neurochem 39: 796-802.

Sherman WR, Munsell LY, Gish BG, Hanchor MP (1985) Effects of systemically administered lithium on phosphoinositide metabolism in rat brain, kidney and testis. J Neurochem 44: 798-807.

Shiu GK, Nemoto EM (1981) Barbiturate attenuation of brain free fatty acid liberation during global ischemia. J Neurochem 37: 1448-1456.

Shiu GK, Nemmer P, Nemoto EM (1983) Reassessment of brain fatty acid liberation during global ischemia and its attenuation by barbiturate anesthesia. J Neurochem 40: 880-884.

Soukup JF, Friedel RO, Schanberg SM (1978) Cholinergic stimulation of polyphosphoinositide metabolism in brain in vivo. Biochem Pharmacol 27: 1239-1243.

Spagnuolo C, Sautebin L, Galli G, Racagni G, Galli C, Mazzari S, Finesso M (1979) $PGF_{2\alpha}$, thromboxane B_2 and HETE levels in gerbil brain cortex after ligation of common carotid arteries and decapitation. Prostaglandins 18: 53-61.

Stavinova WB, Frazer J, Modak AT (1977) Microwave fixation for the study of acetylcholine metabolism. Adv Behav Biol 24: 169-179.

Suzuki N, Nakamura T, Imabayashi S, Ishikawa Y, Sasaki T, Asano T (1983) Identification of 5-hydroxyeicosatetraenoic acid in cerebrospinal fluid after subarachnoid hemorrhage. J Neurochem 41: 1186-1189.

Tang W, Sun GY (1982) Factors affecting the free fatty acids in rat brain cortex. Neurochem Int 4: 269-273.

Tang W, Sun GY (1985) Effects of ischemia on free fatty acids and diacylglycerols in developing rat brain. Int J Develop Neurosci 3: 51-56.

Wolfe LS (1982) Eicosanoids: Prostaglandins, thromboxanes, leukotrienes and other derivatives of carbon-20 unsaturated fatty acids. J Neurochem 38: 1-3.

Wolfe LS, Coceani F (1979) The role of prostaglandins in the central nervous system. Ann Rev Physiol 41: 669-684.

Wooten MW, Wrenn RW (1984) Phorbol ester induces intracellular translocation of phospholipid/Ca^{2+}-dependent protein kinase and stimulates amylase secretion in isolated pancreatic acini. FEBS Lett 171: 183-186.

Yoshida S, Inoh S, Asano T, Sano K, Kubota M, Shimazaki H, Ueta N (1980) Effect of transient ischemia on free fatty acids and phospholipids in the gerbil brain. J Neurosurg 53: 323-331.

Yoshida S, Abe K, Busto R, Watson BD, Kogure K, Ginsberg MD (1982) Influence of transient ischemia on lipid-soluble antioxidants, free fatty acids and energy metabolites in rat brain. Brain Res 245: 307-316.

Phospholipid research and the nervous system
Biochemical and molecular pharmacology
I.A. Horrocks, L. Freysz, G. Toffano (eds)
Fidia Research Series, vol. 4.
Liviana Press, Padova. © 1986

MEMBRANE BOUND DIACYLGLYCEROL LIPASES IN BOVINE BRAIN: PURIFICATION AND CHARACTERIZATION

Akhlaq A. Farooqui, W. Allen Taylor and Lloyd A. Horrocks

Department of Physiological Chemistry, The Ohio State University,
1645 Neil Avenue, Columbus, OH 43210, USA

It is becoming increasingly evident that diacylglycerols, a minor component of mammalian plasma membranes, regulate the activity of protein kinase C (Berridge, 1984). This enzyme has been implicated in the control of cell division, membrane fusion, differentiation, and signal transduction across the cell membrane (Downes, 1983; Nishizuka 1983, 1984; Das and Rand, 1984; Majerus et al., 1984). Furthermore, diacylglycerols have been reported to stimulate the activities of phospholipases by perturbing the bilayer structure of biological membranes (Dawson et al., 1983; 1984; Watson et al., 1984). In brain, diacylglycerols are either phosphorylated to phosphatidic acid by diacylglycerol kinase (Sun, 1983; Bazan, 1983) or hydrolyzed to free fatty acids and monoacylglycerol by diacylglycerol lipase (Cabot and Gatt, 1976, 1977, 1978; Rousseau et al., 1983; Rousseau and Gatt, 1984; Farooqui et al., 1985a,b).

Earlier studies from this laboratory have described sensitive and continuous spectrophotometric assay procedures using thioester substrate analogs for mono-and diacylglycerol lipases, phospholipases, and lysophospholipases (Cox and Horrocks, 1981; Strosznajder et al., 1984; Farooqui et al., 1984 a). We have also shown (Farooqui et al., 1985) that bovine brain contains two diacylglycerol lipases. One is localized in microsomes and the other is found in plasma membranes. Both enzymes can be solubilized with 0.25% Triton X-100. The microsomal diacylglycerol lipase is markedly stimulated by Triton X-100 and $CaCl_2$, whereas the plasma membrane diacylglycerol lipase is strongly inhibited by this detergent while $CaCl_2$ has no effect on its enzymic activity. This article summarizes our studies on the purification and characterization of bovine brain diacylglycerol lipases.

PURIFICATION OF DIACYLGLYCEROL LIPASE
FROM BOVINE BRAIN MICROSOMES

The isolation of bovine brain microsomes and solubilization of diacylglycerol lipase has been described earlier (Farooqui et al., 1985). The solubilized microsomal fraction was fractionated with solid ammonium sulfate. The diacylglycerol lipase activity precipitated in the 0-30% ammonium sulfate fraction was applied to a Sephadex G-75 column. The enzyme was eluted as a sharp peak which did not coincide with the main protein peak. The Sephadex G-75 fraction was chromatographed on a heparin-Sepharose column (Fig. 1A). About 60-65% of the protein was washed out while diacylglycerol lipase was completely retained. The enzyme was eluted as a sharp peak when the column was washed with 0.5M sodium chloride. This step gave a 4-fold purification of diacylglycerol lipase over the previous step, with 75% recovery of enzymic activity (Table 1).

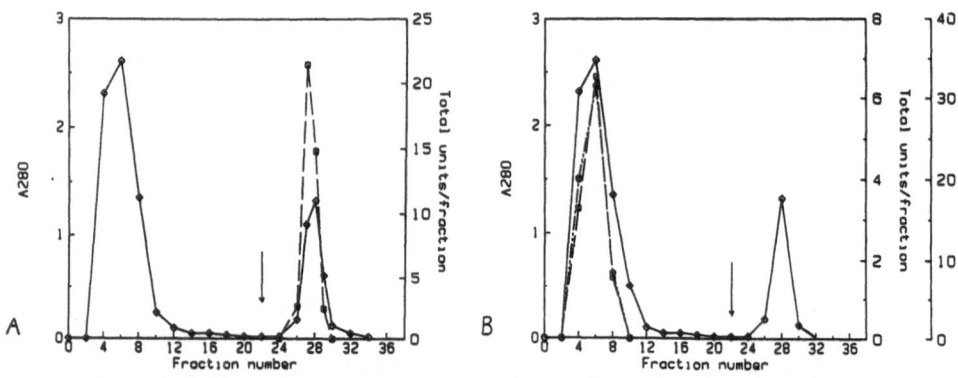

Figure 1. Elution profiles of microsomal mono - and diacylglycerol lipases and lysophospholipase through a heparin-Sepharose column. A) Diacylglycerol lipase activity (□) right ordinate, and A_{280} (◇) left ordinate. B) Monoacylglycerol lipase activity (□) far right ordinate; lysophospholipase activity (□) right ordinate, and A_{280} (◇) left ordinate. 0.5 M NaCl was applied at the arrow.

Washings from the heparin-Sepharose column were assayed for monoacylglycerol lipase and lysophospholipase activities by procedures described elsewhere (Strosznajder et al., 1984; Farooqui et al., 1985; Pendley and Horrocks, unpublished). Both of these enzymes were washed out (Fig. 1B) in yields of 60-70% without any increase in specific activity.

Active fractions from the heparin-Sepharose column were applied to an octyl-Sepharose column. The enzyme was completely retained on the octyl-Sepharose column (Fig. 2) and was eluted as a sharp peak when the column was washed with 0.80% sodium taurocholate. This procedure gave a 3-fold purification of diacylglycerol lipase activity over the previous step with a 73% yield.

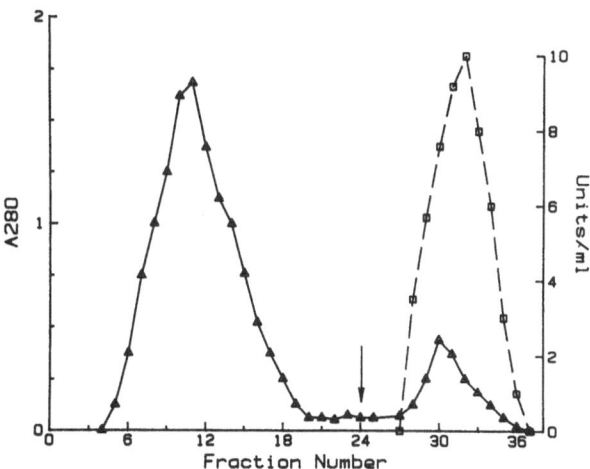

Figure 2. Elution profile of microsomal diacylglycerol lipase through an octyl-Sepharose column. Diacylglycerol lipase activity ($\square$) right ordinate, and A_{280} ($\triangle$) left ordinate. Sodium taurocholate was applied at the arrow.

Table 1. *Purification of microsomal diacylglycerol lipase*

Fraction	Total Protein (mg)	Total Activity (nmol/min)	Sp. Activity (nmol/min/mg)	Yield (%)
12,000 × g supernatant	9972	4216	0.42	100
Solubilized microsomes	1078	2400	2.22	57
0-30% $(NH_4)_2SO_4$	539	1849	3.43	44
Sephadex G-75	150	1560	10.4	37
Heparin-Sepharose	30	1170	39.0	28
Octyl-Sepharose	6	860	143.0	20

The multiple column chromatographic procedures described above gave a 340-fold purification of microsomal diacylglycerol lipase with an overall recovery of 20% (Table 1). The final preparation was not homogeneous as judged by polyacrylamide gel electrophoresis. It showed three bands: one major and two minor. To determine the location of diacylglycerol lipase the unstained gel was cut into 0.5 cm segments and the enzymic activity was determined in these segments after extraction. It was found that the enzymic activity corresponded to the major protein band on the unstained gel.

PROPERTIES OF DIACYLGLYCEROL LIPASE

A comparison of the kinetic parameters of bovine brain diacylglycerol lipases with diacylglycerol lipases from other sources is shown in Table 2. The rate of hydrolysis

184

Table 2. *Kinetic properties of mammalian brain diacylglycerol lipases*

Animal species	pH optimum	K_m	V_{max} (nmol/min/mg)	Reference
Rat microsomes	8.0-8.6	—	8.0	Cabot and Gatt (1976)
Guinea pig synaptosomes	7.0-8.0	500	10.5	Vyvoda and Rowe (1973)
Bovine microsomes (solubilized)	7.5	30 ± 7	7.6 ± 2	Farooqui et al. (1985a,b)
Bovine plasma membrane (solubilized)	7.5	12 ± 2	4.8 ± 0.5	Farooqui et al. (1985a,b)

Results for bovine preparations are mean ± S.D. for three determinations.

of the thioester substrate analog (rac-1,2-S,O-didecanoyl-1-mercapto-2,3-propanediol) of diacylglycerol was quite similar to that of the radiolabeled diacylglycerol. The apparent K_m values for the soluble microsomal and plasma membrane diacylglycerol lipases were 30 and 12 μM respectively, which are 16-40 times lower than the K_m value (Table 2) reported for nerve ending diacylglycerol lipase (Vyvoda and Rowe, 1973).

Heparin, a polysulfated glycosaminoglycan, strongly inhibited both the microsomal and plasma membrane diacylglycerol lipases. No other glycosaminoglycans (hyaluronic acid or chondroitin sulfate) resembled heparin in their effect on diacylglycerol lipases. Monoacylglycerol lipase, which was present in the solubilized microsomal and plasma membrane preparation, was not inhibited by heparin while lysophospholipase was partially inhibited by this glycosaminoglycan (Fig. 3). The differential affect of heparin on mono-and diacylglycerol lipases explains their separation (Farooqui et al., 1984b) by heparin-Sepharose affinity chromatography (Farooqui and Horrocks, 1984).

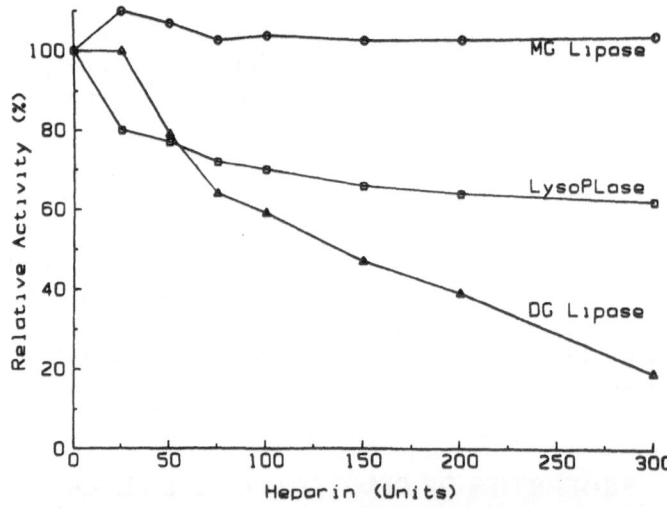

Figure 3. Effect of heparin on microsomal mono- and diacylglycerol lipases and lysophospholipase Monoacylglycerol lipase activity (o), diacylglycerol lipase activity (Δ), and lysophospholipase activity (□).

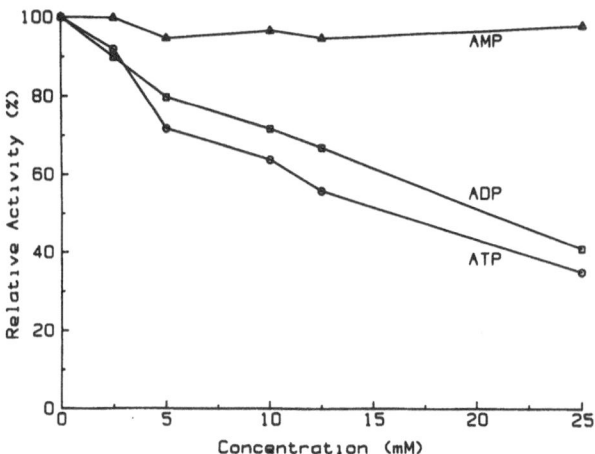

Figure 4. Effects of AMP, and ATP on microsomal diacylglycerol lipase activity. AMP (△), ADP (□), and ATP (o).

The effects of ATP, ADP, and AMP on the microsomal diacylglycerol lipase activity are shown in Fig. 4. Both ATP and ADP caused a concentration-dependent inhibition of diacylglycerol lipase activity. ATP was the most potent inhibitor followed by ADP. AMP had no effect and cyclic AMP (0.5 mM) produced a 20-25% stimulation of the microsomal diacylglycerol lipase. CTP was as effective as ATP in inhibiting the microsomal diacylglycerol lipase. The plasma membrane diacylglycerol lipase was also inhibited by ATP and ADP, but not by AMP. It was suggested (Farooqui et al., 1984b) that the polyanionic nature of the nucleotides was responsible for this inhibition. Similarly, the lysophospholipase activity of bovine brain is also inhibited by ADP and ATP, but the extent of inhibition by these nucleotides was different for diacylglycerol lipase and lysophospholipase (Farooqui et al., 1984b).

Recent studies (Sun, 1985) have also indicated that rat microsomal lysophospholipase is strongly inhibited by ATP when acting on lyso-PtdEtn but ATP did not affect the hydrolysis of lyso-PtdCho.

Polyamines such as spermine and tetracaine, inhibitors of phospholipase A_2 (Cheah and Cheah, 1981), stimulated the partially purified microsomal diacylglycerol lipase (Fig. 5). The stimulation of diacylglycerol lipase by spermine and tetracaine was similar to the stimulatory effect of polyamines on calcium dependent phosphatidylinositol phosphodiesterase (Eichberg et al., 1981). These authors suggested that polyamines could partially mimic the stimulatory effect of $CaCl_2$.

Both diacylclycerol lipases were inhibited by free fatty acids. Palmitate was the most potent inhibitor followed by arachidonate and linoleate (Fig. 6A). The addition of fatty acid free bovine serum albumin to the reaction mixture resulted in reversal of this inhibition (Fig. 6B). Similar results were also obtained by Leibovitz-BenGershon and Gatt (1974) with brain lysophospholipases. Dawson et al. (1982), on the other hand, indicated that the addition of bovine serum albumin to rat brain alkaline phospholipase A[1] resulted in a marked inhibition of this enzyme.

Bovine brain diacylglycerol lipases require no metal ions for their activity. However, the activity of microsomal diacylglycerol lipase was markedly stimulated by $CaCl_2$ (Farooqui et al., 1985). In contrast, the plasma membrane diacylglycerol lipase was not affected by $CaCl_2$. EGTA alone had no effect on the microsomal diacylglycerol lipase but reversed the stimulatory effect of $CaCl_2$ (Farooqui et al. 1985). The calcium binding protein, calmodulin, stimulates the activities of phospholipases from brain (Moskowitz et al., 1983). Diacylglycerol lipases were not affected by calmodulin in the presence or absence of $CaCl_2$. Sodium chloride and sodium azide (up to 100 mM) had no effect, but sodium fluoride at 75 mM produced 50% inhibition of microsomal and plasma membrane diacylglycerol lipases.

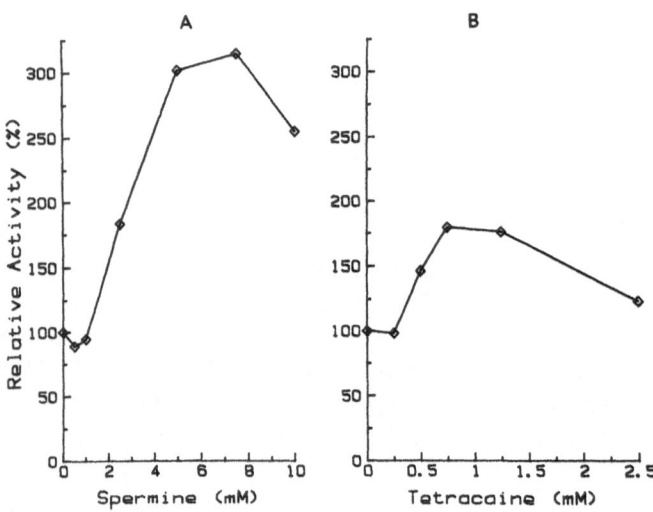

Figure 5. Effects of spermine and tetracaine on microsomal diacylglycerol lipase activity. A) Spermine and B) Tetracaine.

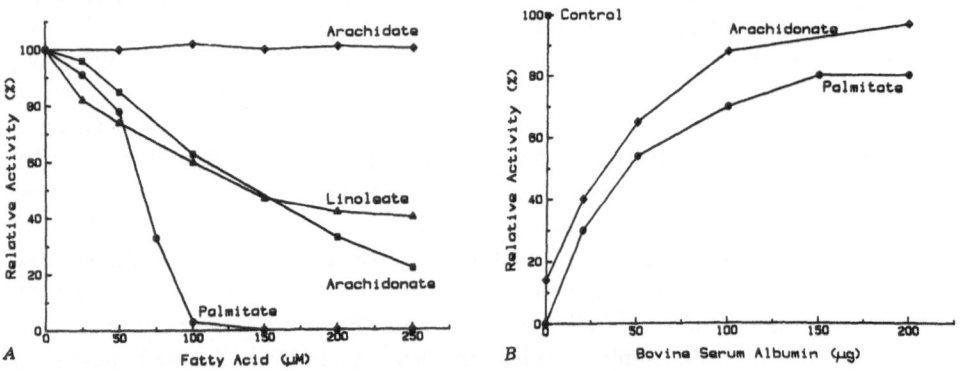

Figure 6. Effects of free fatty acids and bovine serum albumin on microsomal diacylglycerol lipase activity. A) Inhibition by fatty acids, and B) reversal of inhibition by arachidonate (250 μM) and palmitate (150 μM) with bovine serum albumin.

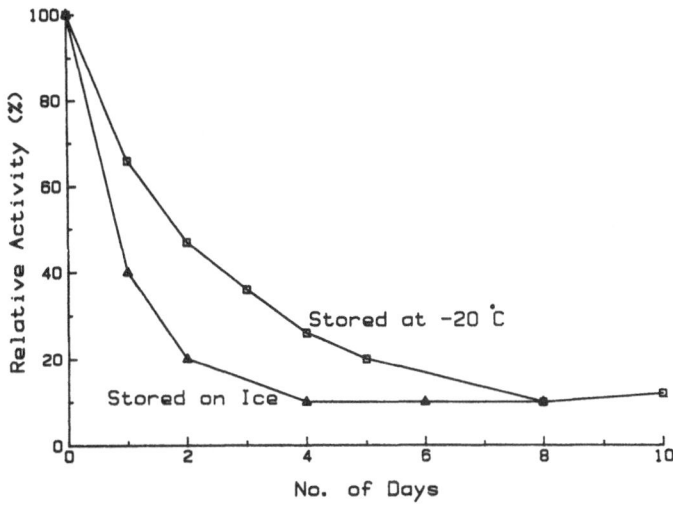

Figure 7. Stability of bovine brain microsomal diacylglycerol lipase. Octyl-Sepharose fraction was dialyzed against 0.05M MOPS buffer, pH 7.4 and stored on ice and at -20°C.

STABILITY OF DIACYLGLYCEROL LIPASE

The final preparation was quite unstable (Fig. 7). It lost 80% of its activity in 2 days when stored on ice. The storage of purified diacylglycerol lipase at -20°C resulted in a slower loss of enzymic activity, 80% of its activity in 5 days. The cause of this instability is not known. However, it is well known that many enzymes require lipid molecules for their activity and the removal of these lipids results in a loss of enzymic activity (Gazzotti and Peterson, 1977; Hidalgo et al., 1978; Basu and Glew, 1984; Farooqui and Horrocks, 1985). It may thus be possible that the loss of diacylglycerol lipase activity in the final preparation is due to the loss of bound lipids which may have been removed during octyl-Sepharose chromatography. Addition of various phospholipids including phosphatidylcholine, phosphatidylethanolamine, or neutral lipids (mono-and diacylglycerols) separately or in combination in 0.1% Triton X-100 did not restore the lost enzymic activity. Attempts to stabilize the diacylglycerol lipase by the addition of 25% sucrose, 25% glycerol, or 0.2M $CaCl_2$ were not successful. It must be noted here that the diacylglycerol lipase purified with heparin-Sepharose was quite stable at -20°C. At this stage, the enzyme can be stored for several months without any appreciable loss of activity.

GLYCOPROTEIN NATURE OF MICROSOMAL DIACYLGLYCEROL LIPASE

The retention of microsomal diacylglycerol lipase on a concanavalin-A Sepharose column (Fig. 8) and its elution with methyl α-D-mannoside indicated that the binding of this enzyme to concanavalin-A Sepharose was through the carbohydrate moiety.

188

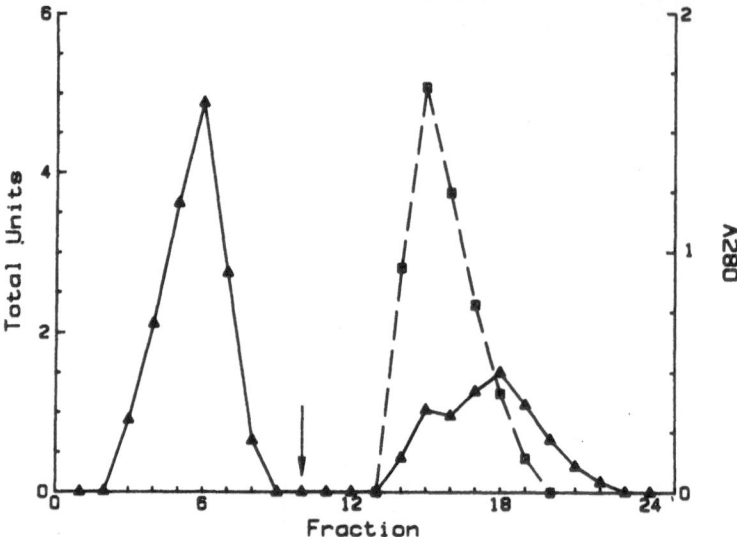

Figure 8. Elution profile of diacylglycerol lipase through a concanavalin-A Sepharose column. A_{280} (▲) right ordinate and diacylglycerol lipase activity (□) left ordinate. Methyl α-D-mannoside (0.5M) was applied at the arrow.

However, our data do not rule out the possibility that microsomal diacylglycerol lipase exists as a complex with the membrane receptor for concanavalin-A and that this complex was cochromatographed without being dissociated by the detergent. Concanavalin-A Sepharose chromatography gave a 5-fold purification of diacylglycerol lipase with 45-50% recovery of enzymic activity. The cause of this low recovery of enzyme from the concanavalin-A Sepharose column is not understood. It may be due either to the irreversible binding of diacylglycerol lipase to concanavalin-A Sepharose, or to inactivation during the chromatographic procedure.

Diacylglycerols have attracted considerable attention. They have been reported to regulate the activities of protein kinase C and phospholipases (Nishizuka, 1983, 1984; Dawson et al., 1983, 1984). The amounts of diacylglycerols in brain may be regulated by diacylglycerol lipases (Cabot and Gatt, 1979; Farooqui et al., 1985). At the same time, their action on diacylglycerol may provide free fatty acids, particularly arachidonic acid, which may be used for the production of prostaglandins and their metabolites.

SUMMARY

Bovine brain contains two membrane bound diacylglycerol lipases. One was found in the microsomal fraction (endoplasmic reticulum) and the other in the plasma membrane fraction. The microsomal diacylglycerol lipase was purified 340-fold using ammonium sulfate fractionation, Sephadex G-75 gel filtration, and chromatography on heparin-Sepharose and octyl-Sepharose. The purified enzyme preparation was not homogeneous but showed one major and two minor bands on polyacrylamide gel electrophoresis. The enzymic activity was localized in the major protein band.

Microsomal and plasma membrane diacylglycerol lipases were strongly inhibited by heparin, nucleotides, and free fatty acids. The inhibition by free fatty acids was reversed by fatty acid free bovine serum albumin. Spermine and tetracaine stimulated microsomal diacylglycerol lipase. The retention of microsomal diacylglycerol lipase on a concanavalin-A Sepharose and its elution by methyl α-D-mannoside indicated the glycoprotein nature of this enzyme.

ACKNOWLEDGEMENTS

This work was supported by grants NS-08291 and NS-10165 from the National Institutes of Health, U.S. Public Health Service.

REFERENCES

Basu A, Glew RH (1984) Characterization of the phospholipid requirement of a rat liver β-glucosidase. Biochem J 224:515-524.

Bazan NG (1983) Metabolism of phosphatidic acid. In: Lajtha A (ed): Handbook of Neurochemistry, Vol 3, 2nd ed, Plenum Press, New York, pp. 17-39.

Berridge MJ (1984) Inositol trisphosphate and diacylglycerol as second messengers. Biochem J 220:345-360.

Cabot MC, Gatt S (1976) Hydrolysis of neutral glycerides by lipases of rat brain microsomes. Biochim Biophys Acta 431: 105-115.

Cabot MC, Gatt S (1977) Hydrolysis of endogenous diacylglycerol and monoacylglycerol by lipases in rat brain microsomes. Biochemistry 16: 2330-2334.

Cabot MC, Gatt S (1978) The hydrolysis of triacylglycerol and diacylglycerol by rat brain microsomal lipase with acidic pH optimum. Biochim Biophys Acta 530:508-512.

Cabot MC, Gatt S (1979) Lipases of rat brain microsomes. Adv Expt Med Biol 101:101-111.

Cheah KS, Cheah AM (1981) Mitochondrial calcium transport and calcium-activated phospholipase in porcine malignant hyperthermia. Biochim Biophys Acta 634:70-84.

Cox JW, Horrocks LA (1981) Preparations of thioester substrates and development of continuous spectrophotometric assays for phospholipase A_1 and monoacylglycerol lipase. J Lipid Res 22:496-505.

Das S, Rand RP (1984) Diacylglycerol causes major structural transitions in phospholipid bilayer membranes. Biochem Biophys Res Commun 124: 491-496.

Dawson RMC, Irvine RF, Hemington N, Hirasawa K (1982) The stimulation of the brain alkaline phospholipase A_1 attacking phosphatidylethanolamine by various salts and metal chelators. Neurochem Res 7: 1149-1161.

Dawson RMC, Hemington NL, Irvine RF (1983) Diacylglycerol potentiates phospholipase attack upon phospholipid bilayers: possible connection with cell stimulation. Biochem Biophys Res Commun 117:196-201.

Dawson RMC, Irvine RF, Bray J, Quinn PJ (1984) Long-chain unsaturated diacylglycerols cause a perturbation in the structure of phospholipid bilayers rendering them susceptible to phospholipase attack. Biochem Biophys Res Commun 125: 836-842.

Downes CP (1983) Inositol phospholipids and neurotransmitter-receptor signalling mechanisms. Trends Neuro Sci 6:313-316.

Eichberg J, Zetusky WJ, Bell ME, Cavanagh E (1981) Effect of polyamines on calcium-dependent rat brain phosphatidylinositol-phosphodiesterase. J Neurochem 36:1868-1871.

Farooqui AA, Horrocks LA (1984) Heparin Sepharose affinity chromatography. Adv Chromatog 23:127-148.

Farooqui AA, Taylor WA, Pendley II CE, Cox JW, Horrocks LA (1984a) Spectrophotometric determination of lipases, lysophospholipase, and phospholipase. J Lipid Res 25:1555-1562.

Farooqui AA, Taylor WA, Horrocks LA (1984b) Separation of bovine brain mono-and diacyglycerol lipases by heparin Sepharose affinity chromatography. Biochem Biophys Res Commun 122:1241-1246.

Farooqui AA, Horrocks LA (1985) Metabolic and functional aspects of neural membrane phospholipids. In: Horrocks LA, Kanfer JN, Porcellati GA (eds): Phospholipids in the nervous system, Vol 2: Physiological roles, Raven Press, New York, pp. 341-348.

Farooqui AA, Pendley II CE, Taylor WA, Horrocks LA (1985) Studies on diacylglycerol lipases and lysophospholipases of bovine brain. In: Horrocks LA, Kanfer JN, Porcellati GA (eds): Phospholipids in the nervous system, Vol 2: Physiological roles, Raven Press, New York, pp. 179-192.

Gazzotti P and Peterson SW (1977) Lipid requirement of membrane bound enzymes. J Bioenerg Biomembr 9:373-386.

Hidalgo C, Thomas DD, Ikemoto N (1978) Effect of the lipid environment on protein motion and enzymatic activity of the sarcoplasmic reticulum calcium ATPase. J Biol Chem 253:6879-6887.

Leibovitz-BenGershon Z, Gatt S (1974) Lysolecithinase of rat brain: further analysis of the effect of substrate on the particulate and microsomal enzyme. J Biol Chem 249:1525-1529.

Majerus PW, Neufeld EJ, Wilson DB (1984) Production of phosphoinositide-derived messengers. Cell 37:701-703.

Moskowitz N, Puszkin S, Schook W (1983) Characterization of brain synaptic vesicle phospholipase A_2 activity and its modulation by calmodulin, prostaglandin E_2, prostaglandin $F_{2\alpha}$, cyclic AMP and ATP. J Neurochem 41:1576-1586.

Nishizuka Y (1983) Phospholipid degradation and signal translation for protein phosphorylation. Trends Biochem Sci 8:13-16.

Nishizuka Y (1984) Turnover of inositol phospholipids and signal transduction. Science 225:1365-1370.

Rousseau A, Dubois G, Gatt S (1983) Subcellular distribution of diacylglycerol lipase in rat and mouse brain. Neurochem Res 8:417-422.

Rousseau A, Gatt S (1984) Utilization of membranous lipid substrates by membrane-bound enzymes. FEBS Lett 167:42-46.

Strosznajder J, Singh H, Horrocks LA (1984) Monoacylglycerol lipase: regulation and increased activity during hypoxia and ischemia. Neurochem Path 2:139-147.

Sun GY (1983) Enzymes of lipid metabolism. In: Lajtha A (ed): Handbook of Neurochemistry, Vol 4, 2nd ed, Plenum Press, New York, pp. 367-383.

Sun GY (1985) Metabolism of lyso-PC and lyso-PE in rat brain subcellular membranes. Trans Am Soc Neurochem 16:105.

Vyvoda OS, Rowe CE (1973) Glyceride lipases in nerve ending of guinea-pig brain and their stimulation by noradrenaline, 5-hydroxytryptamine and adrenaline. Biochem J 132: 233-248.

Watson SP, Ganong BR, Bell RM, Lapetina EG (1984) 1,2-Diacylglycerols do not potentiate the action of phospholipase A_2 and C in human platelets. Biochem Biophys Res Commun 121:386-391.

Phospholipid research and the nervous system
Biochemical and molecular pharmacology
L.A. Horrocks, L. Freysz, G. Toffano (eds)
Fidia Research Series, vol. 4.
Liviana Press, Padova. © 1986

ROLES OF LIPASES IN THE DEVELOPMENT
OF AUTONOMIC NEURONS

Fusao Hirata, Toshio Hattori, Yoshitada Notsu
and Bernd Hamprecht[1]

Laboratory of Cell Biology, National Institute of Mental Health, Bethesda,
Maryland 20205 and [1]Physiologisch-Chemisches Institut der Universität,
87 Wurzburg, West Germany

Glucocorticoids are widely used as therapeutics for immunological and inflammatory diseases (Sayers and Travis, 1970). This anti-inflammatory action of glucocorticoids is now proposed to be associated with the inhibition of the release of arachidonic acid, a precursor of inflammatory agents such as prostaglandins and leukotrienes (Blackwell et al., 1978). Arachidonic acid is mostly present in an esterified form rather than as a free fatty acid. Thus, the rate limiting step of the arachidonate release from intact cells is the activation of phospholipase A_2 or phospholipase C. Furthermore, this anti-inflammatory action of glucocorticoids can be blocked by the previous treatment with cycloheximide and actinomycin D, inhibitors of protein and RNA syntheses, respectively (Hirata et al., 1985b). From these observations, one can assume that glucocorticoids induce the synthesis of a protein(s) which inhibits cellular phospholipases. Recently, the four groups (Flower's group in England, Russo-Marie's group in France, and Goldberg's group and our group in USA) have partially purified and characterized such a factor(s) and named this protein as lipocortin (DiRosa et al., 1984). Isolated lipocortin can suppress the carageenin-induced paw edema, an animal model of acute inflammation, and can inhibit the release of arachidonic acid from various intact cells. Thus, it has been proposed that the inhibition of phospholipases A_2 by lipocortin might be the main mechanism of the anti-inflammatory action of glucocorticoids (Hirata et al., 1985b).

Glucocorticoids are known to affect the growth and differentiation of various cells including neuronal cells (McEwin, 1979). They are necessary for neuronal development and for enhancement of the expression of the adrenergic phenotype (Patterson, 1978). Therefore, it would be of great interest to investigate the mechanism of the phenotypic determination by glucocorticoids and to test effects of lipocortin on development of the autonomic neurons. In this communication, we describe that lipocortin is a possible mediator of glucocorticoids to make these developing neurons adrenergic.

DIFFERENTIATION OF NH15CA2 CELLS BY GLUCOCORTICOIDS

NH15CA2, a neuroblastoma-glioma hybrid cell line, has several characteristics of neuronal cells (Heumann et al., 1979). Different from NG108 (a neuroblastoma-glioma hybrid cell line which is often used as a model system for synaptogenesis), this cell line can synthesize both neurotransmitters, catecholamines and acetylcholine (Nirenberg et al., 1983). Thus, the NH15CA2 cell line provides a better model system to study the neurotransmitter plasticity in developing autonomic neurons.

When NH15CA2 cells were cultured in the presence of 10 μM dexamethasone (Dx), a potent synthetic glucocorticoid, together with 1 mM dibutyryl-cyclic AMP, the majority of cells developed short and multibranched processes. Without Dx, most cells retained a round shape and only a few of them extended neurites after 4 or 7 days in culture. These observations suggest that Dx induces the cellular differentiation of NH15CA2 cells. To study the biochemical mechanism of differentiation of NH15CA2 cells, the activities of tyrosine hydroxylase (TH;EC1.14.16.2) and choline acetyltransferase (CAT;EC 2.3.1.6), key enzymes in the syntheses of noradrenaline and acetylcholine, respectively, were measured. Dx increased TH activity in these cells by approximately 3 fold, but did not change CAT activity significantly. This action of Dx was attributed to the synthesis of TH molecules rather than the activation of TH enzyme activity, because an anti-TH antibody stained the cytoplasm of the Dx-treated cells but not that of the nontreated cells.

INDUCTION OF LIPOCORTIN BY DEXAMETHASONE

To examine whether Dx induces the synthesis of TH directly or indirectly mediating through the synthesis of lipocortin, we treated NH15CA2 cells with monoclonal anti-lipocortin antibody. The treatment with anti-lipocortin antibody resulted in blocking the Dx's effects on the increase in TH activity. Because the control antibody (obtained from mice bearing P3/X63-Ag8 cells) or the anti-lipocortin antibody preabsorbed with partially purified lipocortin had no significant effects on TH activity in the Dx-treated cells, we assumed that the Dx treatment increases TH activity by inducing the synthesis of lipocortin in NH15CA2 cells.

To support this interpretation, we cultured NH15CA2 cells with purified lipocortin. Lipocortin was partially purified from rabbit neutrophils treated with 1 μM fluocinolon acetonide (Hattori et al., 1983) or from HL60 cells, a human promyeloleukemia cell line (unpublished). The morphological changes of these cells induced by lipocortin were essentially identical to those stimulated by the Dx treatment (Fig. 1). Purified lipocortin (2 μg/ml) increased TH activity in these cells approximately 4 fold. Bovine serum albumin (10 mg/ml) or mouse nerve growth factor (100 ng/ml) had no detectable effects on the TH activity. Thus, we concluded that lipocortin is a

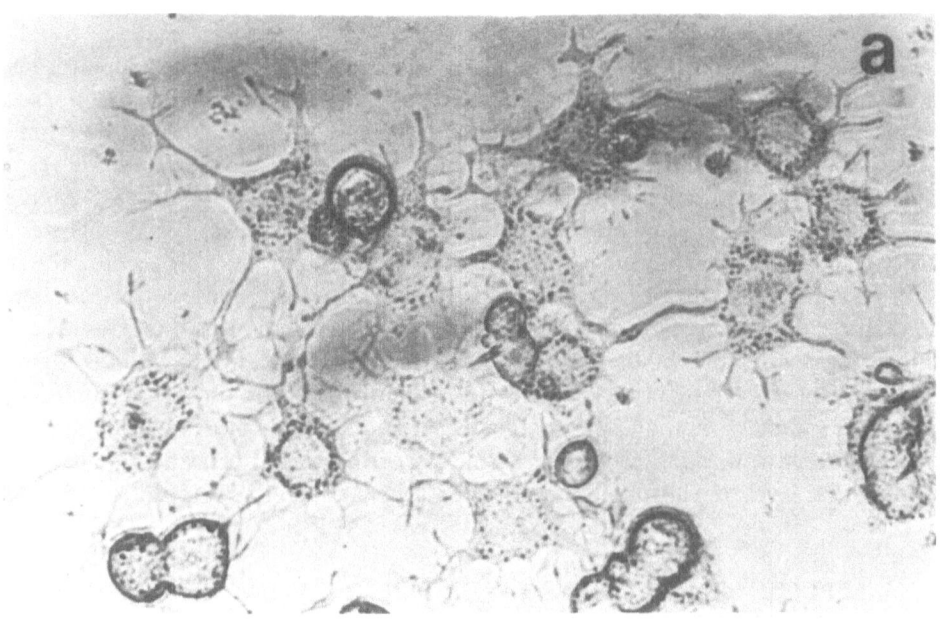

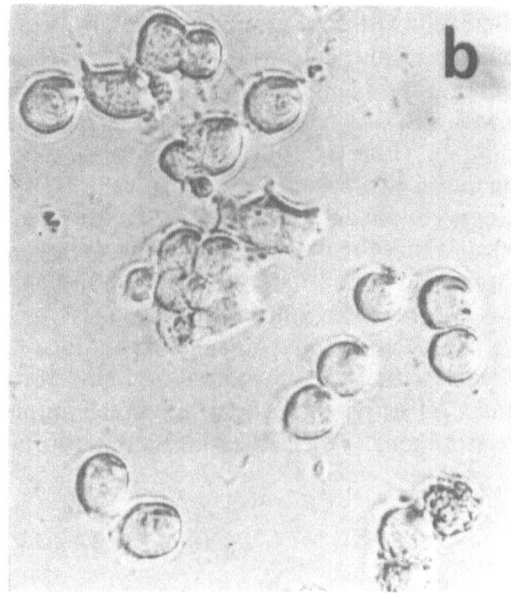

Figure 1. Morphological changes of NH15CA2 cells by lipocortin. NH15CA2 cells were cultured in RPMI 1640 media containing 109 fetal calf serum and 1 mM dibutyryl-cyclic-AMP with (a) or without (b) 4 μg/ml lipocortin for 4 days.

possible mediator of Dx for the induction of TH activity in NH15CA2 cells. To obtain further evidence supporting the hypothesis that Dx induces the synthesis of lipocortin in these cells, we assayed the cellular levels of lipocortin in the Dx-treated and nontreated cells by measuring the binding of an antilipocortin antibody. Nonspecific binding as measured by binding of the control antibody was subtracted. The Dx (10 μM) treatment caused an approximately 2.5 fold increase of the endogenous lipocortin in NH15CA2 cells.

PROPERTIES OF LIPOCORTIN

Lipocortin isolated from the glucocorticoid-treated rabbit neutrophils has a molecular weight around 40,000. This protein is more specific for phospholipase A_2 rather than for phospholipase C, and inhibits the release of arachidonic acid from intact fibroblasts and neutrophils stimulated with bradykinin and fMetLeuPhe, respectively. Lipocortin forms a stoichiometric complex in vitro with porcine pancreas phospholipase A_2. This protein is hydrophobic and has a site(s) that recognizes N-acetylglucosamine, a main sugar residue on the cell surfaces; such properties probably help its reinsertion into plasma membranes to inhibit phospolipase A_2 (Hirata et al., 1985b). At least, some parts of lipocortin are known to reside on cell surfaces, because lipocortin is accessible by the treatment with antilipocortin antibody and is digested by treatment with some proteases. The digestion of lipocortin on the cell surfaces by proteases generally results in phospholipase A_2 activation as measured by the release of arachidonic acid. Thus, cellular levels of lipocortin appear to play an important role in the basal level of phospholipase A_2 activity in the intact cells (Hirata, 1985).

EFFECTS OF PROTEASES ON DIFFERENTIATION OF NH15CA2

To increase the basal activiy of phospholipase A_2 in NH15CA2 cells, we treated these cells with various proteases. β-Chymotrypsin was most effective in decreasing the binding of anti-lipocortin antibody to NH15CA2 cells without affecting the binding of the control antibody. When NH15CA2 cells were cultured in the presence of cyclic-AMP with β-chymotrypsin (50 μg/ml), neurites were extended after 7 days of culture (Fig. 2). The level of lipocortin in these cells in one day after treatment decreased to one-third of those in the control cells. The CAT activity after 4 days in culture increased 3 fold and the TH activity decreased to one half. Taken together, these results suggest that the basal activity of phospholipase A_2 in NH15CA2 cells determines the direction of neurotransmitter plasticity. It should be noted that CAT enhancing factor partially purified from ConA-treated spleen cells (a factor which has properties similar to that isolated from C_6 glioma cells or heart muscle cells) appears to have an activity simillar to that of β-chymotrypsin. Lipocortin and CAT enhancing factor increased TH and CAT activities, respectively, in the primary cultured neuronal cells from superior cervical ganglia of new born rats (Hattori and Hirata, unpublished).

THE MECHANISM OF CELLULAR DIFFERENTIATION

The biochemical mechanism as to how the basal activity of phospholipase A_2 determines the phenotype expression in developing neuronal cells remains to be elucidated. Recently, Ishizaka and his associates have proposed that the suppressor or helper activities of T cells in IgE synthesis are determined by lipocortin and a kallikrein-like protease, respectively (Iwata et al., 1984). Lipocortin inhibits phospholipase A_2 activity in the target cells, whereas the kallikrein-like protease enhances it. HL60 cells, a promyeloleukemia cell line, are also developmentally bipotential for acquiring the

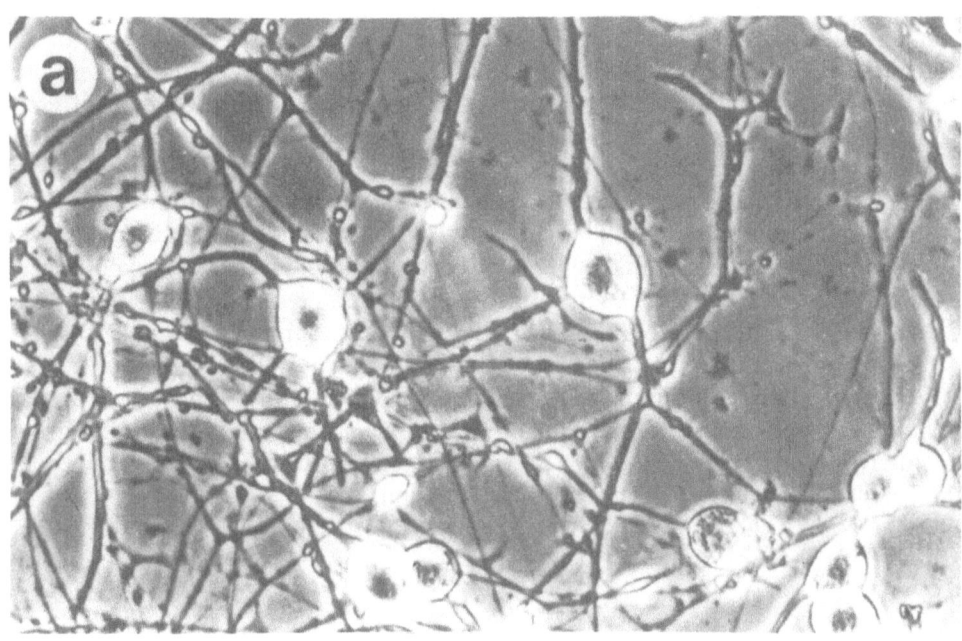

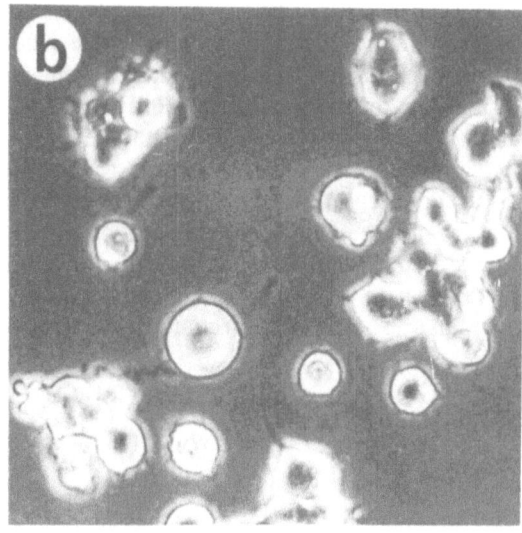

Figure 2. Morphological changes of NH15CA2 cells by β-chymotrypsin. NH15CA2 cells were cultured in RPMI 1640 media containing 10% fetal calf serum and 1 mM dibutyryl-cyclic-AMP with (a) or without (b) 50 μg/ml β-chymotrypsin for 7 days.

characteristics of mature macrophages or neutrophils, depending on the kind of chemical inducer. These cells are differentiated towards macrophage by melittin, a bee venom phospholipase A_2 activator, whereas they obtain the neutrophil properties by the C-MT peptide, a phospholipase inhibitor (Notsu et al., 1985). Amiloride, an inhibitor of the Na^+/H^+ antiporter, and ouabain, an inhibitor of Na^+, K^+-ATPase, can block the differentiation of HL60 cells induced by melittin and C-MT peptide, respectively. Since levels of lysophosphatidylcholine, a product of phosphatidylcholine by phospholipase

A_2, affect the activities of the Na^+/H^+ antiporter and Na^+, K^+-ATPase, we believe that intracellular Na^+ and/or K^+ sequestration regulates gene expression in developmentally bipotential cells (Hirata et al., 1985a). In keeping with this interpretation, monensin and gramicidin, Na^+ ionophores, can promote the cellular differentiation of HL 60 cells towards macrophages. It has been recently reported that the neurotransmitter plasticity of neurons can be affected by depolarization-hyperpolarization, a process which accompanies the fluxes of Na^+ and K^+ (Black et al., 1984). These changes in cellular concentrations of cations often accompany those in intracellular pH. On the other hand, hypo-methylation of DNA is now hypothesized to be involved in active gene expression. Phorbol esters, retinoic acid, melittin, and C-MT peptide, differentiators of HL60 cells, can increase ratios of S-AdoHcy to S-AdoMet to different extents; thus they inhibit DNA methylation to various degrees (Hirata et al., 1985a). Intracellular pH and methylations are also proposed to play an important role in neurite formation of developing neurons (Hattori and Hirata, manuscript in preparation). However, it would require further investigations to explain the detailed mechanism of neuronal differentiation.

SUMMARY

Treatment of NH15CA2, a neuroblastoma-glioma hybrid cell line, and autonomic ganglion cells from superior cervical ganglia with glucocorticoids results in enhancement of the adrenergic phenotype. The glucocorticoid-induced increase in tyrosine hydroxylase activity can be blocked by the simultaneous treatment together with antilipocortin antibody. Consequently, isolated lipocortin mimicked such action of glucocorticoids. Decrease in the cellular level of lipocortin by the treatment with choline acetyltransferase enhancing factor or selected proteases directed these developing neurons to the cholinergic phenotype. Since lipocortin is a naturally occuring phospholipase inhibitory protein that regulates the cellular activities of phospholipases, our results have suggested that the basal activities of phospholipases in developing neuronal cells are associated with cell competence by which the direction of cellular differentiation is determined in these cells. The mechanism of phenotype determination in neurons was briefly discussed.

REFERENCES

Black IB, Adler JE, Dreytus CF, Jonakeit GM, Katz DM, LaGamma EF, Markey KM (1984) Neurotransmitter plasticity at the molecular level. Science 225: 1266-1270.

Blackwell GJ, Flower RJ, Nikamp FP, Vane JR (1978) Phospholipase A_2 activity of guinea pig isolated perfused lungs: Stimulation and inhibition by anti-inflammatory steroids. Brit J Pharmacol 62: 78-79.

DiRosa M, Flower RJ, Hirata F, Parente L, Russo-Marie F (1984) Letter to the editor; Antiphospholipase proteins, Nomenclature announcement. Prostaglandins 28: 441-442.

Hattori T, Hirata F, Hoffman T, Hizuta A, Herberman RB (1983) Inhibition of human natural killer (NK) activity and antibody dependent cellular cytotoxicity (ADCC) by lipomodulin, a phospholipase inhibitory protein. J Immunol 131: 662-665.

Heumann R, Ocalan M, Hamprecht B (1979) Factors from glial cells regulate choline acetyltransferase and tyrosine hydroxylase activities in a hybrid-hybrid cell line. FEBS Letters 107: 37-41.

Hirata F (1985) Molecular mechanisms on the modulation of phospholipid metabolism by glucocorticoids. In: Bailey M (ed): Prostaglandins, Leukotrienes and Lipoxins; Biochemistry, Mechanism of action and clinical application, Plenum, New York; pp. 119-123.

Hirata F, Matsuda K, Wano Y, Hattori T (1985a) The biochemical mechanism of cellular activation. Int J Immunopharmacol, in press.

Hirata F, Notsu Y, Yamada R, Ishihara Y, Wano Y, Kunos I, Kunos G (1985b) Isolation and characterization of lipocortin (lipomodulin). Agents and Action, 17: 263-266.

Iwata M, Akasaki M, Ishizaka K (1984) Modulation of the biological activities of IgE binding factor. VI The activation of phospholipase by glycosylation enhancing factor. J Immunol 133: 1505-1512.

McEwin BS (1979) Influences of adreno-cortical hormones on pituitary and brain functions. In: Baxter JD, Rosseau GG (eds): Glucocorticoid hormone action. Springer-Verlag, Berlin, Heidelberg and New York, pp. 449-465.

Nirenberg M, Wilson S, Higashida H, Rotter A, Krueger K, Busis N, Ray R, Kenimer JF, Adler M (1983) Modulation of synapse formation by cyclic adenosine monophosphate. Science 222: 794-799.

Notsu Y, Namiuchi S, Hattori T, Matsuda K, Hirata F (1985) Inhibition of phospholipases by Met-Leu-Phe-Ile-Leu-Ile-Lys-Arg-Ser-Arg-His-Phe, C terminus of middle-sized tumor antigen. Arch Biochem Biophys 236: 195-204.

Patterson PH (1978) Environmental determination of autonomic neurotransmitter functions. Ann Rev Neurosci 1: 1-17.

Sayers G, Travis RH (1970) Adrenocorticotropic hormones: Adrenocortical steroids and their synthetic analogs. In: Goodman LS, Gilman A (eds): Pharmacological basis of therapeutics. McMillan, New York, pp. 1604-1642.

Phospholipid research and the nervous system
Biochemical and molecular pharmacology
L.A. Horrocks, L. Freysz, G. Toffano (eds)
Fidia Research Series, vol. 4.
Liviana Press, Padova. © 1986

PHOSPHOLIPIDS AND MEMBRANE PHENOMENA INVOLVED IN THE RELEASE OF MEDIATORS OF ANAPHYLAXIS

Y. Pacheco, P. Fonlupt, A.F. Prigent, N. Biot, M. Perrin-Fayolle, H. Pacheco

Centre Hospitalier Universitaire Lyon-Sud, Unité INSERM 205,
Lyon, Villeurbanne, France

The anaphylactic model is of particular interest in the study of inflammatory processes, but it can also inform us of some biological mechanisms in the central or peripheral nervous system. Histamine is now considered as a neuromediator (Schwartz et al., 1979a): histaminergic pathways have been studied in rat brain (Garbarg et al., 1980) and histamine H_1 and H_2 receptors have been described in the brain (Schwartz et al., 1979b). Non neuronal histamine-storing cells having properties strongly resembling those of typical mast cells have also been identified by biochemical and pharmacological approaches (Kruger, 1974). The functional role of mast cells in the central nervous system as in peripheral tissues is far from clear. In peripheral tissues it is generally assumed that they are involved in vascular control under particular pathophysiological conditions, i.e. during immune and inflammatory processes. It can be hypothezised that cerebral mast cells are involved in similar processes. Anaphylactic phenomena involved in tissues are closely interwoven with peripheral nervous system endings disturbances (Boushey et al., 1980). Interactions between inflammatory mediators and neuromediators have already been described (Besedovski et al., 1983). The immune system is now thought to be a sensory organ (Blalock, 1984). It seems probable that the sensory function of the immune system may mimic the neuroendocrine system in terms of a given stimulus eliciting a particular set of hormones and thus a physiological response. Molecular targets of neuromediators have been particularly studied in brain tissue; but these studies benefited from other studies on immune cells or adrenal medulla for example.

MEDIATOR RELEASE DURING ANAPHYLACTIC REACTION

Various mediators are released during an anaphylactic reaction: vasoactive, bronchostrictor and chemotactic mediators as well as active enzymes and structural proteoglycans. Histamine, prostaglandins, leukotrienes, and lysophospholipids are the essential vasoactive, smooth muscle reactive mediators. They also play an important role in chemotaxis. Oxygenation and further transformations of arachidonic acid and certain other polyunsaturated fatty acids result in the formation of biologically active compounds. This was first demonstrated for the prostaglandins and additional studies of the mechanism of prostaglandin biosynthesis led to the detection of new biologically active molecules (prostaglandin endoperoxides, thromboxanes, and prostacyclin) (Kerr and Ganderton, 1982). Recently a new group of biologically active derivatives of polyunsaturated fatty acids, the leukotrienes, was discovered. The pronounced biological effects of these compounds are related to immediate hypersensitivity reactions and inflammation. Various cells are involved in the release of these mediators: mast cells, leukocytes, and macrophages. In IgE mediated hypersensitivity reactions, IgE molecules bind to specific receptors on mast cells and basophil granulocytes with high affinity and the reaction of cell bound IgE molecules with multivalent ligand initiates the release of a variety of preformed or newly generated mediators (Ishizaka and Ishizaka, 1978).

These Fc IgE receptors have been shown on mast cells, basophils, and macrophages. Stimulation of Fc IgE receptors induces several biochemical modifications, which can modify the plasticity and permeability of the plasma membrane (Ishizaka et al., 1980). But membrane enzymatic activation can occur independently of Fc IgE receptor stimulation: for example after stimulation by Ca^{2+} ionophore compounds (Imai et al., 1984). No matter what chemical signals start an anaphylactic reaction, membrane phenomena play an important role in mediator release.

ROLE OF MEMBRANES IN ANAPHYLAXIS

The molecular structure of the membranes is complex. The structure of the membrane lipid bilayer proposed by Nicolson is still accepted, however it seems to be a mobile structure. The membrane is an important storage depot of phospholipids which can liberate fatty acids and fatty acid derived mediators. It also plays an important role in plasticity.

Membranes as a Reservoir of Anaphylactic Mediators

Membrane phospholipids are a source of anaphylactic mediators. Phospholipase A_2 (PLA_2) activation plays an important role in the release of these mediators. This enzyme hydrolyzes esters at carbon 2 of glycerol.

Arachidonic acid (AA) is transformed into leukotrienes by way of the lipooxygenase pathway and into prostacyclin, prostaglandins, and thromboxanes by way of the cyclooxygenase pathway. All of these compounds participate in inflammatory processes (Kerr and Ganderton, 1982). But the membrane is not the only source of these mediators. Lichtenstein and coworkers demonstrated that mast cell and macrophage cytoplasmic lipid bodies appear to represent the major site of intracellular storage and metabolism of products of AA and perhaps other fatty acids taken up from the external milieu (Dvorak et al., 1983).

Lysophospholipids such as lysophosphatidylcholine, which is known to increase membrane permeability, also probably play a role in anaphylaxis (Nath et al., 1983).

Acetylated alkyl phosphoglycerides, a new class of phospholipids, have been shown to possess a remarkable spectrum of extremely potent biological activities categorizing them as potential acute allergic and inflammatory mediators (Benveniste, 1980).

Membrane Plasticity

Membranes are composed primarily of lipids and proteins; the lipids serve the dual purpose of forming a permeability barrier to ions and most polar molecules and providing a fluid solvent for membrane proteins, while the proteins perfom the membrane functions, such as transport, communication and energy transduction. It is generally supposed that most proteins float free in the lipid sea, their motion retarded only by the viscosity of the lipids and by protein-protein collisions and interactions.

Biochemical modifications of phospholipids, particularly their methylation, play an important role (Hirata and Axelrod, 1980). The topology of these enzymes makes possible the translocation of phospholipids from the cytoplasmic side to the outer surface of membranes by successive methylations. This leads to increased mobility of receptors and in several cell types to calcium influx. Phospholipid methylation is closely coupled to phospholipase A_2, a calcium requiring enzyme.

THE ROLE OF PHOSPHOLIPIDS (PL)
IN THE PROCESS OF HISTAMINE RELEASE

PL with Histamine Release Effect

Lysophospholipids [lysophosphatidylcholine (Nath et al., 1983), lysophosphatidylserine (Bruni et al., 1984), and 2-acetyl phosphoglycerides (PAF acether) (Benveniste et al., 1972)] can induce histamine release from mast cells or basophils.

PL Biological Modifications during Histamine Release

— *PL turnover* increases during histamine release. For anti-IgE and Con A stimulation, histamine release and [^{32}P]Pi incorporation into PtdH, PtdIns, and PtdCho were potentiated by passively sensitizing the mast cells with rat IgE or by addition of PtdSer (Schellenberg, 1980).

— *PL methylation.* Crews et al. (1981) demonstrated the close relationship between immune Fc IgE receptor stimulation and PL methylation. Two methyltransferases involved in the methylation of PtdEtn to form PtdCho were demonstrated in cytoplasmic membrane of various cells (red blood cells, mast cells, basophil) (Hirata and Axelrod, 1978). The first methyltransferase (PMT$_1$) catalyzes the methylation of PtdEtn to form phosphatidyl monomethylethanolamine (PtdMeEtn). The second methyltransferase catalyzes the two successive methylations of PtdMeEtn to PtdCho.

The stimulation of phospholipid methylation in the plasma membrane appears to be intrinsic to the processes leading to Ca^{2+} influx and histamine release. Inhibitors of adenosylmethionine dependent methyltransferases (DZA: 3-deaza-adenosine) reduce Ca^{2+} influx and histamine release from basophil (Morita et al., 1981) or enzymatically dispersed human lung mast cells activated with either antihuman IgE or calcium ionophore A 23187 (Benyon et al., 1984). But the hypothesis that phospholipid methylation is obligatory for receptor mediated Ca signals is not supported by some authors (Moore et al., 1984). Chien and Ashman (1984) failed to demonstrate any significant changes in lipid methylation either in unfractionated human peripheral blood mononuclear cells, purified T-cells, B-cells, monocytes, or in combinations of these populations. Other metabolic pathways may operate.

— *Involvement of phospholipid diacylglycerol and PtdIns pathways.*

Recent studies by Sullivan (1982) demonstrated that activation of rat mast cells by antigen, anti-IgE and other stimulants accelerated the conversion of phospholipid to DAG resulting in the accumulation of this potent fusogen. A vast amount of literature has been accumulated which demonstrates that a wide variety of ligand receptor interactions in many cell types results in acceleration of PtdIns turnover. Schellenberg (1980) demonstrated that various stimulators of histamine release increased [^{32}P] phosphate incorporation into PtdH, PtdIns, and PtdCho.

Addition of calcium ionophore A 23187 to mast cells prelabeled with [^{3}H] glycerol induced a rapid and progressive increase in PtdH and DAG which was concomitant with the small rise in PtdIns. A 23187 stimulates arachidonate release, perhaps from PtdCho and triacylglycerol, whereas 48/80 liberates arachidonate from PtdEtn.

The Role of Phospholipase A$_2$ and Arachidonic Acid Metabolism in the Control of Phospholipid Metabolism and Histamine Release

Phospholipase A$_2$ plays also an important role in immediate hypersensitivity reactions. Two agents, p-bromophenacyl bromide (BPB) and mepacrine, which inhibit PLA$_2$ activity in several tissues, cause a dose dependent inhibition of histamine release from human basophils (Marone et al., 1981). The activation of cell surface receptors and/or the intracellular translocation of Ca^{2+} by the ionophore A 23187 activates a basophil PLA$_2$.

ETYA (5, 8, 11, 14-eicosatetraynoic acid, a known inhibitor of cyclooxygenase and lipoxygenase pathways of AA metabolism) inhibits incorporation of [^{32}P]Pi into PtdH, PtdIns, and PtdCho in both unstimulated mast cells and mast cells stimulated by cross linking of surface IgE molecules (Marquardt et al., 1981).

AA inhibits PL labeling in resting mast cells and renders the cells less responsive to secretory signals. Preincubation of the cells with indomethacin blocks AA effects. When AA was added to stimulated mast cell, both PtdCho labeling and mediator release were enhanced (Marquardt et al., 1981).

PL Modifications and Membrane Receptors

The lipids surrounding membrane proteins play an important regulatory role in the activity of some enzymes and in the expression of some membrane antigens. PL methylation influences Ca^{2+}-ATPase activity, beta-adrenoreceptor coupling with adenylate cyclase, and the activity of the receptor for chemotactic factors. Shinitzki (1979) demonstrated that treating cells with cholesterol or hydrophilic cholesterol esters changes the cholesterol phospholipid molar ratio and also alters the expression of a variety of plasma membrane proteins of different types of cells, decreasing membrane lipid packing density by PtdCho treatment, and increasing the expression of MHC determinants.

RELATION BETWEEN ENZYMES OF PL METABOLISM AND PHARMACOLOGICAL COMPOUNDS MODULATING HISTAMINE RELEASE

Indirect Relation

Many substances which can modulate histamine release act indirectly on PL enzymes through an activation or an inhibition of membrane receptors or other enzymes. For example beta-agonist (Hirata et al., 1979) and histamine (Tolone et al., 1982) potentiate PtdCho methylation after binding to beta-receptor or H_2-histamine receptor. Corticosteroids induce lipomodulin an endogenous inhibitor of phospholipase A_2 (Hirata, 1983).

Some compounds modify PL methylation, but we do not know exactly how they act. For example, Ketotifen which is used for treating allergic asthma inhibits PL methylation (Pacheco et al., 1985). Mequitazin (H_1-antihistaminic compound) potentiates PL methylation (Fonlupt et al., 1983).

The addition of EHNA (an inhibitor of adenosine deaminase) together with adenosine and L-homocysteine has been shown to result in accumulation of intracellular AdoHcy in a number of systems and indirectly to inhibit PL methylation. EHNA blocks the IgE mediated histamine release. 3-Deazaadenosine (DZA) a structural analogue of adenosine is both a substrate and an inhibitor of AdoHcy hydrolase. This compound inhibits PL methylation and IgE mediated histamine release (Banyon et al., 1984).

Direct Relation

Two substances are of particular interest in membrane phospholipid metabolism regulation: S-adenosyl-methionine and S-adenosyl-L-homocysteine.

S-ADENOSYL-METHIONINE (AdoMet)

AdoMet seems to be the most versatile methyldonor in mammalian systems, whereas 5-methyltetrahydrofolic acid, methylcobalamine and betaine function as methyl donors only in a few instances (Cantoni, 1952). This compound induces the methylation of PtdEtn to PtdMeEtn and of PtdMeEtn to PtdCho (Hirata and Axelrod, 1980). It inhibits $Na^+K^+ATPase$ (Hattori and Kanfer, 1984). AdoMet plays a role in regulating the uptake of neuromediators in the brain (Fonlupt et al., 1979) and potentiate spontaneous or specific allergen induced histamine release from human leukocytes (Pacheco et al., 1984b).

S-ADENOSYL-L-HOMOCYSTEINE (AdoHcy)

Methylation of proteins, nucleic acid and phospholipids can be inhibited by AdoHcy (Ueland, 1982).

AdoHcy is a potent inhibitor of transmethylation reactions in cell free systems (Gibson et al., 1961). Extracellular AdoHcy does not cross the cell membrane as an intact molecule. However extracellular AdoHcy is hydrolyzed to adenosine and L-homocysteine by AdoHcy hydrolase leaking out of the cells. In addition AdoHcy binds to acceptors on the cell surface (Ueland, 1982).

P Fonlupt demonstrated in vitro and in vivo binding of AdoHcy to membranes from rat cerebral cortex. He demonstrated that the binding sites for AdoHcy in rat brain membranes have a methylation activity which uses PtdEtn as a substrate (Fonlupt et al., 1984) and that the inhibition of PMT_1 by AdoHcy analogues is correlated with their affinity for the [^{3}H]AdoHcy binding sites on rat cortical membranes (Fonlupt et al., 1982).

The physiological role of membrane acceptors for AdoHcy and their relation to methyltransferases localized in the membrane offer a great field for new investigations. AdoHcy has sedative and anticonvulsive properties in the rabbit, rat and cat (Fonlupt et al., 1980a). This observation suggests that AdoHcy has an effect on the central nervous system and is related to effects of AdoHcy on the metabolism of catecholamines.

We have recently shown (Pacheco et al., 1984b) that AdoHcy and its structural analogues (Pacheco et al., 1984a) inhibit human leukocyte histamine release. At the same time AdoHcy inhibits membrane PL methylation. In contrast AdoMet enhances spontaneous and specifically induced histamine release from leukocytes in vitro at a concentration of 10^{-4}M in allergic subjects and spontaneous histamine release at 10^{-5}M concentration in normal subjects. This difference between allergic and normal patients can be explained by different levels in leukocyte membrane PL methylation activity. In fact we demonstrated a decreased activity of PMT_1 in leukocytes from allergic subjects (Fonlupt et al., 1984). This decreased activity depends on the low PMT_1 activity in monocytes and lymphocytes.

By contrast PMT_1 stimulated by noradrenaline and levels of cyclic AMP phosphodiesterase levels are higher in allergic asthmatic patients than in controls. Prigent et al. (1984) demonstrated a significant correlation between cyclic AMP phosphodiesterase and noradrenaline stimulated methyltransferase with both control and allergic subjects, which suggests a close relationship between these two enzymes and indicates the importance of lipids and/or phospholipids in the microenvironment of phosphodiesterase.

CONCLUSION

Membrane phospholipids play an important role both as a reservoir of anaphylactic mediators and as compounds determining membrane plasticity. Pharmacological modulation of the enzymes active in PtdCho metabolism seems of particular interest in some diseases such as anaphylaxis in which membrane mechanisms are important.

S-adenosyl-homocysteine, presently in clinical phase I investigation, is an interesting molecule which can stabilize the membrane and which might have a therapeutic use in insomnia, some neuropsychiatric disorders, and functional diseases such as asthma.

The finding of disturbances in enzymes metabolizing PL in allergic subjects at the same time as modification in enzymes metabolizing nucleotides could explain certain dysfunctions of hormonal receptors and also introduces a new concept in dysfunction of the receptor enzyme complex.

ACKNOWLEDGMENTS

This work was granted by the French Committee of Respiratory Diseases.

REFERENCES

Benveniste J (1980) In: Oehling A (ed) Advances in Allergology. Pergamon Press. Oxford and New York, pp. 195-202.

Benveniste J, Henson PM and Cochrane CG (1972) J Exp Med 136: 1356-1377.

Benyon RC, Church MK and Holgate S (1984) Biochem Pharmacol 33: 2881-2886.

Besedovski HO, Delrey AE and Sorkin E (1983) Immunology Today 4:342-346.

Blalock JE (1984) J Immunology 132: 1067-1070.

Boushey HA, Holtzman MJ, Sheller JR and Nadel JA (1980) Am Rev Respir Dis 121: 389-413.

Bruni A, Bigon E, Battistella A, Boarato E, Mietto L, and Toffano G (1984) Agents and Actions 14: 619-625.

Cantoni G (1952) J Am Chem Soc 74: 2942-2943.

Chien MM and Ashman RF (1984) Molec Immunol 21: 621-625.

Crews FT, Morita Y, Givney AMC, Hirata F, Siraganian RP and Axelrod J (1981) Arch Biochem Biophys 212: 561-571.

Dvorak AM, Dvorak HF, Peters SP, Shulman ES, MacGlashan DW, Pyne K, Harvey VS, Gallis J and Lichtenstein LM (1983) J Immunol 131: 2965-2975.

Fonlupt P, Barailler J, Roche M, Cronenberger L and Pacheco H (1979) CR Acad Sci Paris 288: 283-286.

Fonlupt P, Dubois M, Gallet H, Biot N and Pacheco H (1983) CR Acad Sci Paris 296: 1005-1007.

Fonlupt P, Pacheco Y, Macovschi O, Dubois M, Biot N, Pacheco H (1984a) Clin Chim Acta 136: 13-18.

Fonlupt P, Rey C, Comte N, Dubois M and Pacheco H (1984b) CR Soc Biol 178: 45-51.

Fonlupt P, Rey C and Pacheco H (1980a) J Neurochem 36: 165-170.

Fonlupt P, Rey C and Pacheco H (1982) Life Sci 3: 655-170.

Fonlupt P, Roche M, Cronenberger L and Pacheco H (1980b) Can J Pharmacol Physiol 58: 160-166.

Fonlupt P, Roche M and Pacheco H (1980c) Can J Pharmacol Physiol 58: 493-498.

Garbarg M, Barbin G, Llorens C et al (1980) In: Essman B (ed): Neurotransmitters, Receptors and Drug Action. Spectrum Publications, New York, pp. 179-202.

Gibson KD, Wilson JD and Udenfriend S (1961) J Biol Chem 236: 673-679.

Hattori H and Kanfer JN (1984) J Neurochemistry 42: 204-208.

Hirata F (1983) Biochimie Inflammation. pp. 61-66, Masson ed., Paris.

Hirata F and Axelrod J (1978) Proc Natl Acad Sci. USA 75: 2348-2352.

Hirata F and Axelrod J (1980) Science 209: 1082-1090.

Hirata F, Strittmatter WJ and Axelrod J (1979) Proc Natl Acad Sci. USA 76: 368-372.

Imai A, Ishizuka Y, Nakashima S and Nozawa Y (1984) Arch Biochemistry Biophysics 232: 259-268.

Ishizaka T, Hirata F, Ishizaka K and Axelrod J (1980) Proc Natl Acad Sci. USA 77: 1903-1907.

Ishizaka T and Ishizaka K (1978) J Immunol 120: 800-805.

Kerr JW and Ganderton MA (1982) Mediators of allergic tissue reactions. Symposium in London 82 ICACI. p. 3-46.

Kruger PG (1974) Experientia 30: 810-811.

Marone G, Kagey-Sobotka A and Lichtenstein LM (1981) Clin Immunol Immunopathol 20: 231-239.

Marquardt DL, Nicolotti RA, Kennerly DA and Sullivan TJ (1981) J Immunology 127: 845-849.

Moore JP, Johannsson A, Hesketh TR, Smith GA and Metcalfe JC (1984) Biochem J 221: 675-684.

Morita Y, Chiang PK and Siraganian RP (1981) Biochem Pharmacol 30: 785-791.

Nath P, Josui AP and Agraval KP (1983) J Allergy Clin Immunol 72: 351-358.

Pacheco Y, Bensoussan P, Biot N, Duprat P, Fonlupt P, Prigent AF, Rey C, Dubois M, Pacheco H and Perrin-Fayolle M (1985) Prog Resp Res 19: 106-112.

Pacheco Y, Bensoussan P, Biot N, Fonlupt P, Pacheco H, Perrin-Fayolle M and Gero M (1984a) Am Rev Respir Dis 129 (suppl.), A25 (abstract).

Pacheco Y, Macovschi O, Biot N, Fonlupt P, Perrin-Fayolle M and Pacheco H (1984b) Clin Allergy 14: 37-43.

Prigent AF, Fonlupt P, Dubois M, Nemoz G, Pacheco H, Pacheco Y, Biot N and Perrin-Fayolle M (1984) Clin Chim Acta 143: 225-223.

Schellenberg RR (1980) Immunology 41: 123-129.

Shinitzky M and Souroujon M (1979) Proc Natl Acad Sci USA 76: 4438-4440.

Schwartz JC, Barbin G, Baudry M et al (1979a) In: Essman B and Valzeili L (eds): Current Developments in Psychopharmacology vol. 5, pp. 173-261. Spectrum Publications. New York.

Schwartz JC, Palacios JM, Barbin G et al. (1979b) In: Yellin TO (ed): Histamine Receptors. Spectrum Publications. New York. pp. 161-183.

Sullivan TJ (1982) In: Becker EL and Austen KF (eds): Fourth Internat Symposium. Liss, New York, p. 229.

Tolone G, Bonasera L and Pontieri GM (1982) Experientia 38: 966-968.

Ueland PM (1982): Pharmacol Rev 34: 223-253.

Phospholipid research and the nervous system
Biochemical and molecular pharmacology
L.A. Horrocks, L. Freysz, G. Toffano (eds)
Fidia Research Series, vol. 4.
Liviana Press, Padova. © 1986

ROLE OF LIPIDS DURING FUSION OF MODEL AND BIOLOGICAL MEMBRANES

A.J. Verkleij

Institute of Molecular Biology, University of Utrecht
Padualaan 8, 3584 CH Utrecht, The Netherlands

INTRODUCTION

Membrane fusion is an ubiquitous event in cell biology. Some of the important biological phenomena in which membrane fusion is involved are: (i) fusion of the sperm and the egg membrane which leads to fertilization, (ii) the secretion of neurotransmitters, insulin and other hormones, and digestive enzymes from their respective storage vesicles inside the gland cells, referred to as exocytosis, and (iii) the uptake of viruses and removal of receptor ligands from the surface (receptor-mediated endocytosis). In fact, every biological membrane has the potential to fuse, but this potentiality may be revealed more in one membrane than in another. In most types of intracellular membranes, such as endoplasmic reticulum, coated vesicles, endosomes, lysosomes, and Golgi cisternae, fusion takes place continuously.

Many studies have been undertaken with the aim of understanding the basic principles of the membrane fusion process itself and the factors involved and/or actively modulating this process. In recent years this basic interest in membrane fusion has been further stimulated. It has been realized that the application of artificially induced fusion is a powerful tool for hybridization of cells for the production of monoclonal antibodies, for introducing membrane components into the cell membrane, and also in relation to the introduction of drugs into cells (targeting).

At present many effectors are known to trigger membrane fusion, such as Ca^{2+} in exocytosis, hormones, growth factors, and antibodies in receptor-mediated endocytosis and pH during fusion of endocytotic vesicles and lysosomes. Moreover, the involvement of many other substances like ATP, cAMP, GTP, drugs and Ca^{2+}-binding proteins has been reported. Also, membrane proteins appear to be involved in the fusion. The state of the cytoskeleton, the extent of glycosylation and the distribution of membrane-spanning proteins all can affect the process (see Schramm et al., 1982). In line with the latter aspect, it is generally assumed that membrane proteins, both the extrinsic or membrane skeleton proteins and intrinsic membrane-spanning proteins have to be cleared from the fusion site before actual fusion can start. Clearance of membrane proteins or lateral reorganization of the latter proteins could not be visualized using rapid freezing. No alterations of intramembranous particles could be seen using reliable fast freezing fixation (Chandler and Hauser, 1980; Schmidt et al., 1983), indicating that only a small area of lipid bilayers appears to be sufficient for fusion (local point fusion). Of great importance to note is also that membrane fusion is extremely fast in the order of msec and under strict control.

Whereas it may be clear that proteins and many factors are crucial in the modulation of membrane fusion, it is lipids that actually fuse. This implies that the lipids of the fusing membranes have to come into close apposition, which requires a reduction in electrostatic repulsion and in the hydration forces of the lipids. Subsequently, the lipids of the two bilayers have to join, which implies that they temporarily and locally have to leave the bilayer configuration at the fusion point. Finally, the bilayers fuse with each other, which includes bilayer destabilization.

It has been postulated that special fusogenic lipids such as lysophosphatidylcholine (Lucy, 1970), monoglycerides (Ahkong et al., 1973), or phosphatidylserine play a pivotal role in membrane fusion. If there is a universal mechanism for the involvement of lipids in membrane fusion, it is highly unlikely that one particular lipid is required in this process; this in relation with the high variability of lipid compositions in biological membranes. Phosphatidylserine, for instance, is not present in all membranes (compare bacterial, chloroplast, and eukaryotic membranes). Moreover, in light of the lipid asymmetry, the role of phosphatidylserine in membrane fusion can only be valid for exocytosis or other fusion events from the cytoplasmic site, since this phospholipid appears to be exclusively located in the cytoplasmic monolayer of the plasma membrane (Op den Kamp, 1979) and coated vesicles (Altstiel and Branton, 1983).

In an alternative hypothesis it has been proposed that the property of phospholipids to adopt the hexagonal II phase and/or inverted micelle conformation is crucial for membrane fusion (Cullis and Hope, 1978; Verkleij et al., 1979a). This hypothesis is based on the fact that: (i) in any membrane lipids are present which upon isolation can adopt the hexagonal II phase or the analogous non-bilayer phase of inverted micelles at physiological conditions, (ii) fusion occurs between artificial membranes in which part of the lipids like to adopt the hexagonal II phase, (iii) all effectors known as fusion factors, including Ca^{2+}, pH, T and apolar peptides can trigger bilayer to hexagonal II and inverted micelle configurations, and (iv) the fusion intermediate to be expected in this concept, the inverted micelle, has actually been visualized by electron microscopy (the so-called lipidic particle). In this contribution all these points will be summarized and discussed in the light of biologically and artificially induced membrane fusion.

HEXAGONAL II PHASE AND INVERTED MICELLES

Lipids which prefer to adopt the hexagonal II (H_{II}) phase at physiological conditions can be found in almost any biological membrane. In this conformation, the lipids are organized in hexagonally arranged cylinders in which the polar headgroups of the lipid molecules surround a narrow aqueous channel. The organization of lipids in the

H_{II} phase can be detected with X-ray diffraction (Luzzati et al., 1968), freeze-fracturing (Deamer et al., 1970; Verkleij, 1984) and ^{31}P nuclear magnetic resonance (^{31}P NMR) (Cullis and De Kruijff, 1979).

Examples of lipids which adopt the H_{II} phase at physiological conditions are unsaturated phosphatidylethanolamines (PtdEtn), monogalactosyldiglycerides from chloroplasts, and monogalactosyldiglycerides from *Acholeplasma laidlawii*. The tendency of other naturally occurring lipids, like cardiolipin and phosphatidic acid, to form the H_{II} phase is dependent on pH and the presence of divalent cations. These negatively charged phospholipids can also adopt the H_{II} phase in the presence of the local anaesthetics dibucaine and chlorpromazine. Cardiolipin adopts the H_{II} phase in the presence of cytochrome c. Moreover, the H_{II} phase can be induced by gramicidin, can be modulated in the presence of cholesterol, appears in complex mixtures of synthetic lipids and lipid extracts of *E. coli*, mitochondria, and rod outer segment membranes (see for reviews Cullis et al., 1982; Verkleij et al., 1984). Recently and worth mentioning in the light of the phosphatidylinositol response and membrane fusion, it has been reported that diacylglycerols can induce the H_{II} phase in phosphatidylethanolamine, phosphatidylcholine and phosphatidylserine bilayers (Das and Rand, 1984).

The transition from bilayer to hexagonal H_{II} phase in systems of purified phosphatidylethanolamines is remarkably abrupt and occurs within a temperature range of only a few degrees. The latter point can be deduced from the fact that one has to use fast-freezing rates (> 1000 degrees/sec) to prevent the H_{II} bilayer transition in order to preserve the H_{II} phase for freeze fracturing (Van Venetië et al., 1981). Differential scanning calorimetry also showed that the enthalpy change involed in the structural rearrangement is very small compared to the heat uptake needed to melt the solid bilayer. Because the polymorphic transition occurs above the gel-liquid crystalline transition of the bilayers it can be concluded that the H_{II} phase is a liquid-crystalline or fluid phase (Cullis and De Kruijff, 1979).

Next to the pure H_{II} phase also an alternative analogous non-bilayer configuration can be found, i.e. the inverted micelle. This inverted micelle is visible as a well-defined lipidic particle (Verkleij et al., 1979b; Verkleij, 1984). Such lipidic particles have been found in pure phosphatidylethanolamines and cardiolipin as well as in lipid mixtures in which one of the lipids prefers the hexagonal H_{II} phase, e.g. phosphatidylethanolamine-containing systems, cardiolipin/phosphatidylcholine in the presence of Ca^{2+}, and cardiolipin in the presence of local anaesthetics and cardiotoxin (Gulik-Krzwycki et al., 1981; Batenburg et al., 1985; for a review see Verkleij, 1984). Inverted micelles or lipidic particles can in fact be seen as intermediary structures between bilayer and H_{II} phase and vice versa and also as fusion intermediates in bilayer fusion.

FUSION IN ARTIFICIAL MEMBRANES

The hypothesis that lipids preferring the hexagonal H_{II} phase and/or inverted micelle configuration play a crucial role in membrane fusion has been strongly supported by experiments with model membrane systems. This was first demonstrated with vesicles composed of an equimolar mixture of cardiolipin and egg-phosphatidylcholine (Fig. 1). These vesicles in fact fuse in a non-leaky fashion (Wilschut et al., 1982). More

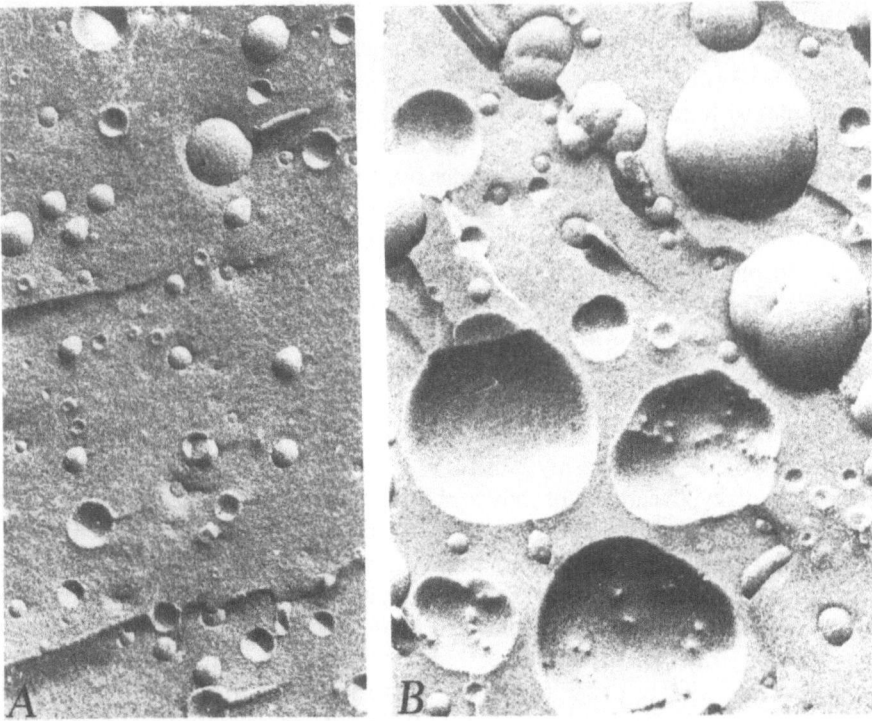

Figure 1. Freeze-fracture preparations of lipid vesicles composed of an equimolar mixture of cardiolipin and lecithin before (A) and after (B) addition of Ca^{2+}. Magnification 120,000 ×.

relevant for biological fusion events are mixtures of phosphatidylethanolamines and negatively charged phospholipids. Ca^{2+}, which is well-known to promote biological membrane fusion, triggers bilayer to non-bilayer (H_{II} and/or inverted micelle) transitions isothermally in these mixtures (Cullis et al., 1982). An increase in temperature also promotes such a transition. It is therefore logical to suppose that the presence of Ca^{2+} or an increase in temperature allows the non-bilayer tendency of endogenous lipids to be expressed, thus promoting the fusion event. These predictions hold for a variety of mixtures when Ca^{2+} is added (see for reviews Cullis et al., 1982; Verkleij, 1984), such as phosphatidyl ethanolamine/phosphatidylserine, phosphatidylethanolamine/phosphatidylinositol, phosphatidylethanolamine/phosphatidic acid, phosphatidylethanolamine/cardiolipin, phosphatidylethanolamine/phosphatidylcholine/cholesterol/phosphatidylserine and when the temperature is raised in phosphatidylethanolamine/ phosphatidylcholine/cholesterol. In all these cases the incubations result in the formation of larger structures showing that net fusion has occurred.

Lipidic Particles

In all these mixtures undergoing fusion this event is associated with the appearance of lipidic particles sometimes localized in regions corresponding in the fusion interface.

In a recent review (Verkleij, 1984), it has been demonstrated that the different particle types, which confused the issue in the early days of the discovery of these lipidic particles, can be interpreted as a reflection of different stages during the fusion event.

Fig. 2 shows a tentative drawing and the corresponding particles at the different fusion stages. In the 'adhesion' stage, two neighbouring bilayers form polar contact points requiring local dehydration and charge neutralization which is characteristic for H_{II}-preferring lipids. The bilayers are still intact but the contact points rise to deflections in the freeze-fractured membranes because of the local high curvature. Such deflections are most likely not homogenous in size (diameter) and are not well-defined. Cusp-like particles may reflect this stage and indeed represent intermembrane attachment sites (IMAS model and cusp model; Miller, 1980; Hui and Stewart, 1981). In the second stage the bilayers 'join', which enables intermixing of their lipids. This may either proceed with an inverted micelle or extended micelle as an intermediate. Fracturing of a joining stage containing of well-defined particles and complementary pits. In the last step there will be a fission which can give rise either to the two original vesicles or to real fusion by intermixing of the two aqueous compartments of both vesicles. In the former case the bilayer will temporarily pass a stage similar to that of intermembrane attachment sites, whereas in the latter case the fracturing passing through the newly formed aqueous channel will show a marked change from cusp shapes to volcanoes with flat tops. The flat tops arise as a result of fracturing through ice in aqueous pores.

Although all these lipidic particles have been found in systems which undergo fusion, it has to be mentioned that these particles could not be found during the initial rounds of fusion and are seen at later stages (where these structures have a longer lifetime; Bearer et al., 1982; Verkleij et al., 1984). The reason for not seeing the lipidic particles during the initial fusion rounds is likely due to the fact that fusion, which is not arrested at the joining stage but proceeds as expected, is so fast that the transient intermediates (the inverted micelles) escape detection by freeze-fracture electron microscopy.

It has to be noted that the presence of non-bilayer lipids is not a prerequisite for fusion of artificial membranes.

For instance, dimyristoyl phosphatidylserine vesicles (Van Dijck et al., 1978), as well as phosphatidylserine vesicles upon the addition of Ca^{2+} (Papahadjopoulos et al., 1978) fuse upon repeatedly passing the gel to liquid-crystalline phase. However, in light of the facts that H_{II}-preferring lipids occur in any biological membrane (which is not the case for these lipids) and that the formation of solid domains is not preceding membrane fusion, the involvement of non-bilayer lipids is made more likely and attractive as a general mechanism for biological membrane fusion.

FUSION IN BIOLOGICAL MEMBRANES

All the notions discussed above that: (i) biological membranes contain lipids which can adopt H_{II} phase or inverted micellar organizations, (ii) both fusion and bilayer/H_{II} phase transitions can be modulated by similar factors including Ca^{2+}, pH, and also apolar peptides, etc., (iii) bilayer/H_{II} transitions are at the same time scale as membrane fusion (order of msec), and (iv) biological membrane fusion is a local point fu-

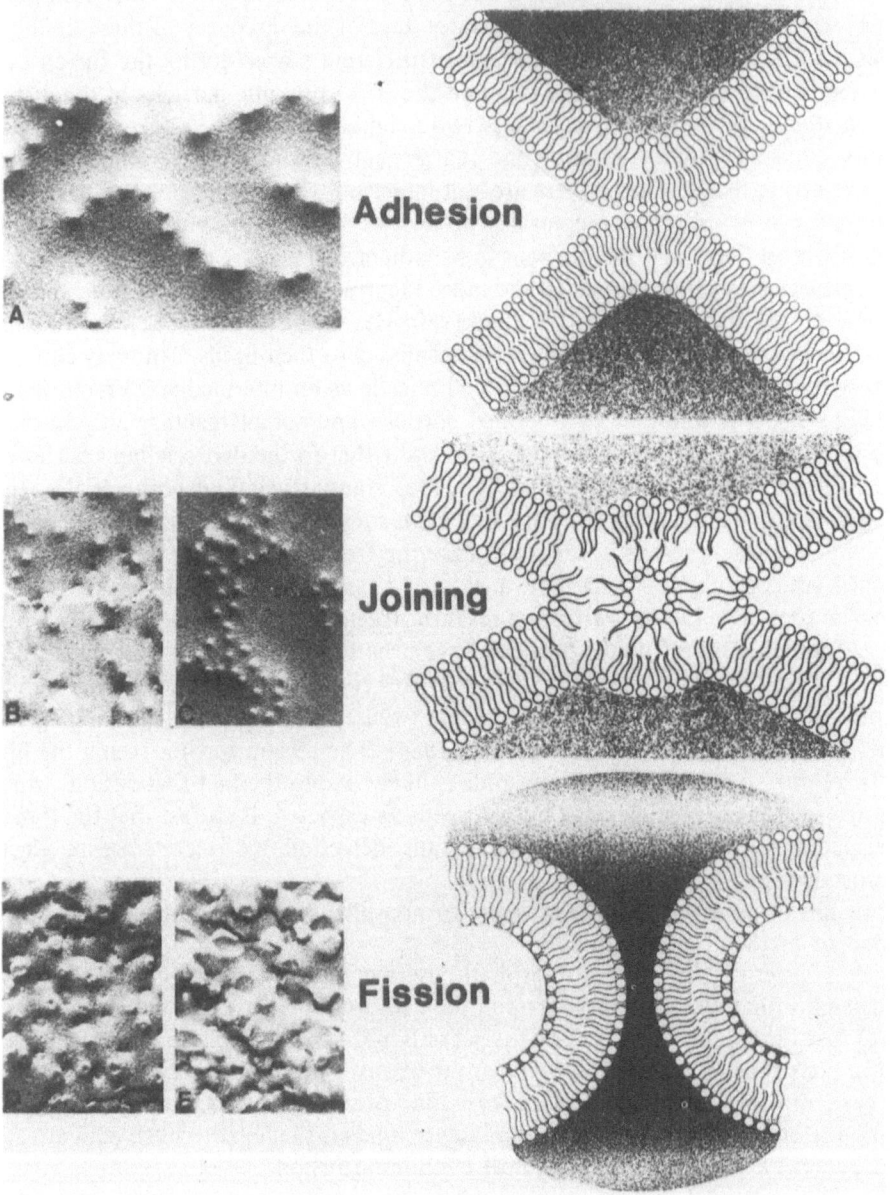

Figure 2. Tentative drawings of membrane fusion intermediates and their possible corresponding features as visualized by freeze fracturing. A-D are micrographs taken from the sample DOPE/DOPC containing dioleoyl species of PtdEtn and PtdCho with cholesterol (molar ratio 3:1:2) heated up to 60°C for 10 min, cooled down to 4°C, and subsequently frozen at 4°C. Magnification 75,000 x.

sion at which site the lipids have to leave temporarily the bilayer configuration, corroborate the universal hypothesis, although not a direct proof, that H_{II} phase-preferring lipids by virtue of their ability to adopt non-bilayer structures (inverted micelles) are pivotal for fusion of biological membranes.

As has been discussed, lipidic particles seen in artificial systems represent fusion intermediates at different stages of the fusion but might escape detection during real membrane fusion. In biological systems it is theoretically even more difficult to trap such intermediary structures, even with fast-freezing, because: (i) of the short life-time, (ii) the problem to synchronize the fusion event, and (iii) the amount of fusion events per surface. Even with exocytosis only a few hundred vesicles at most per cell are involved in fusion. However, recently Schmidt et al. (1983) have been able to catch exocytosis in chromaffin cells using fast-freezing devices. After stimulation of the exocytotic activity in a controlled way by adding carbachol, the plasma membrane displayed structural features which strongly resemble or better which are identical to the 'lipidic particles' found in model systems.

Figure 3 (see also Schmidt et al., 1983) shows some of the structural intermediates of this biological fusion. One can observe undefined semi-spherical protrusion or cusp-like particles, a well-defined lipidic particle (arrow-head) and volcano-like particles with a flat top (arrows), which likely resemble the 'adhesion', the 'joining', and the 'fission' stage, respectively. From this study it is also clear that no particle aggregation (clearance) is visible before or during fusion and that the fusion is a local event. On the other hand, one has to realize that it is very hard, or maybe even impossible, to make a direct correlation between fusion and these morphological features in other systems where fusion cannot be controlled.

How is fusion then modulated in biological systems? There are in principle different possibilities. It could be that intrinsic proteins may prevent fusion if they are laterally homogenously distributed over the surface. Support for this idea is coming from Tarashi et al. (1982). In these studies it was shown that glycophorin keeps phosphatidylethanolamine in the lamellar phase rather than in the H_{II} phase which it prefers. Addition of a lectin which induces aggregation of the glycophorin (although not visible) allows fusion and the formation of H_{II} phases. This phenomenon may be of relevance for receptor-mediated fusion.

Alternatively, extrinsic proteins may prevent fusion of membranes which undergo fusion upon a local increase of Ca^{2+}. Examples are the cytoskeleton of the erythrocyte membrane, clathrin coats in the fusion of coated vesicles and the endosome (Alstiel and Branton, 1983), and probably cytoskeleton elements in exocytosis. These membrane skeletons have to leave the fusion site. As a result of Ca^{2+} binding to phosphatidylserine (which is exclusively in the cytoplasmic leaflet of the membrane; Op den Kamp, 1979), the bilayer stabilization is lost (Cullis et al., 1982) and fusion will occur.

Fusion may also be induced by a local accumulation of a non-bilayer lipid. Of special interest is diacylglycerol, which has been shown to induce the H_{II} phase in phosphatidylethanolamine, phosphatidylcholine, and phosphatidylserine bilayers (Das and Rand, 1984). Exocytosis in many cell types (Baker and Knight, 1984) and myoblast fusion (Wakelam, 1983) are closely linked to the phosphatidylinositol response of which diacylglycerol is a product.

Lastly it has been proposed that a class of peptides of some viruses become apolar upon a pH drop in the lysosome which induces fusion. This has led Lucy (1984) to

214

suggest that hydrophobic peptides possibly induce fusion. In general, hydrophobic peptides should be cleaved off either from intrinsic or extrinsic proteins by activation of endogenous cellular proteinases. Other examples mentioned in his review are melittin and alamethicin. At first, this may appear to be contradictory with our hypothesis, but in fact it is not because peptides, including cytochrome c, gramicidin (Verkleij, 1984) melittin (Batenburg and coworkers, unpublished observations), cardiotoxin (Gulik-Krzywicki et al., 1981; Batenburg et al., 1985) and possibly alamethicin (as suggested by Lau and Chan, 1975) can induce non-bilayer structures like inverted micelles or H_{II} cylinders.

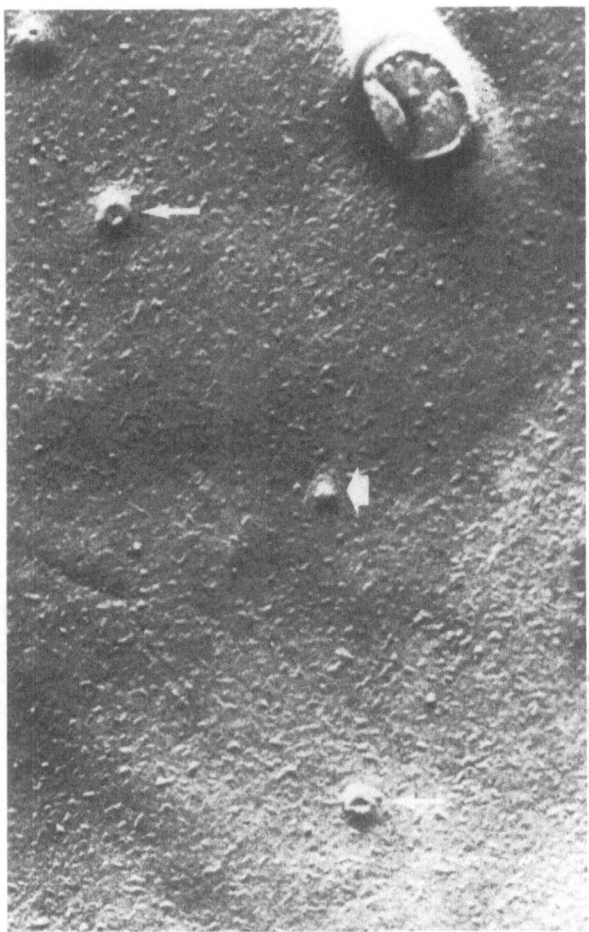

*Figure 3.*Freeze-fractured plasma membrane of chromaffin cells during stimulation with carbachol. This EF aspect exhibits features which are associated with exocytosis. Arrow head indicates a 'lipidic particle'-like structure. Arrows indicate some of the exocytotic openings similar to the volcano-like protrusion found in model systems which represents a fission stage. EF: exoplasmic fracture face. Magnification 140,000 × . This micrograph was provided by Prof. Dr. H. Plattner and permission for pubblication has been given by the authors.

In summary, fusion may proceed after taking away the limiting factors, which do not allow fusion to take place, or by local changes of the membranes, i.e. local accumulation of lipids or hydrophobic peptides, which allows adhesion and induces non-bilayer configuration lipids locally, resulting in membrane fusion.

REFERENCES

Ahkong QF, Fisher D, Tampion W, Lucy JA (1973) The fusion of erythrocytes by fatty acids, esters, retinol and α-tocopherol. Biochem J 136: 147-155.

Altstiel L, Branton D (1983) Fusion of coated vesicles with lysosomes: measurement with a fluorescence assay. Cell 32: 921-929.

Baker PF, Knight DE (1984) Calcium control of exocytosis in bovine adrenal medullary cells. Trends Neurol Sci 7: 120-127.

Batenburg AM, Rochat H, Verkleij AJ, de Kruijff B (1985) The penetration of a cardiotoxin into cardiolipin model membranes and its applications on lipid organization. Biochemistry 24: 7102-7110.

Bearer EL, Düzgünez N, Friend DS, Papahadjopoulos D (1982) Fusion of phospolipid vesicles arrested by quick freezing. The question of lipidic particles as intermediates in membrane fusion. Biochim Biophys Acta 693: 93-102.

Chandler DE, Hauser JE (1980) Arrest of membrane fusion events in mast cells by quick freezing. J Cell Biol 86: 666-674.

Cullis PR, Hope MJ (1978) Effects of fusogenic agents on membrane structure of erythrocyte ghosts and the mechanism of membrane fusion. Nature (Lond) 271: 672-674.

Cullis PR, de Kruijff B (1979) Lipid polymorphism and the functional roles of lipids in biological membranes. Biochim Biophys Acta 559: 399-420.

Cullis PR, de Kruijff B, Hope MJ, Verkleij AJ, Nayar R, Farren SB, Tilcock C, Madden TD, Bally MB (1982) Structural properties of lipids and their functional roles on biological membranes. In: Aloia RC (ed): Membrane Fluidity Vol 2, Academic Press, New York, pp 40-79.

Das S, Rand RP (1984) Diacylglycerol causes major structural transitions in phospholipid bilayer membranes. Biochem Biophys Res Commun 124: 491-496.

Deamer DW, Leonard R, Tardieu A, Branton D (1970) Lamellar and hexagonal lipid phases visualized by freeze etching. Biochim Biophys Acta 219: 47-60.

Gulik-Krzywicki T, Balerna M, Vincent JP, Luzdanski M (1981) Freeze-fracture study of cardiotoxin action on axonal membrane and axonal membrane lipid vesicles. Biochim Biophys Acta 643: 101-114.

Hui SW, Steward TP (1981) 'Lipidic particles' are intermembrane attachment sites. Nature (Lond) 290: 427-428.

Lau ALY, Chan SJ (1975) Alamethecin-mediated fusion of lecithin vesicles. Proc Natl Acad Sci USA 72: 2170-2174.

Lucy JA (1970) The fusion of biological membranes. Nature (Lond) 227: 814-817.

Lucy JA (1984) Do hydrophobic sequences cleaved from cellular polypeptides induce membrane fusion reactions in vivo? FEBS Lett 166: 223-231.

Luzzati V, Gulik-Krzywicki T, Tardieu A (1968) Polymorphism of lecithins. Nature (Lond) 218: 1031-1034.

Miller RG (1980) Do 'lipidic particles' represent intermembrane attachment sites? Nature (Lond) 287: 166-167.

Op den Kamp JAF (1979) Lipid asymmetry in membranes. Ann Rev Biochem 48: 47-71.

Papahadjopoulos D, Portis A, Pangborn W (1978) Calcium-induced lipid phase transitions and membrane fusion. Ann NY Acad Sci 308: 50-66.

216

Schmidt A, Patzak A, Lingg G, Winkler H, Plattner H (1983) Membrane events in adrenal chromaffin cells during exocytosis: a freeze-etching analysis after rapid cryofixation. Eur J Cell Biol 32: 31-37.

Schramm M. Oates J, Papahadjopoulos D, Loyter A (1982) Fusion and implantation in biological membranes. Trends Pharmacol Sci 3: 221-229.

Taraschi TF, van der Steen ATM, de Kruijff B, Tellier C, Verkleij AJ (1982) Lectin-receptor interactions in liposomes: evidence that binding of wheat-germ agglutinin to glycoprotein phosphatidylethanolamine vesicles induces non-bilayer structures. Biochemistry 21: 5756-5764.

Van Dijck, PWM, de Kruijff B, Aarts PAMM, Verkleij AJ, de Gier J (1978) Phase transitions in phospholipid model membranes of different curvature. Biochim Biophys Acta 506: 183-191.

Van Venetië R, Hage WJ, Bleumink JG, Verkleij AJ (1981) Propane jet-freezing: a valid ultra-rapid freezing method for preservation of temperature-dependent lipid phases. J Microsc 123: 287-292.

Verkleij AJ (1984) Lipidic intramembraneous particles. Biochim Biophys Acta 779: 43-63.

Verkleij AJ, Mombers C, Gerritsen WJ, Leunissen-Bijvelt J, Cullis PR (1979a) Fusion of phospholipid vesicles in association with the appearance of lipidic particles as visualized by freeze fracturing. Biochim Biophys Acta 555: 358-361.

Verkleij AJ, Monbers C, Leunissen-Bijvelt J, Ververgaert PHJ (1979b) Lipidic intramembraneous particles. Nature (Lond) 279: 162-163.

Verkleij AJ, Leunissen-Bijvelt J, de Kruijff B, Hope M, Cullis PR (1984) Non-bilayer structures in membrane fusion. In: Cell Fusion, Ciba Foundation Symposium 103, Pitman London, pp 45-59.

Wakelam MJO (1983) Inositol phospholipid metabolism and myoblast fusion. Biochem J 214: 77-82.

Wilschut J, Holsappel M, Jansen R (1982) Ca^{2+}-induced fusion of cardiolipin/phosphatidyl-choline vesicles monitored by mixing of aqueous contents. Biochim Biophys Acta 690: 297-301.

Phospholipid research and the nervous system
Biochemical and molecular pharmacology
L.A. Horrocks, L. Freysz, G. Toffano (eds)
Fidia Research Series, vol. 4.
Liviana Press, Padova. © 1986

SERINE PHOSPHOLIPIDS IN CELL COMMUNICATION

A. Bruni, L. Mietto[1], A. Battistella[1], E. Boarato[1], P. Palatini and G. Toffano[1]

Department of Pharmacology, Univerisity of Padova,
Largo Meneghetti 2, 35131 Padova, Italy;
[1]Department of Pharmacology, Fidia Research Laboratories,
Via Ponte della Fabbrica 3/A, 35031 Abano Terme, Italy

INTRODUCTION

Communication between distant cells depends on the release of signals in the intercellular space. These signals (first messengers) are collected by receptors located in the external cell membrane of a target cell a and then transduced in the cell interior through a perturbation of the second messenger system. Phospholipids, as components of plasma membrane, participate in this signalling system. Cycles of polyphosphoinositide degradation and resynthesis in response to the activation of membrane receptors generate the second messengers diacylglycerol and inositol trisphosphate (Berridge and Irvine, 1984). Since 1976, when the first pharmacological effects of phosphatidylserine (PdtSer) in vivo were described (Bruni et al., 1976), our efforts have aimed at exploring whether serine phospholipids are involved in a system of intercellular communication. As shown in Figure 1, the basic assumption of this hypothesis is that PtdSer is held in the inner side of plasma membrane in order to avoid the exposure of the serine headgroup to the external environment. Since the first observations of Verkleij et al. (1973) and Gordesky and Marinetti (1973), many studies have confirmed that this hidden position is the preferred distribution of PtdSer in the plasma membrane of eukaryotic cells. When a cell is broken, PtdSer is exposed and lysoPtdSer may be generated. These two phospholipids, either as components of a damaged membrane (Ptd Ser) or as free monomers in solution (lysoPtdSer) may reach responsive cells signalling that adjustment or repair is requested. To establish a first messenger effect of serine phospholipids at least three criteria must be fulfilled: (a) the generation of lysoPtdSer upon cell damage, (b) the existence of binding sites for serine phospholipids in sensitive cells and (c) the functional response following the interaction of these compounds with the target cell.

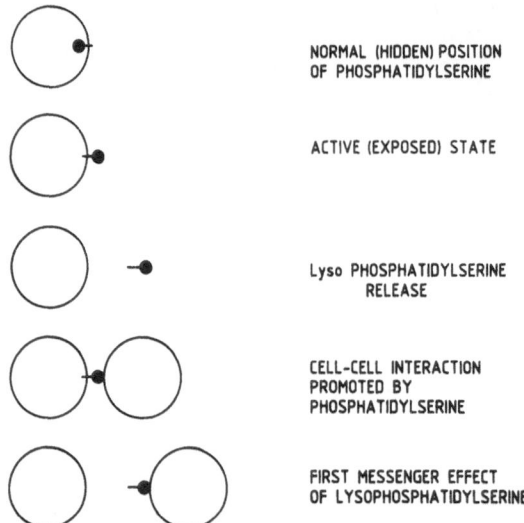

NORMAL (HIDDEN) POSITION
OF PHOSPHATIDYLSERINE

ACTIVE (EXPOSED) STATE

Lyso PHOSPHATIDYLSERINE
RELEASE

CELL-CELL INTERACTION
PROMOTED BY
PHOSPHATIDYLSERINE

FIRST MESSENGER EFFECT
OF LYSOPHOSPHATIDYLSERINE

Figure 1. Role of serine phospholipids in cell communication. Exposed phosphatidylserine may promote cell-cell interaction (Tanaka and Schroit, 1983). Alternatively, the soluble lysophosphatidylserine is released to transmit a signal to target cells (first messenger effect).

GENERATION OF LYSOPHOSPHATIDYLSERINE

LysoPtdSer is not found in the blood and tissues of mammalians. However, plasma from several animal species (Bruni and Toffano, 1985) and human inflammatory synovial fluids (Punzi et al., 1986) contain a phospholipase A_2 active on PtdSer. The studies of Billah et al. (1980) have shown that horse platelets do not produce lysoPtdSer upon stimulation but they do so after deoxycholate-induced damage. In our experiments we used a suspension of rat leukocytes. These are fragile, short-lived cells that migrate in a large number in cavities where an appropriate stimulus is applied. Furthermore, phospholipase A_2 has been found to be associated with the granules of these cells (Franson et al., 1974). In order to detect the release of serine phospholipids we took advantage of their peculiar activity in rat mast cells. Both PtdSer and lysoPtdSer are able to initiate histamine release provided the incubation medium is supplemented with minute amounts of nerve growth factor (Bruni et al., 1982). As shown in Table 1, the incubation of rat lenkocytes in a medium where mast cells were subsequently added, resulted in the release of a factor inducing histamine secretion in the presence of nerve growth factor. Erythrocytes were without effect. The activity of the leukocyte suspension was retained in the supernatant after centrifugation. The leukocyte-derived factor was stable when heated 20 min at 60°C and was extractable by chloroform methanol 2:1. Radiolabeled serine was included in the incubation medium to see whether lysoPtdSer was formed under these conditions. At the end of the incubation the leukocyte suspension was extracted with chloroform-methanol and the phospholipids separated by two dimensional thin layer chromatography. Five phospholipid species were labeled, including lysoPtd Ser. The distribution of serine among these phospholipids is shown in Table 2. As ex-

Table 1. *Leukocyte-induced histamine release*

Cell	% histamine release	
	Without β-NGF	With β-NGF
Mast cells alone	3.9±0.7	4.7±0.7
Mast cells plus 0.05 μM lysoPtdSer	4.9±1.1	27.9±2.9
Leukocyte plus mast cells	5.1±1.4	24.7±1.9
Erythrocytes plus mast cells	6.2±0.7	5.4±1.3

10^7 cells (rat erythrocytes, casein-elicited rat leukocytes) were incubated 120 min at 37°C in 0.9 ml of a buffered saline solution (Lagunoff, 1972) containing 1 mM calcium chloride, 10 mM glucose, 1 mg/ml serum albumin. After this incubation, 0.1 ml of the medium containing 5×10^5 mast cells (3-5 μg of histamine base) were added and the incubation was continued for 30 more min. The cells were centrifuged and the histamine content measured in the supernatant by a fluorimetric procedure. β-NGF denotes the addition of 10 ng of the β-subunit of mouse nerve growth factor together with the mast cells. Mean±S.E.M. of six experiments.

Table 2. *Incorporation of radiolabeled serine into leukocyte phospholipids*

Phospholipid	Per cent of the radioactivity found in the phospholipid fraction	Cell to medium ratio
Phosphatidylethanolamine	18.3 (15.2-20.3)	14.5 (7.1- 28)
Phosphatidylserine	61.3 (60.0-62.7)	44.8 (7.4-104)
Sphingomyelin	18.1 (17.1-19.4)	48.2 (6.6-133)
Lysophosphatidylethanolamine	0.5 (0.3- 0.8)	
Lysophosphatidylserine	1.8 (0.8- 3.1)	1.5 (0.6-2.7)

10^7/ml of casein-elicited rat leukocytes were incubated 3 h at 37°C in a buffered saline solution (Lagunoff, 1972) containing 1 mM calcium chloride, 10 mM glucose, 1 mg/ml serum albumin and U-[^{14}C] serine (0.3 μM, 110000 DPM/ml of medium). After 10 min centrifugation at 2500 RPM, the supernatant and the sediment were extracted separately. The data are given as per cent of the radio-activity recovered in the phospholipid fraction (4760 ± 360 DPM/10^7 cells). The cell to medium ratio denotes the ratio between the labeled phospholipid found in the pellet and in the supernatant. At the end of incubation the determination of released lactic dehydrogenase indicated 10-15% cell mortality. Mean values of 5 experiments. Ranges in parenthesis.

pected, the greatest fraction of radioactivity was recovered in PtdSer, followed by phosphatidylethanolamine and sphingomyelin. LysoPtdSer comprised approx. 2% of the radioactivity found in the phospholipid fraction (2.8% of the total pool of serine phospholipids). Since it was found that PtdSer comprised 8% of the total leukocyte phospholipids and that 10^7 leukocytes had 80 nmol phospholipids, it could be calculated that the lysoPtdSer production amounted to 0.18 nmol/10^7 cells. As shown by the ratio between the phospholipids in the cell and in the supernatant (Table 2), lysoPtdSer was the only phospholipid formed in a large proportion in the extracellular medium. From these data it is apparent that lysoPtdSer may reach outside the cells a sufficient concentration to activate the mast cells in the presence of nerve growth factor.

PHOSPHATIDYLSERINE-CELL INTERACTION

Studies along several lines have shown that certain cells have binding sites for PtdSer, eliciting a functional response upon their occupancy. The first indication came from experiments in which PtdSer liposomes were shown to interact with rat mast cells

220

and to enhance the antigen-induced histamine release (Goth et al., 1971). Subsequently, PtdSer-containing liposomes were tested on phagocytic cells. Both rodent macrophages (Schroit and Fidler, 1982) and human monocytes (Mehta et al., 1982) were able to bind and incorporate substantial amounts of these liposomes. More recently, mouse erythrocytes have been enriched in PtdSer in the external monolayer of the plasma membrane. The presence of PtdSer promoted the interaction of erythrocytes with autologous macrophages and their subsequent ingestion (Tanaka and Schroit, 1983). Our studies confirm and extend these observations. As shown in Figure 2 mouse macrophages and

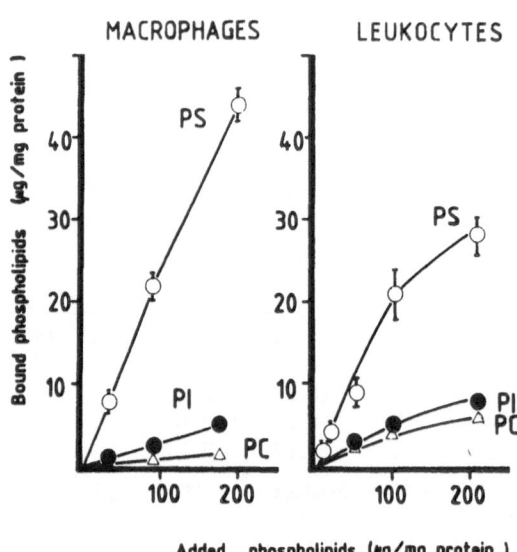

Figure 2. Phosphatidylserine uptake by phagocytic cells. Mouse macrophages, or rat leukocytes (300 μg protein) were incubated 1 h at 37°C with the indicated amount of radiolabeled phospholipid vesicles (20000 DPM) in 1 ml of a buffered saline solution (Lagunoff, 1972) containing 1 mM calcium chloride, 10 mM glucose, 1 mg/ml serum albumin. Macrophages were incubated as a monolayer adherent to 35 mm diameter tissue culture dishes, leukocytes as a cell suspension. After the incubation, the macrophage monolayer was washed 5 times. The leukocyte suspension was washed twice by centrifugation and finally passed through a layer of 20% serum albumin. PS, phosphatidylserine; PC, phosphatidylcholine; PI, phosphatidylinositol.

rat leukocytes bind cosiderable amounts of radiolabeled PtdSer vesicles. Phosphatidylcholine and phosphatidylinositol were bound only at a low extent. The binding of PtdSer to the macrophage monolayer was massive and showed no tendency to saturation. Time course experiments revealed that it was completed within 2 h at 37°C. At this time, more than 90% of bound radioactivity could be recovered as unchanged PtdSer after extraction of the macrophage-monolayer with chloroform methanol 2:1. Trypsin treatment of monolayers (1 mg/ml, 30 min) slightly inhibited the uptake of PtdSer which was instead strongly prevented (80%) at 0°C. Although these results indicated that PtdSer vesicles were selectively taken up by mouse macrophages, the quantitative assessment of the process was hampered by non-specific liposome binding or aggregation to the cell monolayer, especially when large PtdSer amounts were employed. More satisfactory results were obtained with rat leukocytes, likely due to the greater efficiency of the washing procedure which was devised. In these cells the PtdSer uptake displayed an apparent saturation starting at input levels of 100-200 μg PtdSer/mg protein. The uptake was rapid in the first hour of incubation then declined in the second hour, it was inhibited by low temperatures (62% at 0°C), by a pretreatment of leukocytes with trypsin (75% at 0.1 mg/ml for 30 min), by deoxyglucose and cytochalasin B (60% at 10 mM and 10 μg/ml, respectively). These results show that the rat leukocyte membrane bears binding sites for serine phospholipids and that their occupancy is followed by the ingestion of the PtdSer vesicles.

TARGET CELLS

A list of the preparations responsive to serine phospholipids is shown in Table 3. A very sensitive group of cells are the mast cells from rodents in which serine phospholipids, chiefly lysoPtdSer, induce histamine release. Rat serosal mast cells are

Table 3. *Target cells for serine phospholipids*

Cell	Ligand	Response	References
Rat mast cells	PtdSer, lysoPtdSer	histamine release	Boarato et al., 1984
Mouse mast cells	lysoPtdSer	histamine release	ibid.
Gerbil mast cells	lysoPtdSer	histamine release	ibid.
Hamster mast cells	lysoPtdSer	histamine release	Leung and Pearce, 1984
Rodent macrophages	PtdSer	endocytosis	Schroit and Fidler, 1982
Human monocytes	PtdSer	endocytosis	Mehta et al., 1982
Rat leukocytes	PtdSer	endocytosis	this paper
Rat pituitary cells	PtdSer	inhibition of PtdIns metab.	Bonetti et al., 1985

peculiar within this group since histamine release is not induced directly but through a permissive effect on other secretagogues (e.g. antigen, nerve growth factor). By contrast, in mouse, gerbil and hamster mast cells lysoPtdSer acts as a direct agonist. However, significant differences can be also found within these cells. For example, the effect of lysoPtdSer in the mouse mast cells is not dependent on the presence of calcium, whereas in gerbil mast cells extracellular calcium is needed (Boarato et al., 1984). To explain the activation of mast cells by serine phospholipids several possibilities have been considered. The phospholipids were at first regarded as calcium ionophores (Morgan and Svec, 1972), and later as activators of phosphatidylethanolamine methylation (Ishizaka et al., 1980). The existence of receptors for these compounds in the mast cell membrane has been also considered (Mietto et al., 1984). In rat mast cells, the hypothesis that serine phospholipids promote the mechanism of calcium influx is supported by the calcium-dependent synergism observed between lysoPtdSer and the phorbol ester, tetradecanoylphorbolacetate (Battistella et al., 1985). A second group of sensitive cells are the phagocytic cells, macrophages and leukocytes which are able to interact and internalize PtdSer vesicles. Since endocytosis depends on the "circumferential interaction mediated by multiple bridge formation between the surface of the phagocytes and the particles" (Goldman and Bar-Shavit, 1982), it is likely that these cells bear receptors for serine phospholipids. This view is consistent with the trypsin-induced inhibition of PtdSer endocytosis in rat leukocytes. Whether these receptors are specific for the serine head-group of these phospholipids remains to be established.

Recent evidence that PtdSer induces inhibition of phosphatidylinositol metabolism in rat pituitary cells opens this field to a new development (Bonetti et al., 1985). In this preparation the effect of PtdSer is similar to that induced by dopamine. Consistently, antagonism by anti-dopaminergic agents is observed. Further experiments are required to establish whether this effect is direct or mediated by an interaction with stimulatory compounds present in the preparation.

SUMMARY

Cell damage leads to the exposure of phosphatidylserine to the extracellular environment. Our work aims at exploring whether this drastic change in the membrane structure is used as a message, signalling that tissue injury has occurred and that a defense reaction is required. Two possibilities have been considered. The first is the generation of lysophosphatidylserine which is suitable to become a soluble messenger of the cell injury, the second is a direct role of phosphatidylserine, bound to the damaged cell. Experimental evidence shows that both mechanisms are possible. Rat leukocytes upon incubation in a buffered saline medium produce lysophosphatidylserine from damaged membranes. The lysoderivative may reach and activate added mast cells. On the other hand, model membranes made of phosphatidylserine vesicles interact with leukocytes and macrophages and are incorporated. On these bases a system of cell communication begins to emerge in which there are cells producing lysophosphatidylserine or exposing phosphatidylserine to release signals calling for defensive reactions. At the present stage of investigation the cells shown to be target for serine phospholipids are the rodent mast cells, the phagocytic cells and, possibly, the rat pituitary cells.

REFERENCES

Battistella A, Mietto L, Toffano G, Palatini P, Bigon E, Bruni A (1985) Synergism between lysophosphatidylserine and the phorbol ester tetradecanoylphorbolacetate in rat mast cell. Life Sci 36: 1581-1587.

Berridge MJ, Irvine RF (1984) Inositol trisphosphate, a novel second messenger in cellular signal transduction. Nature (London) 312: 315-321.

Billah MM, Lapetina EG, Cuatrecasas P (1980) Phospholipase A_2 and phospholipase C activities of platelets. Differential substrate specificity, calcium requirement, pH dependence and cellular localization. J Biol Chem 255: 10227-10231.

Boarato E, Mietto L, Toffano G, Bigon E, Bruni A (1984) Different responses of rodent mast cells to lysophosphatidylserine. Agents and Actions 14: 613-618.

Bonetti AC, Canonico PL, MacLeod RM (1985) Brain cortex phosphatidylserine inhibits phosphatidylinositol turnover in rat anterior pituitary glands. Proc Soc Exp Biol Med 180: 79-83.

Bruni A, Bigon A, Boarato E, Mietto L, Toffano G (1982) Interaction between nerve growth factor and lysophosphatidylserine on rat peritoneal mast cells. FEBS Letters 138: 190-192.

Bruni A, Toffano G (1985) Influence of serine phospholipids on biogenic amine secretion in vivo and in vitro. In: Horrocks LA, Kanfer JN, Porcellati G (eds): Phospholipids in the Nervous System. Vol. 2: Physiological Roles, Raven Press, New York; pp. 21-29.

Bruni A, Toffano G, Leon A, Boarato E (1976) Pharmacological effects of phosphatidylserine liposomes. Nature, London 260: 331-333.

Franson R, Patriarca P, Elsbach P (1974) Phospholipid metabolism by phagocytic cells. Phospholipase A$_2$ associated with rabbit polymorphonuclear leukocyte granules. J Lipid Res 15: 380-389.

Goldman R, Bar-Shavit Z (1982) Phagocytosis-Modes of particle recognition and stimulation by natural peptides. In: Karnovsky ML and Bolis L (eds): Phagocytosis, past and future. Academic Press, New York; pp. 259-281.

Gordesky ES, Marinetti GV (1973) The asymmetric arrangement of phospholipids in the human erythrocyte membrane. Biochem Biophys Res Comm 50: 1027-1031.

Goth A, Adams HR, Knoohuizen V (1971) Phosphatidylserine: selective enhancer of histamine release. Science 173: 1034-1035.

Ishizaka T, Hirata F, Ishizaka K, Axelrod J (1980) Stimulation of phospholipid methylation, Ca^{2+} influx and histamine release by bridging of IgE receptors on rat mast cells. Proc Natn Acad Sci USA 77: 1903-1906.

Lagunoff D (1972) The mechanism of histamine release from mast cells. Biochem Pharmacol 21: 1889-1896

Leung KBP, Pearce FL (1984) A comparison of histamine secretion from peritoneal mast cells of the rat and hamster. Br J Pharmacol 81: 693-701.

Mehta K, Lopez-Berestein G, Hersh EM, Juliano RL (1982) Uptake of liposomes and liposome-encapsulated muramyl dipeptide by human peripheral blood monocytes. J Reticuloend Soc 32: 155-164.

Mietto L, Battistella A, Toffano G, Bruni A (1984) Modulation of lysophosphatidylserine-dependent histamine-release. Agent and Actions 14: 376-378.

Morgan JL, Svec P (1972) The effect of phospholipids on anaphylactic histamine release. Br J Pharmacol 46: 741-752.

Punzi L, Todesco S, Toffano G, Catena R, Bigon E and Bruni A (1986) Phospholipids in inflammatory synovial effusion. Rheumatol Int 6: 7-11

Schroit AJ, Fidler IJ (1982) Effects of liposome structure and lipid composition on the activation of the tumoricidal properties of macrophages by liposomes containing muramyl dipeptide. Cancer Res 42: 161-167.

Tanaka Y, Schroit AJ (1983) Insertion of fluorescent phosphatidylserine into the plasma membrane of red blood cells. Recognition by autologous macrophages. J Biol Chem 258: 11335-11343.

Verkleji AJ, Zwaal RF, Roelofsen B, Confurius P, Kastellin D, Van Deenen LLM (1973) The asymmetric distribution of phospholipids in the human red cell membrane. A combined study using phospholipases and freeze-etch electron microscopy. Biochim Biophys Acta 323: 178-193.

Phospholipid research and the nervous system
Biochemical and molecular pharmacology
L.A. Horrocks, L. Freysz, G. Toffano (eds)
Fidia Research Series, vol. 4.
Liviana Press, Padova. © 1986

PHOSPHOLIPID METABOLISM IN AGING BRAIN

A. Gaiti, C. Gatti, M. Puliti, M. Brunetti

Istituto di Chimica Biologica, Perugia, Italy

INTRODUCTION

The problem of aging may be considered from different points of view. Based on the main tendencies in the theoretical approach to aging, such points can be summarized as follows: a) predetermination (genetical?) of the maximal extent of life; b) result of random negative influences upon cells (Hocman, 1979). However, leaving aside the theoretical approach to the problem, valuable insight may be gained by studying the variety of metabolic changes during aging even though in these types of studies it is difficult to determine the causes and effects. Nevertheless, it is known that some biological functions are strictly connected with the membrane and that each modification of them may influence cellular functions. On this basis, we have studied some aspects of the metabolism of phospholipids in the brain, as these compounds influence the membrane status. In particular, we have determined the rate of the main phospholipid biosynthesis and turnover in aged rat brain with the final target of trying to connect some alterations of physiological functions to the variations of this metabolism. The knowledge of these metabolic changes connected with aging may also be useful for a pharmacological approach to the problem.

PHOSPHOLIPID BIOSYNTHESIS

We tested the choline and ethanolamine phosphoglycerides (CPG and EPG) biosynthesis both in vitro and in vivo in the rat brain, its subfractions, cellular types and areas, by using different precursors (Brunetti et al., 1979; Gaiti et al., 1982a; 1982b). The results obtained (summarized in Fig. 1) indicate a general decline of the brain structure capacity to synthesize CPG and EPG. The studies indicate that the decrease is related to the *age of animals and is progressive up to the 18th month of age* (Brunetti et al., 1979;

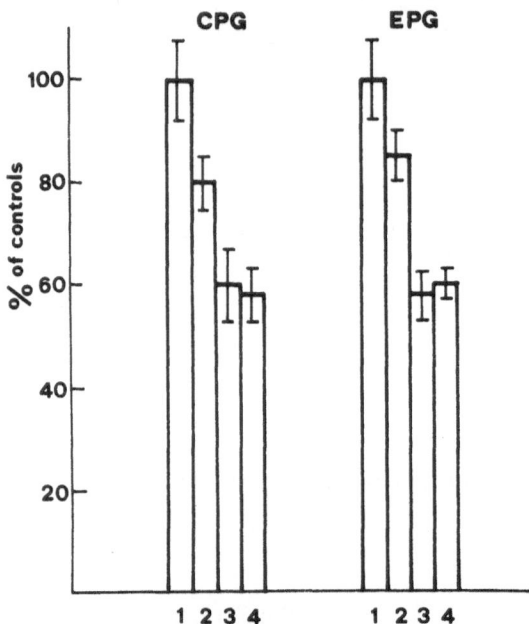

Figure 1. Decrease of in vitro phospholipid biosynthesis in rat brain microsomes (precursor CDP-base). Results are expressed as % of controls (100%). Age of rats: 1, 2, 3, 4 = 6, 12, 18, 24 months respectively (2).

Gaiti et al., 1982a). The effect of age is evident in neurons only and not in glial cells (Gaiti et al., 1982b). By using CDP-choline or CDP-ethanolamine as precursors of in vitro biosynthesis and crossing in a suitable way the microsomes from adult and aged rat brain (used as an enzyme source) in the presence of exogenous diglycerides with a different unsaturation degree, we were also able to suggest that the unsaturation degree of diglycerides may be one of the possible causes of biosynthesis decline (Brunetti et al., 1979).

There have also been reports of another way to synthesize phospholipid, which is called base-exchange reaction (Porcellati et al., 1971). This enzymatic system is able to change the hydrophilic head of phosphoglyceride molecules by exchanging, without energy requirement, its nitrogenous base with ethanolamine, choline, or serine (and others) present in the cells as free bases. The above metabolic pathway represents, up to now, the main way of synthesizing serine phosphoglycerides (SPG) in the nervous system, even though a different additional mechanism has been suggested (Pullarkat and Sbaschnig-Agler, 1981).

The in vitro tests in the presence of choline and ethanolamine free bases indicated that no variation of the base-exchange activity occurs in microsomes from whole brain with age (Brunetti et al., 1979). Recently, we also tested the base-exchange using serine as free base and microsomes from different brain areas of 4-month and 24-month old rats. The results are reported in Table 1.

Table 1. *Serine-exchange reaction in different brain areas from adult and aged rats*

	AD	AG	
Cortex	17.3 ± 1.5	19.3 ± 2.1	+ 11%
Striatum	8.2 ± 1.2	7.6 ± 0.6	—
Hippocampus	14.0 ± 1.0	15.7 ± 0.7	+ 8%

The results are expressed as nmol/mg protein/h and represent the average of 6 tests (2 experiments).
AD: 4-month old and AG: 24-month old rats.
Incubation conditions: tubes were incubated at 37°C for 20 min, at pH 8.1 (HEPES 40mM), in the presence
of 1.5mM L-3[^{3}H] Serine (SA 1.53 Ci/mol) and 2.5mM Ca^{++}.

As it is possible to see in Table 1, the base-exchange activity differs in accord to
the area examined and a slight increase is evident in the cortex and hippocampus from
aged brain. This increase is presently of little significance due to the small number of
tests (2 experiments in triplicate), but is significant (p < 0.05) by considering the two experiments separately.

The in vivo studies provided a good support to the above in vitro results. We
injected 3[^{3}H]serine into the lateral ventricle of the brain of 4-month and 24-month
old rats. At increasing time intervals from the injection, the animals were sacrificed
by decapitation and the brain quickly dissected in different areas. The phosphoglycerides
extracted from different areas were divided into P-base (phospholipase C), fatty acids
(as methyl esters) and glycerol, in order to determine the distribution of the isotopes
into different portions of the phospholipids.

Table 2 shows the specific activity (SA) of the P-serine which we found in the cortex of adult and aged rats at different time intervals. It is apparent that the rate of serine
incorporation into corresponding phospholipids of aged cortex is quite similar to that
of adult cortex. Thus, the base-exchange activity seems unaffected by age also in vivo,
as well as the decarboxylation of SPG into EPG (Butler and Morell, 1983), as shown
in Table 3.

Table 2. *In vivo incorporation of free serine into serine phosphoglycerides from cortex*

Time	25	60	120	360
Adult	3.8	4.0	3.9	6.5
Aged	5.6	3.0	4.5	4.5

The values are expressed as SA (nCi/nmol of P-serine) × 10^{-3}. Time is expressed in minutes. AD: 4-month
old and AG: 24-month old rats. For details, see the text.

Table 3. *Labeling of cortex ethanolamine phosphoglycerides after intraventricular injection of free serine*

Time	25	60	120	360
Adult	3.0	1.5	1.3	2.5
Aged	4.7	1.7	1.1	1.3

The values are expressed as SA (nCi/nmol of P-Etn) × 10^{-3}.
Time is expressed in minutes. For details, see the text.

PHOSPHOGLYCERIDES TURNOVER

We determined the turnover of EPG in six different cerebral areas of 4-month and 22-month old rats, by injecting a solution contàining [³H] glycerol together with [¹⁴C] ethanolamine into the brain lateral ventricle. The animals were sacrificed at increasing time intervals from the injection (up to 14 hrs). The areas examined behave quite differently in respect to their utilization of the precursors. In aged rats the utilization of the water-soluble precursors of EPG synthesis decreases in all of the six brain areas examined and the apparent half-lives of the two precursors are longer than in adults (in most areas from aged brain it was impossible to calculate the half-lives of glycerol and ethanolamine in the fast lipid pools). These last data suggest that aging may have a different effect on the catabolic activities as well as phospholipid biosynthesis (Gaiti et al., 1984).

Following these suggestions, we examined the turnover of two different fatty acids (palmitic and arachidonic acids) in five brain areas of adult and aged rats (Gatti et al., 1985). As can be seen from Figure 2, the utilization of both fatty acids, injected together

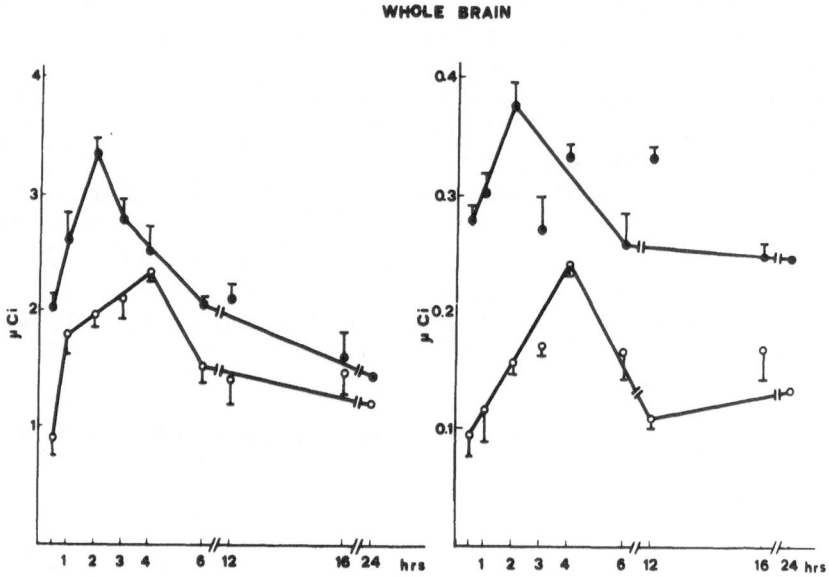

WHOLE BRAIN

Figure 2. Radioactivity content of whole brain after intraventricular injection of 2 μCi of 1[¹⁴C]arachidonic acid, SA 59 mCi/mmol together with 20 μCi of [9,10-(n)-³H] palmitic acid, SA 500mCi/mmol. Left: ³H; right: ¹⁴C.

into the lateral ventricle of the brain, is decreased by aging. However, the radioactivity content slopes draw together to a similar value at 24 hours after the injection or shortly later. This behavior is more evident in the SA slopes of total particulate lipids from the different areas. By considering that at the time interval adopted a relevant percentage of the ¹⁴C is still present in the molecule of the arachidonic acid (Gatti et al., 1985), we may suggest that the aged brain retains the unsaturated fatty acid longer than the adult one.

PHOSPHOLIPID DEACYLATION AND ACYLATION REACTIONS

As postulated by Webster and Alpern (1964), in brain tissue a cycle of deacylation-acylation of diacyl glycerophospholipids is active, analogous to that suggested by Lands (1960) for liver. Three enzymes are directly involved, namely phospholipase A, acyl-CoA synthetase and acyl-CoA: lysophosphatide acyl transferase. Two other enzymes, namely acyl-CoA hydrolase and lysophospholipase, could conceivably regulate the cycle by controlling the substrate concentration.

Age shows a different effect on the phospholipase A activities according to the substrate (PC or PE) utilized and area examined (Gaiti et al., 1985). In particular, the hydrolysis of both the fatty acids of PE is decreased in most areas from aged rats (about 50% of decrease), while the phospholipase A_2 activity against PC is stimulated in some areas, for example + 150% in striatum (Gaiti et al., 1985).

We also tested the enzymic activities connected with acylation process. The tests of activity by using free fatty acids, lysophospholipids and CoA-SH, ATP as substrates and cofactors, indicated that in aged brain areas no variation of the acylation takes place in the presence of palmitic acid (data not shown). On the contrary, an increase of the activity is evident in some areas by using arachidonic acid and 2-lyso-glycerophosphoethanolamine (lyso-PE) as substrate (Table 4).

Table 4. *Arachidonic acid incorporation into lyso phosphoglycerides in vitro*

Substrate	Lyso-PC			Lyso-PE		
	AD	AG	%	AD	AG	%
Cortex	0.15	0.13	—13	0.09	0.13	+ 44
Striatum	0.11	0.14	+ 27	0.10	0.12	+ 20
Hippocampus	0.12	0.12	—	0.11	0.13	+ 18
Rest	0.11	0.09	—18	0.06	0.06	—
Cerebellum	0.11	0.08	—27	0.08	0.07	—

Lyso-PC: 2-lyso-1-acyl glycerophosphocholine
Lyso-PE: 2-lyso-1-acyl glycerophosphoethanolamine
AD: adult, 4-month old rats; AG: aged, 24-month old rats.
Each value represents an average of 5 points with a deviation lesser than 10% and is expressed in nmol/mg protein/min.
Incubation conditions: tubes were incubated 20 min at 37°C at pH 7.4 (50 mM Tris-HCl), 20 mM $MgCl_2$ in the presence of ATP 5mM, CoA 1mM, DTT 0.1mM, BSA 5 mg/ml and [1-^{14}C] arachidonic acid 80 μM (SA 1.25 Ci/mol).

CONCLUSIONS

From the data which we have been collecting up to now, we may draw some conclusions as follows. During aging the brain loses progressively its ability to synthesize the main phospholipid by the ex novo pathway. This loss is due to different causes such as, for example, the reduced capacity to utilize the most simple precursors of the phosphoglyceride biosynthesis, the variation of diglycerides unsaturation, as well as other causes. This phenomenon, however, is not general because, as we were able to demonstrate, some areas (see hippocampus) are relatively unaffected by age as well as some cellular types (see glial cells). The reasons for that are actually unknown and, probably, are related to the particular functions for these areas or cells.

While the ex novo phospholipid biosynthesis is affected by aging, from our results it appears that the brain structures counteract this decline with the enzymatic mechanisms involved in the modification of only a portion of phospholipid molecule.

The base-exchange is either unaffected or slightly stimulated by age. This means that the cerebral structure retains the capacity to modify phospholipid hydrophilic head. Moreover, by considering the synthesis of SPG and its conversion, via decarboxylation, to EPG (which also seems unaffected by aging, see Table 3), we may hypothesize an important contribution of the base-exchange to phospholipid turnover in aged brain. Butler and Morrell (1983) calculated a minimum contribution of SPG to EPG over 7%, even though the authors say that this contribution is largely underestimated. Based on this observation we may suppose that this pathway may recycle in aged rat brain a significant pool of phospholipid molecules. This recycle of phospholipids starts from a base-exchange reaction between free serine and EPG or CPG molecules, decarboxylation of synthesized SPG to EPG, and, then, again base-exchange.

In addition, the deacylation-acylation cycle may provide an additional way to renew phospholipid molecules acting on their hydrophobic tail. The increased incorporation of arachidonic acid into lyso-PE (see Table 4) is significant in this regard.

Both mechanisms are energy-saving, the first because it does not require energy and the second because it needs the activation of only one fatty acid and, in aged brain, they may together ensure a quite sufficient turnover of phospholipid molecules to membrane structures, counteracting in part the strong decline of ex novo biosynthesis without waste of energy, which aged brain seems to be relatively lacking (Leong et al., 1981).

An additional fascinating hypothesis may also be suggested, even though we have not yet obtained conclusive evidence. By considering that, in spite of the high degree of EPG unsaturation, arachidonic acid is rapidly incorporated into position 2 of CPG (Gatti et al., 1985; Yau and Sun., 1974), which reaches one of the highest SA (together with inositol phosphoglycerides) briefly after the intraventricular injection of radioactive fatty acid (Gatti et al., 1985), we may suppose that in some areas of aged rat brain the increased phospholipase A_2 activity against PC may serve to ensure a good availability of unsaturated fatty acid pool. The unsaturated fatty acids, which become available in this way, could be incorporated directly into phospholipid molecule by deacylation-acylation cycle (see above).

ACKNOWLEDGMENTS

This work was supported in part by a grant from the Consiglio Nazionale delle Ricerche, Rome; contribution no. 83.02657.56/115.07444, Progetto Finalizzato "Medicina preventiva e riabilitativa", SP2.

REFERENCES

Brunetti M, Gaiti A, Porcellati G (1979) Synthesis of phosphatidylcholine and phosphatidylethanolamine at different ages in the rat brain in vitro. Lipids 14: 925-931.

Butler M and Morell P (1983) The role of phosphatidylserine decarboxylase in brain phospholipid metabolism. J Neurochem 41: 1445-1454.

Gaiti A, Brunetti M, Piccinin GL, Woelk H, and Porcellati G (1982a) The synthesis in vivo of choline and ethanolamine phosphoglycerides in different brain areas during aging. Lipids 17: 291-296.

Gaiti A, Sitkiewicz D, Brunetti M, Porcellati G (1981) Phospholipid metabolism in neuronal and glial cells during aging. Neurochem Res 6: 11-20.

Gaiti A, Brunetti M, Gatti C, Porcellati G (1984) Turnover of ethanolamine phosphoglycerides in different brain areas of adult and aged rats. Neurochem Res 9: 1549-1558.

Gaiti A, Gatti C, Brunetti M, Teolato S, Calderini G, Porcellati G (1985) Phospholipase activities in rat brain areas during aging. In: Horrocks LA, Kanfer JN, Porcellati G (eds): Phospholipids in the nervous system. Vol 2: Physiological roles. Raven Press, New York, pp. 155-162.

Gatti C, Noremberg K, Brunetti M, Teolato S, Calderini G, Gaiti A (1986) Turnover of palmitic and arachidonic acids in the phospholipids from different brain areas of adult and aged rats. Neurochem Res 11: 241-252.

Hocman G (1979) Biochemistry of aging. Int J Biochem 10: 867-876.

Lands WEM (1960) Metabolism of glycerolipids. II. The enzymatic acylation of lysolecithin. J Biol Chem 235: 2233-2237.

Leong SF, Lai JCK, Lim L, Clark JB (1981) Energy metabolising enzymes in brain regions of adult and aging rats. J Neurochem 37: 1548-1556.

Porcellati G, Arienti G, Pirotta M, Giorgini D (1971) Base-exchange reaction for the synthesis of phospholipids in nervous tissue: The incorporation of serine and ethanolamine into the phospholipids of isolated brain microsomes. J Neurochem 18: 1395-1417.

Pullarkat RK, Sbaschnig-Agler M (1981) Biosynthesis of phosphatidylserine in rat brain microsomes. Biochim Biophys Acta 663: 117-123.

Webster GR and Alpern RJ (1964) Studies on the acylation of lysolecithin by rat brain. Biochem J 90: 35-42.

Yau TM, Sun GY (1974) The metabolism of 1-^{14}C arachidonic acid in the neutral glycerides and phosphoglycerides of mouse brain. J Neurochem 23: 99-104.

Phospholipid research and the nervous system
Biochemical and molecular pharmacology
L.A. Horrocks, L. Freysz, G. Toffano (eds)
Fidia Research Series, vol. 4.
Liviana Press, Padova. © 1986

PHARMACOLOGICAL PROPERTIES OF PHOSPHATIDYLSERINE IN THE AGING BRAIN: BIOCHEMICAL ASPECTS AND THERAPEUTIC POTENTIAL

G. Calderini, F. Bellini, A.C. Bonetti, E. Galbiati, R. Rubini, A. Zanotti and G. Toffano

Fidia Neurobiological Research Laboratories,
Via Ponte della Fabbrica 3/A, 35031 Abano Terme, Padova, Italy

INTRODUCTION

Our interest in phospholipids as pharmacological agents for the central nervous system (CNS) arose many years ago when a phospholipid mixture extracted from bovine brain was shown to affect a number of cerebral biochemical and functional activities (Maniero et al., 1973; Toffano et al., 1974; Leon et al., 1975; Toffano et al., 1976). Soon after it was realized that, of the various phospholipids, only phosphatidylserine (PtdSer) was able to mimic the effects observed with the whole mixture (Bruni et al., 1976; Leon et al., 1978; Bigon et al., 1979; Toffano and Bruni, 1980). Our attention was then focused on the pharmacological properties of PtdSer and its effects at the molecular level. In young animals biochemical, behavioral and electroencephalographic observations indicated an effect on catecholaminergic and cholinergic systems (Toffano et al., 1978; Casamanti et al., 1979: Pepeu et al., 1980; Toffano and Bruni, 1980; Canonico et al., 1981; Toffano et al., 1981; Aporti et al., 1982). Because these neurotransmitter pathways are deeply involved in the aging-dependent deterioration of cerebral functions, it was important to assess the PtdSer pharmacological properties in the aging animals. A multiple approach was adopted including the assessment of parameters ranging from EEG and behavior to the molecular mechanisms involved in information processing.

PHARMACOLOGICAL EFFECTS OF PtdSer TREATMENT IN AGED ANIMALS: ELECTROPHYSIOLOGICAL AND BEHAVIORAL PARAMETERS

Among cerebral functions, memory seems to be extremely sensitive to the aging process, both in man and in experimental animals (Bartus et al., 1982). Male Sprague-

Dawley rats, tested on the passive avoidance response, exhibit a progressive loss in the retention of an aversive experience, such as the electrical foot shock, as a function of age (Calderini et al., 1985b; Aporti et al., 1985). In this strain of animals impairment of memory function seems to be linked to the appearance in the EEG recordings of bursts of spontaneous, asymptomatic spike-wave discharges. The abnormal EEG activity occurs in about 15% of 10-12 month and 90% of 20-24 month old rats (Calderini et al., 1985b; Aporti et al., 1985). In middle aged animals (15-19 months) the link between these two parameters is particularly evident because memory dysfunction is present only in animals with abnormal EEG discharges. Moreover the extent of memory deficit seems to correlate with the number of bursts present in the EEG recordings (Calderini et al., 1985a; Aporti et al., 1985).

In this subpopulation of animals prolonged treatment with PtdSer (15 mg/kg i.p.) consistently reduces the number and duration of abnormal bursts (Calderini et al., 1985a; 1985b; Aporti et al., 1985). Neither phosphatidylcholine (PtdCho) at equimolar dose nor the major phosphatidylserine constituents, namely serine and oleic acid, possess a similar pharmacological activity, thus suggesting a structure-activity relationship (Fig. 1). In the same group of animals PtdSer treatment leads also to a parallel improvement in memory function, as suggested by the increased retention observed in the passive avoidance test (Fig. 2). Interestingly no pharmacological effect could be observed in animals of the same age which do not reveal electrical abnormalities in the EEG recordings (Fig. 2). The behavioral effects observed after PtdSer treatment are in agreement with previous findings obtained in aged Wistar (Drago et al., 1981) and Fisher 344 (Corwin et al., 1985) rat strains.

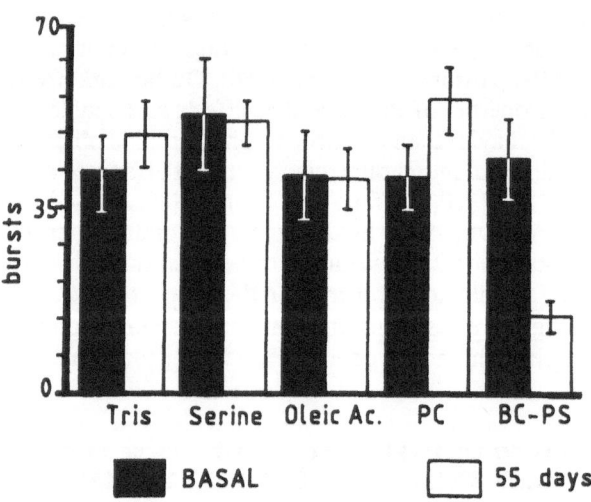

Figure 1. Number of bursts present in the EEG recordings after chronic treatment with PtdSer (BC-PS), PtdCho (PC), serine and oleic acid at equimolar dose.

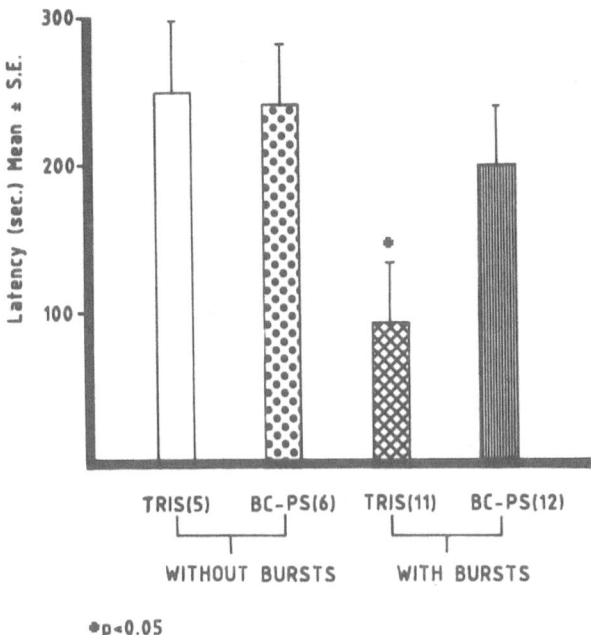

Figure 2. Effect of chronic PtdSer (BC-PS) treatment on passive avoidance performance of 17 month old rats with and without bursts of spike-wave discharges. Rats were treated with PtdSer 15 mg/kg i.p. for 60 days.

PtdSer TREATMENT MODIFIES AGING EFFECTS
ON SOME MECHANISMS INVOLVED IN INFORMATION PROCESSING

The overall activity of the brain depends on the efficiency of the systems special-ized in transmission and transduction of biological messages. The continuous interac-tion of the brain with the environment depends on the delicate balance between signal input and output. This heavy traffic of impulses relies on the rapid and reliable decodification of electrical and humoral signals. The decodification mechanism is located in cell plasma membranes and utilizes various devices (Triggle, 1980; Rasmussen and Barret, 1984; Takai et al., 1984). Among these, phosphatidylinositol (PtdIns) metabolism and protein kinase C activity have recently received considerable attention (Downes, 1983; Nishizuka, 1984).

PtdIns Metabolism

The involvement of PtdIns metabolism in the control of prolactin (PRL) secretion is indicated by the capacity of dopamine (DA) to inhibit both the incorporation of ^{32}P into PtdIns and spontaneous PRL secretion from isolated pituitary glands (Bonetti et al., 1985). In old animals the incorporation of ^{32}P into PtdIns is significantly reduced in spite of normal PRL plasma levels (Bonetti et al., 1985). An increased number of DA binding sites have been detected in pituitary glands of aged rats (Arita et al., 1984).

236

These modifications have been explained as compensatory mechanisms aiming to maintain the efficiency of the hypothalamus-hypophysis-prolactin system otherwise diminished by the age-dependent reduction of DA turnover in the hypothalamus (Meites et al., 1978; Gudelsky et al., 1981).

Prolonged PtdSer treatment (15 mg/kg i.p. for 30 days) decreases both the ^{32}P incorporation into PtdIns and the plasma prolactin levels in young and old animals (Fig. 3). The PtdSer effect on PtdIns metabolism is lower in aged than in young rats and is not equivalent to the effect on PRL plasma levels. This is presumably due to the low basal incorporation of ^{32}P into PtdIns and to the fact that the system can hardly be inhibited more than 50% with respect to the level of young rats. In parallel studies it was found that at variance from the case of DA, the PtdSer inhibition of PtdIns metabolism in vitro is not associated with a decrease of PRL secretion (Fig. 4). The

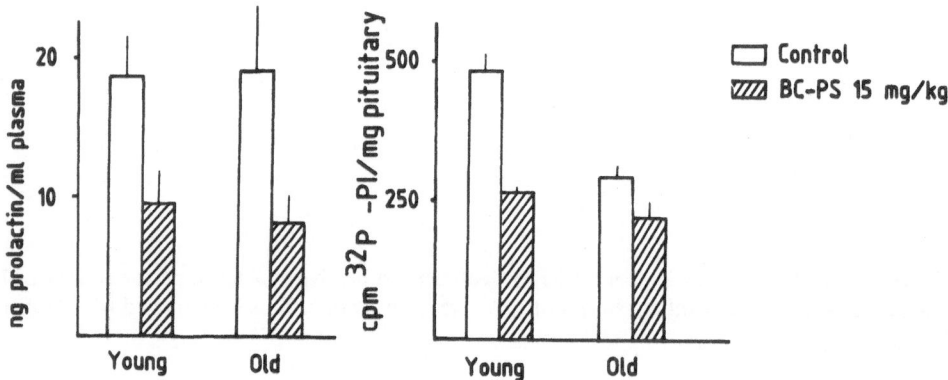

Figure 3. Effect of PtdSer (BC-PS) administration on plasma prolactin levels and ^{32}P incorporation into PtdIns (PI) of the pituitary in young and old male rats. Chronic administration (30 days)of PtdSer resulted in a significant ($p < 0.05$ Dunnet test) decrease of plasma PRL levels both in young (3 months) and old (25 months) rats. The treatment caused a significant ($p < 0.01$ Dunnet test) inhibition of ^{32}P incorporation into PtdIns in young rats only. Incorporation into other phospholipids was not affected.

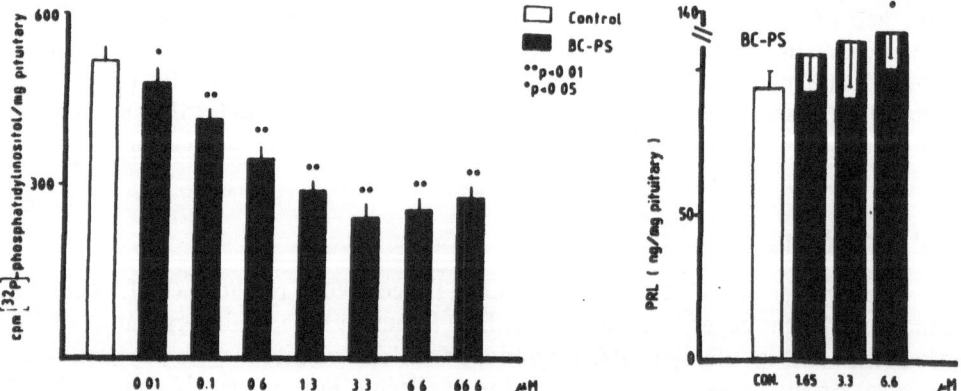

Figure 4. Effects in vitro of various PtdSer (BC-PS) concentrations on pituitary ^{32}P incorporation into PtdIns and PRL secretion.

disconnection between these two events suggests that PtdSer is not acting at the pituitary level through DA receptors and that other steps are presumably involved in the cascade of events leading to the cellular response. This is also supported by the fact that in vitro PtdSer potentiates the inhibitory effect of a submaximal dose of DA on ^{32}P incorporation into PtdIns (Fig. 5).

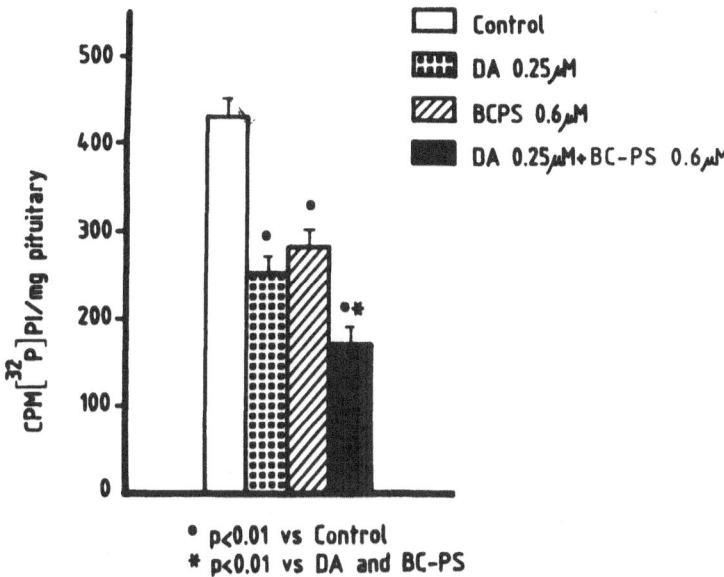

Figure 5. Synergism between PtdSer (BC-PS) and dopamine on ^{32}P incorporation into pituitary phosphatidylinositol (PI).

Protein Kinase C

Since the discovery of a specific phospholipid-sensitive, Ca^{++}-dependent protein kinase, much interest has developed on the role of this enzyme in the process of transduction of biological signals. Part of the interest arose from the observation that diacylglycerol enhances the reaction velocity of this protein kinase and increases its affinity for Ca^{++} down to the nanomolar range (Takai et al., 1979; Kishimoto et al., 1979). We have recently found that aging can also modify the activity and distribution of this enzyme within the cell. In particular, in the brain of aged rats the activity of protein kinase C in the soluble fraction is significantly reduced (Fig. 6). Because protein kinase C activity is not changed in the membrane bound fraction, an imbalance of the activity of this enzyme between the cellular compartments may occur with aging.

PtdSer counteracts the age-induced changes of protein kinase C activity and distribution (Fig. 7). This effect is probably linked to the specific requirement of protein kinase C for PtdSer in order to achieve complete activation.

238

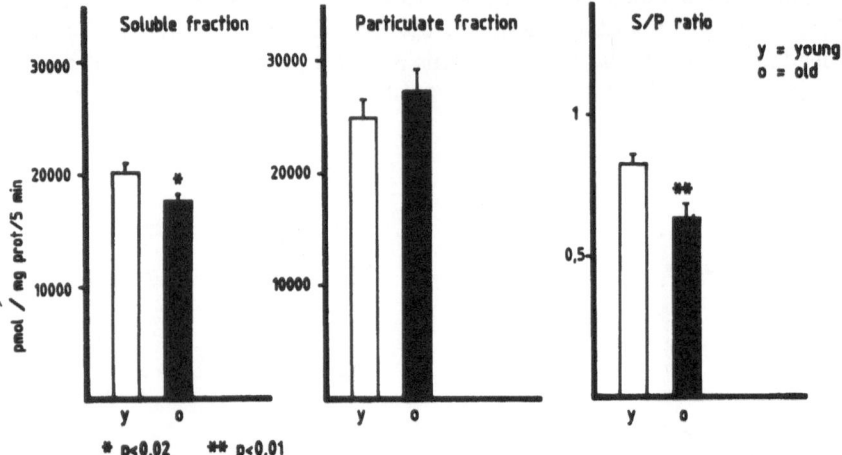

Figure 6. Age-dependent changes in protein kinase C activity in rat brain cortex. Data were obtained with different preparations, obtained from at least 10 animals. When the determinations were performed in preparations obtained from pooled cortices, the decrease in protein kinase C activity as a function of age was much more evident reaching sometimes 50% of the young control values.

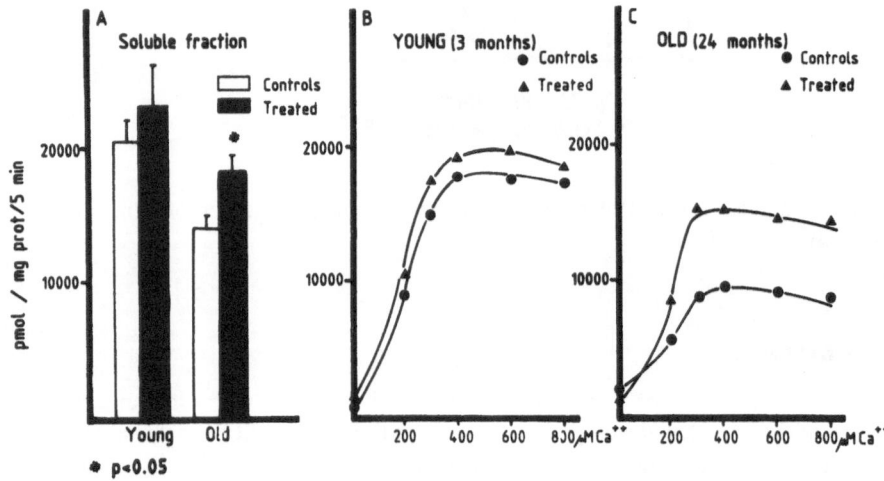

Figure 7. Effect of long-term treatment with PtdSer (BC-PS) (15 mg/kg/i.p. for 1 month) on protein kinase C activity in the cortices of young and old animals. A. Protein kinase C activity was assayed in standard conditions (500 μM Ca^{++}: 250μg/ml PtdSer) in preparations obtained from cortices of different animals.
B-C. Ca^{++} curves were assayed in presence of 250 μg/ml PtdSer in samples obtained pooling together the single animal preparations.

CONCLUSION

In agreement with previous results (Drago et al., 1980; Corwin et al., 1985; Aporti et al., 1985) we report here that PtdSer treatment can improve the memory function altered by the aging process. The increase of memory function is paralleled by a normalization of the electroencephalographic profile suggesting a close relationship between the two events. These results seem to us very promising for the use of this phospholipid as a new pharmacological tool for memory dysfunctions in old people. The biochemical results obtained at the molecular level clearly indicate that PtdSer can interfere with the mechanisms involved in signal decodification and transduction. Such effects can be responsible for the improvement observed in behavioral and electroencephalographic parameters. These results suggest a direct pharmacological effect of PtdSer at cerebral levels. This view is consistent with previous results on the cholesterol to phospholipids molar ratio and on the Na^+, K^+-ATPase activity in purified synaptic plasma membranes of old rats (Calderini et al., 1985a). However other effects should also be considered for a complete evaluation of the mechanism of action of this phospholipid. In fact phosphatidylserine can be converted in the blood stream into its lysoderivative, which is very active in releasing chemical mediators from peripheral cells (Bruni and Toffano, 1982). Recently lysophosphatidylserine has been shown to stimulate the incorporation of unsaturated fatty acids into PtdIns and PtdCho, thus playing an important role in the renewal of membrane phospholipid aliphatic chains (Sbaschnig-Agler and Pullarkat, 1985). Furthermore both PtdSer and lysoPtdSer are known to interact in vitro with nerve growth factor for mast-cell activation (Bruni et al., 1985).

All this evidence supports a beneficial effect of PtdSer treatment of aged people which, in our opinion, deserves careful clinical investigation.

REFERENCES

Aporti F, Nelsen JM, Goldstein L (1982) A quantitative EEG study at cortical and subcortical levels of the effects of brain cortical phosphatidylserine (BC-PS) in rats, and of its interaction with scopolamine. Res Commun Psychol Psychiat Behav 7: 131-143.

Aporti F, Borsato R, Calderini G, Rubini R, Toffano G, Zanotti A, Valzelli L, Goldstein L (1986) Age-dependent spontaneous EEG bursts in rats: effect of brain phosphatidylserine. Neurobiol Aging, 7(2): 115-120.

Bartus RT, Dean RL III, Beer B, Lippa AS (1982) The cholinergic hypothesis of geriatric memory dysfunction. Science 217: 408-416.

Bigon L, Boarato E, Bruni A, Leon A, Toffano G (1979) Pharmacological effects of phosphatidylserine liposomes: regulation of glycolysis and energy level in brain. Br J Pharmacol 66: 167-174.

Bonetti AC, Bellini F, Calderini G, Galbiati E, Toffano G (1985) Age-dependent changes in the molecular mechanisms controlling prolactin secretion in aged male rats: effect of a treatment with phosphatidylserine. Endocrinology, submitted.

Bruni A, Leon A, Boarato E (1976) Effect of polar lipids on cerebral content of free glucose in mice. In: Porcellati G, Amaducci L, Galli C (eds): Function and metabolism of phospholipids in the central and peripheral nervous systems. Plenum Press, New York, pp. 271-283.

Bruni A, Toffano G (1982) Lysophosphatidylserine, a short-lived intermediate with plasma membrane regulatory properties. Pharmacol Res Comm 14: 469-484.

Bruni A, Mietto L, Battistella A, Boarato E, Palatini P, Toffano G (1986) Serine phospholipids in cell communication, this volume.

240

Calderini G, Aporti F, Bellini F, Bonetti AC, Rubini R, Teolato S, Xu C, Zanotti A, Toffano G (1985a) Phospholipids as pharmacological tools in the aging brain. In: Horrocks LA, Kanfer JN, Porcellati G (eds): Phospholipids in the nervous system. Vol. 2: Physiological roles. Raven Press, New York, pp. 11-19.

Calderini G, Aporti F, Bellini F, Bonetti AC, Teolato S, Zanotti A, Toffano G (1985b) Pharmacological effect of phosphatidylserine on age-dependent memory dysfunction. In: Olton DS, Gamzu E, Corkin S (eds): Memory dysfunctions: an integration of animal and human research from preclinical and clinical perspectives. Ann New York Acad Sci 44: 504-506.

Canonico PL, Annunziato L, Toffano G, Bernardini R, Stanzani S, Foti M, Clementi G, Drago F, Scapagnini U (1981) In vivo and in vitro interference of phosphatidylserine liposomes on prolactin secretion in the rat. Neuroendocrinology 33: 358-362.

Casamenti F, Mantovani P, Amaducci L, Pepeu G (1979) Effect of phosphatidylserine on acetylcholine output from the cerebral cortex of the rat. J Neurochem 32: 529-533.

Corwin J, Dean RD III, Bartus RT, Rotrosen J, Watkins DL (1985) Behavioral effects of phosphatidylserine in the aged Fischer 344 rat: amelioration of passive avoidance deficits without changes in psychomotor task performance. Neurobiol Aging 6: 11-15.

Downes CP (1983) Inositol phospholipids and neurotransmitter-receptor signalling mechanisms. Trends Neuro Sci 6: 313-316.

Drago F, Canonico PL, Scapagnini U (1981) Behavioral effects of phosphatidylserine in aged rats. Neurobiol Aging 2: 209-213.

Gudelsky GA, Nansel DD, Porter JC (1981) Dopaminergic control of prolactin secretion in the aging male rat. Brain Res 204: 446-450.

Kishimoto A, Takai Y, Mori T, Kikkawa U, Nishizuka Y (1980) Activation of calcium and phospholipid-dependent protein kinase by diacylglycerol, its possible relation to phosphatidylinositol turnover. J Biol Chem 255: 2273-2276.

Leon A, Toffano G (1975) Possible role of BC-PL in enhancing [32]Pi incorporation into mice brain phospholipids. In: Porcellati G, Amaducci L, Galli C (eds): Function and metabolism of phospholipids in the central and peripheral nervous systems. Plenum Press, New York, pp. 307-313.

Leon A, Benvegnù D, Toffano G, Orlando P, Massari P (1978) Effect of brain cortex phospholipids on adenylate-cyclase activity of mouse brain. J Neurochem 30: 23-26.

Maniero G, Toffano G, Vecchia P, Orlando P (1973) Intervention of brain cortex phospholipids in pyridoxal phosphate-dependent reactions. J Neurochem 20: 1401-1409.

Meites G, Huang HH, Simpkins JW (1978) Recent studies on neuroendocrine control of reproductive senescence in rats. In: Schneider EL (ed): The aging reproductive system. Raven Press, New York, pp. 213-235.

Nishizuka Y (1984) Turnover of inositol phospholipids and signal transduction. Science 225: 1365-1370.

Pepeu G, Gori G, Bartolini L (1980) Pharmacologic and therapeutic perspectives on dementia: an experimental approach. In: Amaducci L, Davison AN, Antuono P (eds): Aging of the brain and dementia. Raven Press, New York, pp. 271-274.

Rasmussen H, Barrett PQ (1984) Calcium messenger system: an integrated view. Physiol Rev 64: 938-984.

Sbaschnig-Agler M, Pullarkat RK (1985) Lysophosphatidylserine dependent incorporation of acyl CoA into phospholipids in rat brain microsomes. Neurochem Int 7: 295-300.

Takai Y, Kishimoto A, Kikkawa U, Mori T, Nishizuka Y (1979) Unsaturated diacylglycerol as a possible messenger for the activation of calcium-activated, phospholipid-dependent protein kinase system. Biochem Biophys Res Comm 91: 1218-1224.

Takai Y, Kikkawa U, Kaibuchi K, Nishizuka Y (1984) Membrane phospholipid metabolism and signal transduction for protein phosphorylation. In: Greengard P, Robison GA (eds): Advances in Cyclic Nucleotide and Protein Phosphorylation Research. Raven Press, New York, vol. 19, pp. 119-158.

Toffano G, Gonzato P, Aporti F, Castellani A (1974) Variations of brain enzymatic activities in experimental atherosclerosis. Atherosclerosis 20: 427-436.

Toffano G, Leon A, Benvegnù D, Boarato E, Azzone GF (1976) Effect of brain cortex phospholipids on catecholamine content of mouse brain. Pharmacol Res Comm 8: 581-590.

Toffano G, Leon A, Mazzari S, Savoini G, Teolato S, Orlando P (1978) Modification of noradrenaline hypothalamic system in rat injected with phosphatidylserine liposomes. Life Sci 25: 1093-1102.

Toffano G, Bruni A (1980) Pharmacological properties of phospholipid liposomes. Pharmacol Res Comm 12: 829-845.

Toffano G, Battistella A, Aporti F, Orlando P (1981) Central pharmacological effects of phosphatidylserine liposomes. In: Bazan NG, Paoletti R, Iacono JM (eds): New trends in nutrition, lipid research and cardiovascular diseases. Alan R. Liss, New York, pp. 91-99.

Triggle DJ (1980) Receptor-hormone interrelationships. In: Bittar EE (ed): Membrane structure and function. Wiley-Interscience, New York, pp. 3-58.

Phospholipid research and the nervous system
Biochemical and molecular pharmacology
L.A. Horrocks, L. Freysz, G. Toffano (eds)
Fidia Research Series, vol. 4.
Liviana Press, Padova. © 1986

ESSENTIAL FATTY ACIDS AND PHOSPHOLIPIDS AS DRUGS IN THE TREATMENT OF ALCOHOLISM

John Rotrosen, David J. Segarnick, Adam Wolkin

Department of Psychiatry, New York University School
of Medicine and Psychiatry Service, New York Veterans
Administration Medical Center, 408 First Avenue,
New York, N Y 10010, USA

INTRODUCTION

This chapter will focus on the potential use of lipids and phospholipids as therapeutic agents in treating alcoholism, and on behavioral, biochemical and physiological interactions between alcohol, essential fatty acids, and prostaglandins. The rationale for these approaches is based on alcohol's effects on membrane fluidity, lipid metabolism, and prostaglandin synthesis.

When administrated in vitro, or given acutely in vivo, ethanol causes lipid membranes to become more fluid and disordered (Hill and Bangham, 1975; Littleton and John, 1977). This effect probably derives from ethanol's ability to become inserted into the lipid membrane matrix. It is likely that it is this biophysical effect on membrane fluidity that mediates many of ethanol's acute actions on enzymatic and receptor-mediated processes. Chronic in vivo exposure to ethanol induces an adaptive reordering of membrane lipid composition (involving increases in cholesterol: phospholipid ratios, and changes in phospholipid-esterified fatty acids) and membrane structure, as well as changes in the sensitivity of receptors for dopamine, norepinephrine, serotonin, and histamine; adenylate cyclase and ATPase activity are also changed in directions opposite to that seen with acute administration (Cicero, 1978; Ellingboe, 1978). This "adaptive" reordering becomes non-adaptive, or pathological, when alcohol is withdrawn; the associated decreases in membrane fluidity are thought to underly many components of the alcohol withdrawal syndrome.

Recently, several laboratories have shown that alcohol enhances synthesis of prostaglandins (PGs) and thromboxanes (TXBs) from their essential fatty acid (EFA) precursors both in vitro (Rotrosen et al., 1980; Manku et al., 1979; Segarnick et al., 1985a) and in vivo (Anton et al., 1983; George and Collins, 1985). In addition, chronic administration of alcohol inhibits the desaturation of cis-linoleic acid (cLA, 18:2) to gamma-linolenic acid (GLA, 18:3), which is in turn the precursor for dihomo-gamma-linolenic acid (DGLA, 20:3), and then arachidonic acid (AA, 20:4) (Nervi et al., 1980; Reitz, 1980). These effects on EFA metabolism mediate at least some of alcohol's behavioral, physiological, and pathological actions. Thus, certain acute effects of alcohol can be attenuated by pretreatment with prostaglandin synthesis inhibiting non-steroidal anti-inflammatory drugs (NSAIDs), and can be potentiated by pretreatment with PG precursor EFAs. Certain chronic effects of alcohol (which may be viewed as being a result of EFA/PG deficiency) are attenuated by administration of EFAs.

There are three major obstacles to be dealt with in conceptualizing and designing lipid therapies:

(1) the complexity of membrane structure and function,

(2) the complexity of lipid metabolism (e.g., EFAs get converted to phospholipids, and vice versa), and

(3) the vast pools of lipids already present in the body (i.e., can pharmacologic doses of lipids effectively alter lipid metabolism?).

THERAPEUTIC GOALS

Given these problems, is it reasonable to attempt to develop lipid drugs to treat the ubiquitous effects of alcohol, and if so, which of these effects? Data from studies with rodents and from a small number of studies in man suggest that indeed it is reasonable, and that a surprisingly broad spectrum of acute and chronic, behavioral, and physiological effects of alcohol may be susceptible to therapy. Targeted therapeutic goals, based on preliminary work which is reviewed below, might include the following:

(1) acute effects on CNS function such as acute intoxication, sedation, and even death from massive acute overdoses,

(2) physical pathology associated with chronic use such as the appearance of fatty liver,

(3) withdrawal symptomatology seen after discontinuation of chronic use,

(4) fetal alcohol syndrome — decreased fetal viability, craniofacial abnormalities, and mental retardation seen in infants born to alcoholic mothers,

(5) cognitive deterioration associated with chronic alcohol use, particularly the reversible cognitive impairment seen in alcohol withdrawal, and

(6) craving for alcohol associated with prolonged abstinence.

These goals should be viewed in the context of the following questions regarding the relationship of underlying pathology to lipid therapies:

(1) does alcohol alter lipid and PG metabolism, and do these changes underly the appearance of pathology?

(2) can lipids administered as drugs reverse or prevent these changes?

(3) are there strategies that are best suited for prevention and others for treatment of already existing pathology? and

(4) given the vast quantities of lipids comprising the total membrane pool, how do lipid drugs work?

BIOCHEMICAL INTERACTIONS

Ethanol enhances platelet synthesis of prostanoids from exogenous precursors (Manku et al., 1979; Rotrosen et al., 1980; Segarnick et al., 1985a), and increases levels of both PGs and TXBs in vivo in brain (Anton et al., 1983; George and Collins, 1985). Both of these effects occur at pharmacologically relevant doses or concentrations of ethanol, and are dose-related. In brain these effects are regionally specific. In addition precursor — and pool — dependent specificity is seen in experiments in vitro with human platelets. Thus, ethanol (100-400 mg%) enhances conversion of exogenous, non-esterified (free pool) DGLA to PGE_1, but has no effects on conversion of AA to PGE_1 (Table 1). In contrast, synthesis of prostanoids from EFAs esterified to phospholipids (membrane pool) is not affected by ethanol either under basal conditions or in the presence of thrombin (Segarnick et al., 1985a).

Table 1.

		Ethanol Concentration			
Substrates	*Product*	*0 mg%*	*100 mg%*	*250 mg%*	*500 mg%*
[^{14}C]AA	TXB	$0.66 \pm .02$	$0.69 \pm .02$	$0.78 \pm .02$	$0.81 \pm .05$
[^{14}C]AA	PGE_2	$1.01 \pm .14$	$1.02 \pm .04$	$1.04 \pm .11$	$0.89 \pm .02$
[^{14}C]DGLA	TXB	$0.10 \pm .01$	$0.11 \pm .01$	$0.11 \pm .02$	$0.12 \pm .02$
[^{14}C]DGLA	PGE_1	$1.45 \pm .04$	$1.69 \pm .05$	$1.65 \pm .05$	$2.08 \pm .21$

Washed resuspended platelets (not pulse-labelled) were incubated for 15 minutes in the presence of ethanol (0-500 mg%); [^{14}C]-AA 50 μmol or [^{14}C]-DGLA 50 μmol were added for the last 5 minutes of the incubation. Prostanoids formed were separated by TLC. Data are expressed as nmol of labelled prostanoid formed/mg protein (mean $\pm$ S.E.M. of triplicates). Similar results were obtained in experiments using platelets from other donors, and confirmed using HPLC or argentation TLC separations. Significant effects of ethanol were seen for TXB formation from [^{14}C]-AA ($F[3,11] = 5.22$, $p = .033$), and for PGE_1 formation from [^{14}C]-DGLA ($F[3,11] = 5.30$, $p = .026$) by one way ANOVA.

These latter findings are consistent with Crawford's (1983) speculations that two discrete lipid pools exist; one, a stable membrane pool protected from phospholipase attack except under circumstances of cell injury or activation (e.g., thrombin), and the second, a metabolically active pool that is directly and readily influenced by dietary EFAs and other lipids. The existence of this latter pool might explain how, in the presence of vast quantities of endogenous lipids, small doses of exogenous lipids can exert potent behavioral and physiological effects. Such an understanding is crucial to proposing lipid therapies.

BEHAVIORAL INTERACTIONS

Pretreatment with PGE_1 and PG precursor EFAs potentiate the acute sedative effects of ethanol. Conversely, aspirin and other nonsteroidal anti-inflammatory drugs attenuate ethanol induced sedation. These effects are alcohol-specific, i.e., sedation induced by other agents such as chloral hydrate or pentobarbital is not influenced by either EFAs or NSAIDs. However, with other alcohols (e.g., t-butanol) both EFA and NSAID • interactions do occur (Table 2).

Table 2. *Effects of pretreatments on ethanol, chloral hydrate, pentobarbital and t-butanol induced sleep time in minutes (mean ± SEM)*

Pretreatment	Treatment	Sleep Time
saline	ethanol 3.6 g/kg	30.9 ± 0.8
PGE$_1$ 4 mg/kg	ethanol 3.6 g/kg	59.2 ± 3.5
GLA 100 μl	ethanol 3.6 g/kg	44.3 ± 1.2
Olive Oil 350 μl	ethanol 3.6 g/kg	33.2 ± 0.9
Aspirin 15 mg/kg	ethanol 3.6 g/kg	20.3 ± 0.7
Indomethacin 5 mg/kg	ethanol 3.6 g/kg	16.7 ± 0.9
Ibuprofen 15 mg/kg	ethanol 3.6 g/kg	21.0 ± 0.6
saline	chloral hydrate 300 mg/kg	36.7 ± 1.5
EPO 350 μl	chloral hydrate 300 mg/kg	35.2 ± 1.2
Aspirin	chloral hydrate 300 mg/kg	35.3 ± 1.3
saline	pentobarbital 40 mg/kg	32.7 ± 1.0
GLA 100 μl	pentobarbital 40 mg/kg	31.0 ± 0.8
Aspirin 15 mg/kg	pentobarbital 40 mg/kg	31.0 ± 0.9
saline	t-butanol 1.6 g/kg	35.5 ± 1.4
Aspirin	t-butanol 1.6 g/kg	22.3 ± 1.2
GLA 200 μl	t-butanol 1.6 g/kg	45.3 ± 1.1

GLA = gamma linolenic acid; EPO = evening primrose oil

There is also evidence in support of the hypothesis that chronic ethanol administration produces a functional EFA/PG deficiency that may at least in part underly the signs and symptoms of ethanol withdrawal. In mice, PGE$_1$, PGE$_2$, and GLA all effectively reduce alcohol withdrawal intensity by approximately 50% (Table 3) (Rotrosen et al., 1980; Sergarnick et al., 1985b). NSAIDs partially attenuate the beneficial effects of precursors, but not those of PGE$_1$ itself. NSAIDs do not affect the intensity of withdrawal when administered alone. Similar findings have now been reported in rats chronically fed an alcohol containing liquid diet (Karpe et al., 1983).

Additional behavioral and physiological EFA/NSAID/ethanol interactions have been reported for open field activity (Ritz et al., 1981), hypothermia (George et al., 1981), and mortality due to very high dose ethanol (George et al., 1982).

In preliminary studies we have seen that phosphatidylserine attenuates alcohol withdrawal and decreases the percentage of mice eliciting audiogenic seizures (Table 4). PtdSer is rapidly cleaved to form lysoPtdSer plus a free fatty acid. Because PtdSer is highly enriched in EFAs it is possible that exogenous PtdSer may provide a metabolically active source of PG precursor EFAs. Alternatively, lyso-PtdSer itself has been proposed by Bruni and Toffano (1982) to be a short-lived intermediate with potent membrane regulatory properties.

EFFECTS OF EFAs ON THE DEVELOPMENT OF HEPATIC PATHOLOGY

Several laboratories have now reported that EFAs (both GLA and AA) as well as PGE$_1$ administered chronically with ethanol prevent or attenuate the development of fatty liver assessed both by classic histological criteria and by the presence of in-

Table 3. *Effects of PGs, EFAs and NSAIDs on withdrawal intensity at 10 hours after discontinuation of ethanol vapor*

Pretreatment	Dose	Treatment	Dose	n	Withdrawal Score
—	—	saline	—	10	6.0 ± 0.3
—	—	PGE_1	4 mg/kg	10	2.5 ± 0.5
—	—	PGE_2	4 mg/kg	10	4.1 ± 0.2
—	—	$PGF_{1\alpha}$	4 mg/kg	10	5.5 ± 0.3
—	—	$PGF_{2\alpha}$	4 mg/kg	10	6.1 ± 0.3
Saline	—	saline	—	10	5.0 ± 0.2
Saline	—	GLA	100 μl	10	2.4 ± 0.4
Aspirin	15 mg/kg	GLA	100 μl	10	3.3 ± 0.2
Indomethacin	5 mg/kg	GLA	100 μl	10	3.9 ± 0.3
Ibuprofen	15 mg/kg	GLA	10 μl	10	4.2 ± 0.2
—	—	Aspirin	15 mg/kg	10	5.7 ± 0.3
—	—	Indo	5 mg/kg	10	4.9 ± 0.4
—	—	Olive Oil	350 μl	10	5.2 ± 0.3

Table 4. *Effects of phosphatidylserine (SO nq/kg) on ethanol withdrawal intensity at 10 hours after discontinuation of ethanol vapor. Withdrawal was assessed using a rating scale different from that shown in Table 3*

Condition	n	Withdrawal Score	Seizures
Saline	10	27.5 ± 1.9	4
PtdSer	10	12.9 ± 1.1	0

creased hepatic triglyceride levels (Wilson et al., 1973; Goheen et al., 1983; Segarnick et al., 1985c). In addition, Cunnane et al. (personal communication) have shown that EFAs prevent the development of fatty liver that is seen after administration of carbon tetrachloride.

FETAL ALCOHOL SYNDROME

Fetal alcohol syndrome occurs in infants born to alcoholic mothers, and is characterized by low birth weight, craniofacial abnormalities, and mental retardation. Alcohol administered to pregnant mice on or around gestation day 10 causes an increase in prenatal mortality and in birth defects in surviving pups; decreased fetal weight is also seen. Randall and Anton (submitted) have shown that NSAIDs given before alcohol reduce prenatal mortality and reduce the number of fetal malformations.

CLINICAL TRIALS

Glen (1984) and we (Segarnick et al., in press; Wolkin et al., in preparation) have conducted placebo-controlled clinical therapeutic trials with prostanoid precursor EFAs in both acute alcohol withdrawal and in prolonged abstinence. These studies have been designed to assess the effects of GLA 300-400 mg/day (administered as evening primrose oil) on withdrawal symptoms, alcohol craving, liver function, cognitive performance (attention, short-term memory, and abstraction), and fatty acid profiles in erythrocytes, plasma phospholipids and plasma triglycerides.

The data reported by Glen suggest that EFA supplementation reduces craving for alcohol during prolonged withdrawal, and hastens both the recovery of liver function (as measured by serum liver enzymes) and the recovery of impaired cognitive function.

Our data, based on analysis of 27 alcoholics admitted to the study in early withdrawal, suggest the following. Alcoholics in early withdrawal had lower levels of 18:2 and 18:3 than did normal non-alcoholic controls, and had higher levels of 20:4. During recovery, both 18:2 and 18:3 increased toward normal control levels. The changes in fatty acids were more pronounced in lipid pools with faster turnover rates (plasma triglycerides and phospholipids) than in the more stable erythrocyte membrane phospholipids.

An unexpected but very strong correlation was seen between aberrant liver enzyme values and decreased levels of 18:2 and 18:3, as well as increased levels of 20:3, 20:4, and 22:4 in all patients at baseline (early withdrawal, before EFA supplementation was begun). These findings are consistent with our findings in rodents and with a hypothesis that high levels of long chain unsaturated fatty acids may protect against alcohol-induced liver damage. Correlation data for liver enzymes and plasma phospholipid fatty acids are shown in Table 5.

Clinical symptoms of alcohol withdrawal, measures of cognitive function, and enzyme measures of liver function showed consistent improvement during treatment. However, on only a small number of assessments were differences observed between the EFA and placebo supplemented groups.

Table 5. *Correlation of liver enzymic activities and plasma phospholipid fatty acids*

| | Plasma Phospholipids | |
	18:2	22:4
Glutamic oxalacetic transaminase	0.54 (p = .01)	—0.54 (p = .01)
Glutamic pyruvic transaminase	0.56 (p = .02)	—0.48 (p = .05)
Alkaline phosphatase	0.47 (p = .03)	—0.47 (p = .04)

Values are Pearson correlation coefficients. Associated two-tailed levels of significance are shown in parentheses (n = 18 to 21).

SUMMARY AND CONCLUSIONS

EFAs, EFA-containing phospholipids, and their prostanoid products appear to play significant roles as mediators or modulators of many effects of ethanol. These include acute and chronic effects, and abstinence effects, and extend to a number of different organ systems. There is substantial evidence now to support the existence of separate structural and metabolic pools of both EFAs and phospholipids which differentially contribute to these interactions. These differential roles should be taken into consideration in the design and assessment of lipid drugs.

REFERENCES

Anton RF, Randall CL, Wallis CJ (1983) Effect of acute ethanol exposure on in vivo brain cortical PGE levels in C3H mice. Alcoholism 7: 104.

Bruni A and Toffano G (1982) Lysophosphatidylserine, a short-lived intermediate with plasma membrane regulatory properties. Pharmacol Res Commun 14: 469-482.

Cicero T (1978) Tolerance to and physical dependence on alcohol: behavioral and neurobiochemical mechanisms. In: Lipton MA, DiMascio A and Killam KF (eds): Psychopharmacology: A Generation of Progress. Raven Press, New York, pp. 1603-1615.

Crawford MA (1983) Background to essential fatty acids and their prostanoid derivatives, Br J Pharmacol 58: 9-16.

Ellingboe J (1978) Effects of alcohol on neurochemical processes. In: Lipton MA, DiMascio A and Killam KF (eds): Psychopharmacology: A Generation of Progress. Raven Press, New York, pp. 1653-1664.

George FR and Collins AC (1985) Ethanol's behavioral effects may be partly due to increases in brain prostaglandin production. Alcoholism, Clin Exp Res 9: 143-146.

George FR, Jackson SJ, Collins AC (1981) Prostaglandin synthesis inhibitors antagonize hypothermia induced by sedative hypnotics. Psychopharmacology 74: 241-244.

George FR, Elemer GI, Collins AC (1982) Indomethacin significantly reduces mortality due to acute ethanol exposure. Substance and Alcohol Actions/Misuse 3: 267-274.

Glen I (1984) Possible pharmacologic approaches to the prevention and treatment of alcohol related CNS impairment: results of a double blind trial of essential fatty acids. In: Edwards G and Littleton J (eds): Pharmacological Treatments for Alcoholism. Crooms-Helm, London, pp. 331-345.

Goheen SC, Larkin EC, Ananda Rao G (1983) Severe fatty liver in rats fed a fat-free ethanol diet and its prevention by small amounts of dietary arachidonate. Lipids 18: 285-289.

Hill M and Bangham AD (1975) General depressant drug therapy: a biophysical aspect. Adv Exp Med Biol 59: 1-10.

Karpe F, Neri A, Änggård E (1983) The effect of dietary primrose oil on ethanol withdrawal in the rat. Acta Pharmacol Toxicol 53: 18P.

Littleton JM and John GR (1977) Synaptosomal membrane lipids of mice during continuous exposure to ethanol. J Pharm Pharmacol 29: 579-580.

Manku MS, Oka M, Horrobin DF (1979) Differential regulation of the formation of prostaglandins and related substances from arachidonic acid and dihomogammalinolenic acid. I. Effects of ethanol. Prostaglandins Med 3: 119-128.

Nervi AM, Peluffo RD, Brenner RR, Leikin AI (1980) Effect of ethanol administration on fatty acid desaturation. Lipids 15: 263-268.

Randall CL and Anton RF: Aspirin reduces alcohol-induced prenatal mortality and malformations in mice, submitted for publication.

Reitz RC (1980) The effects of ethanol administration on lipid metabolism. Lipids 15: 263-268.

Ritz MC, George FR, Collins AC (1981) Indomethacin antagonizes ethanol-induced but not pentobarbitol-induced behavioral activation. Substance and Alcohol Actions/Misuse 2: 289-299.

Rotrosen J, Mandio D, Segarnick D, Traficante LJ, Gershon S (1980) Ethanol and prostaglandin E_1: biochemical and behavioral interactions. Life Sciences 26: 1867-1876.

Segarnick DJ, Rjer H, Rotrosen J (1985a) Precursor- and pool - dependent differential effects of ethanol on human platelet prostanoid synthesis. Biochem Pharmacol 34: 1343-1346.

Segarnick DJ, Mandio Cordasco D, Rotrosen J (1985b) Prostanoid modulation (mediation?) of certain behavioral effects of ethanol. Pharmacol Biochem Behav, 23:71-75.

Segarnick DJ, Mandio Cordasco D, Agura V, Cooper NS, Rotrosen J (1985c) Gamma linolenic acid inhibits the development of the ethanol-induced fatty liver. Prostaglandins Leukotrienes Med 17: 277-282.

Segarnick DJ, Wolkin A, Rotrosen J (1986) Biochemical, behavioral and clinical interactions between essential fatty acids, prostaglandins and ethanol. Advances in Lipid Research, in press.

Wilson DE, Engel J and Wong R (1973) Prostaglandin E_1 prevents alcohol-induced fatty liver. Clinical Research 21:829.

Phospholipid research and the nervous system
Biochemical and molecular pharmacology
L.A. Horrocks, L. Freysz, G. Toffano (eds)
Fidia Research Series, vol. 4.
Liviana Press, Padova. © 1986

AXONALLY TRANSPORTED PHOSPHOLIPIDS AND NEURITE REGROWTH

Mario Alberghina

Institute of Biochemistry, Faculty of Medicine,
University of Catania, Catania, Italy

The growing body of knowledge concerning the axonal transport of neuronal intracellular components is very impressive today even if restricted to the multidisciplinary achievements obtained in investigating the phospholipid transport alone. Despite the large number of investigators who have sought to determine, at a molecular level, the dynamics of phospholipid transport and a great variety of substances and organelles in both directions, we do not actually have a clear idea of the mechanism underlying this process. However, recent work from many laboratories, addressed towards the understanding of how normal cells function in delivering substances to the periphery, is promising in this regard. Interestingly, the recognition of the features of intracellular transport in axotomized neurons capable of regeneration paves the way to a tighter linkage of innovative research between basic neuroscience and clinical concern.

CHARACTERISTICS OF FAST INTRA-AXONAL TRANSPORT OF PHOSPHOLIPIDS

Fast axonal transport of phospholipids is a well established phenomenon in a variety of neuronal pathways, both in PNS and CNS. Included are the rabbit, rat, chick and goldfish optic systems (Grafstein et al., 1975; Toews et al., 1979; Haley et al., 1979a,b; Dziegielewska et al., 1980; Alberghina et al., 1981a; Lee et al., 1982), the rabbit hypoglossal nerve (Miani, 1963; Alberghina et al., 1983a), the rat, mouse and bullfrog sciatic nerve (Abe et al., 1973; Tang et al., 1974; Longo and Hammerschlag, 1980; Gould et al., 1982; Alberghina et al., 1983b,c; Guy and Bisby, 1983), the chicken oculomotor nerve (Droz et al., 1978), as well as the rat nigrostriatal tract in the central nervous system (Toews et al., 1980; Hitzemann and Loh, 1982).

Our understanding of the active bidirectional movement of all the phospholipid classes relies upon experiments in which hydrophilic and hydrophobic radioactive lipid precursors, injected into the vicinity of various neuronal types employed as experimental models, are incorporated into the perikaryal phospholipids. These compounds in their turn are rapidly found from the soma outward along the elongated axonal structures and neuronal constituents of nerve fibres (Grafstein et al., 1975; Currie et al., 1978;

Haley et al., 1979b; Toews et al., 1979; Alberghina et al., 1981a, 1982b,c). Experiments carried out with various transport models indicate that axonally transported phospholipids can exchange with stationary or other moving structures of axon (i.e. smooth endoplasmic reticulum, axolemma, mitochondria, precursors of synaptic vesicles). This might also be the case for plasmalogens (Miani, 1963) and cholesterol (Rostas et al., 1975). Phosphatidylcholine seems committed to transport sooner than phosphatidylethanolamine, and differences among phospholipids in the kinetics of accumulation along the nerves have been observed (Toews et al., 1983). In addition, the glycerophosphoryl moiety of individual phospholipids seems transported as an intact group because, after intravitreal injection of [^{3}H] glycerol and [^{32}P] phosphate, the ^{3}H/^{32}P ratio does not change along the optic pathway (Toews et al., 1979).

Inositol phospholipids constitute an intriguing case in studying the axonal transport of phospholipids. They are axonally transported in the form of PtdIns, PtdIns4P, and PtdIns(4,5)P in the rabbit optic nerve and tract (Alberghina et al., 1981b; 1982b) and can be synthesized in the axons from diffusibile and co-transported free myo-inositol (Kumara-Siri and Gould, 1980; Alberghina et al., 1981b; Gould et al., 1983). While fitting into the general patterns of axonal transport, these compounds do not accumulate in the nerve ending regions. [^{14}C] Dipalmitoylglycerophosphocholine (a very minor lipid molecule seen in the nervous tissue) or metabolically inert analogs, intravitreally injected into the rabbit optic system, did not undergo any modification while moving towards the synaptic terminals (Lee et al., 1982).

Based on these findings, a major function of fast-transported phospholipids may well be to maintain the ongoing process of renewal of membrane-associated components. The observed analogy with the fast-transport of structural proteins (membrane-bound enzymes, synaptic components, glycoproteins etc.) has suggested that the material rapidly moving along the axons is in the form of preassembled membranes rather than individual molecules (Abe et al., 1973; Grafstein et al., 1975; Longo and Hammerschlag, 1980; Grafstein and Forman, 1980; Guy and Bisby, 1983).

In addition, phospholipid molecules, in downward transit along the axons, are translocated into the periaxonal myelin in the nerve (Haley and Ledeen, 1979a; Droz et al., 1981; Alberghina et al., 1982a; Gould et al., 1982; Ledeen and Haley, 1983; Brunetti et al., 1983). Although the translocation appears to be quantitatively a minor event as compared with the large amount of extra-axonal myelin lipids synthesized by glia themselves (Droz et al., 1981; Alberghina et al., 1982a), it seems worthwhile to gain further insight into the mechanism and the possible functional significance of the interaction between axonal process of the neurons and glia, referred to phospholipids.

STRATEGY FOR EXAMINING AXON-MYELIN TRANSFER OF PHOSPHOLIPIDS

We have examined the possibility that the 1,2-diradyl groups of the molecular species of major phospholipids are modified during their transport and insertion into specific compartments of the rabbit optic system.

In order to label phospholipids synthesized in the retinal ganglion cells, we injected intravitreally [2-^{14}C]ethanolamine or [l-^{14}C]arachidonic acid, or a mixture of

[³H]arachidonic acid and [¹⁴C]hexadecanol; then we measured the labeling of [¹⁴C]-ethanolamine-containing phospholipids or the [¹⁴C]-labeled phospholipids, and the ³H/¹⁴C ratios in PtdCho, PtdEtn and their subclasses as well as in other phospholipids (PtdIns, PtdSer) of the retina and subcellular fractions of the optic nerve and tract (Alberghina et al., 1985). One and 8 days after labeling with [2-¹⁴C]ethanolamine, the ¹⁴C radioactivity was chiefly localized in both diacyl and alkenylacyl types of the PtdEtn isolated from the myelin fraction and from axons, whereas the increase of labeling observed in the axons by 8 days was higher than that measurable in the myelin fraction. This suggests that any axonally transported diradyl-GPE type was selectively transferred to myelin. Further to this, higher values of the ³H/¹⁴C ratio measured in position 2 for both diacyl- and alkenylacyl-GPE isolated from retina as well as myelin and axonal fraction of the optic nerve and tract, at day 1 and 8, were observed with exception of alkenylacyl types of retinal PtdEtn. This suggested a preferential participation of [³H]arachidonic acid (or elongated-desaturated analogs) in the acylation of position 2 of the glycerol backbone of PtdEtn committed to transport along the optic pathway and indicated that [1-¹⁴C]hexadecanol was incorporated into 1- and 2-positions of the PtdCho and PtdEtn aliphatic chains, either as O-alkyl and O-alkenyl moieties or as an O-acyl group after oxidation to palmitoyl-CoA.

The ³H/¹⁴C ratios, obtained in double-labeling experiments for total PtdCho and PtdEtn isolated from retina, myelin fraction, and axons of optic nerve and tract, are reported in Figure 1. In the axons these ratios were substantially unchanged comparing the values for the optic tract, at 1 and 8 days after precursor injection. This was also true for fractionated diradyl species of PtdEtn (data not shown). The results could be consistent with a metabolic stability of axonally transported phospholipids in the proximo-distal axonal segments of the optic pathway. But, at both time intervals (1 and 8 days) of labeling, the ratios for the same phospholipid fractions isolated from myelin in both optic structures were substantially different (higher for PtdEtn) in comparison with those found in the axons. From these differences it could be inferred that at least a portion of the diradyl types of the major phospholipids were remodelled by an acyl-transfer mechanism (Lands, 1960; Webster and Alpern, 1964; Baker and Thompson, 1972). In fact, turnover phenomena of labeled phospholipids in the axons can really be ruled out because the calculated apparent half-lives of newly-synthesized phospholipid acyl and alkenyl groups in the rat optic tract are many weeks (Toews and Morell, 1981). It appears also unlikely that local de novo synthesis may significantly contribute to the formation of lipids, transported and exchanged in the axons (Currie et al., 1978; Droz et al., 1979; Gould et al., 1983). Moreover, an axonal localization of acyltransferase enzymes has been found in the trigeminal nerve of neonatal rat during myelination (Benes et al., 1973). Taken together, the results shown in this study strengthen the view that the presence of independent acyl-exchange reactions in the nerve fibers is likely.

The deacylation-reacylation sequence of phospholipid turnover plays an essential role in biochemical events associated with a variety of cell membranes (Bell and Coleman, 1980). Such a view, however, must account for resident enzymes that act on the cytoplasmic side of the axonal membrane. Retailoring in fact must involve local hydrolysis, i.e. a phospholipase A_1 and A_2, plasmalogenase and lysophospholipase activity, and an acyltransferase system, whose presence in the optic nerve and tract is not

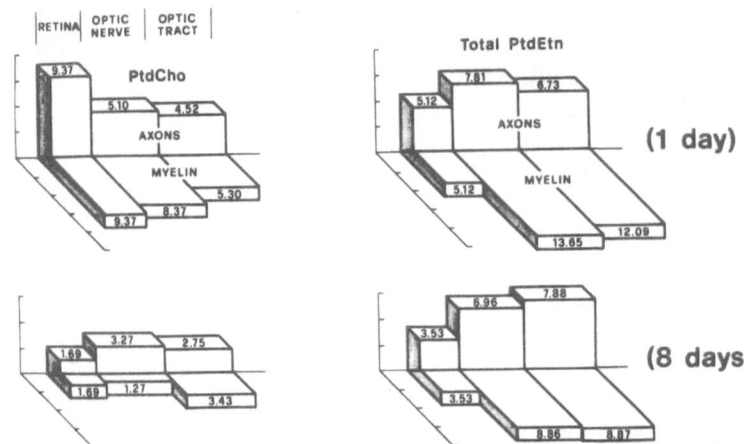

Figure 1. ^{3}H/^{14}C ratios of labeled total PtdCho and PtdEtn in the various optic structures. Rabbits were injected intravitreally with a mixture of 40 μCi of [^{3}H]arachidonic acid and 10 μCi of [l-^{14}C]hexadecanol, and killed 1 and 8 days later. Axons and myelin fractions were separated as previously described (Alberghina et al., 1985). Each value reported in the blocks represents the average from two experiments in which individual values were within 15% of the mean.

proven as yet. In this connection it is of interest that myelin itself contains very little phospholipase A$_1$ and A$_2$ activities (Woelk and Porcellati, 1973), while neither plasmalogenase (Ansell and Spanner, 1968) nor acyltransferase activity (Benes et al., 1973) could be detected.

With respect to other suggestions capable of explaining the phospholipid remodelling process, another possibility is that the relative enzyme systems sustaining the process of acyl group transfer may be operative in some compartments surrounding the axon (Fig. 2). Paranodal regions of the glial cells as well as adaxonal structures of the internodal regions of the nerve fibres, rather than myelin, appear to be in fact alternative loci of retailoring of the specific pool of phospholipids derived from the axon. Autoradiographic techniques could be helpful in checking this hypothesis.

Whichever is the case, the non-identity of axonal and myelin phospholipid composition and, perhaps, fatty acid profiles (Sun and Sun, 1972; DeVries and Norton, 1974; DeVries and Zmachinski, 1980) strongly substantiate the belief that the neuronal phospholipids channelled towards the myelin sheath must be remodelled before they are inserted into the multilamellar periaxonal membrane.

PRESENCE OF PHOSPHOLIPASE AND ACYLTRANSFERASE ACTIVITIES IN CNS AND PNS AXOLEMMA-ENRICHED FRACTIONS

We examined nerve tissues from both CNS and PNS in order to demonstrate the occurrence of phospholipase and lysophospholipid acyltransferase activities in axonal compartments. Selected marker enzymes measured throughout various subcellular fractions isolated following the procedure of DeVries et al., (1983) and Yoshino et al., (1983)

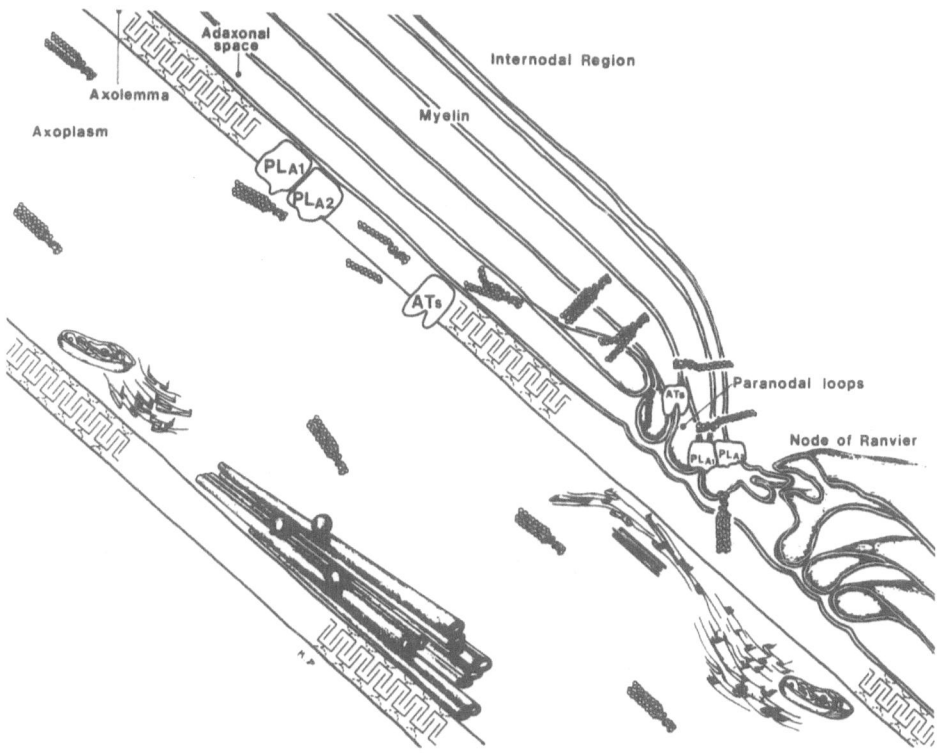

Figure 2. Schematic representation of two possible processes of axon-myelin transfer of intra-axonally conveyed phospholipids. In the section of an idealized nerve fibre, phospholipids are actually depicted as magnified, space-filling molecular models travelling along the axoplasm and axonal structures. A phospholipid molecule is seen to be interacting with hydrolytic enzymatic proteins (PLA1 and PLA2) in the inner surface of the axolemma (upper left) and then deacylated to form a lyso-compound. The subsequent insertion of the aliphatic chain into lysophospholipid by acyltransferases (ATs) might take place in the same membrane. The remodelled lipid molecule is then embedded in the myelin sheath. The same mechanism of deacylation-reacylation might occur in the paranodal loops which run in perinodal region of the fibre (lower right). Phospholipid molecules pass first through the axolemmatic membrane and then enter the glial processes whereby they are retailored.

displayed the highest detectable specific activitiy of AChE, 5'-nucleotidase, Na^+K^+-ATPase in the axolemma (fraction of sucrose concentration from 25 to 30% in the gradient). Concomitantly, all the fractions enriched in axolemma possessed a marked specific activity of phospholipase A_1 and A_2 (Tables 1 and 2) which were in the range of the activity of whole homogenate. The occurrence of a very active 2-acylGro$P$Cho acyltransferase (which has not been tested before in the nervous system) as well as l-acylGro$P$Cho acyltransferase, with palmitoyl-CoA and linoleoyl-CoA as acyl group donor respectively, was also found associated with all membrane fractions; their distribution, however, was clearly different from that of the phospholipases (Tables 1 and 2). The highest specific activity was that of the axolemmatic fractions, and its

Table 1. *Phospholipase and Acyl-CoA: lyso-PtdCho acyltransferase activities measured in tissue fractions prepared from brain stem and optic nerve and tract of the rabbit*

Tissue	Fraction	Phospholipase*		Acyl-CoA:lyso-PtdCho acyltransferase**	
		A_1	A_2	1-acyl-*sn*-GroPCho	2-acyl-*sn*-GroPCho
	Homogenate	46.6	0.073	1.94	0.35
	Myelinated axons	60.2	0.103	—	—
Brain stem	Myelin fraction (15-24% sucrose)	22.1	0.063	1.58	0.31
	Axolemma (25-30% sucrose)	37.7	0.087	2.92	1.00
	Homogenate	42.2	0.020	2.27	0.47
Optic nerve and tract	Myelin fraction (15-24% sucrose)	46.0	0.015	1.32	0.45
	Axolemma (25-30% sucrose)	34.2	0.058	3.74	0.91

The values are the mean of 2-3 experiments. S.E.M. was less than 20%. GroPCho is glycero-3-phosphocholine.
(*) The enzyme activities (expressed as nmol/h/mg protein) were determined in the NaCl extracts of acetone-dried powders. Phospholipase A_1 was assayed with 1-[9,10(n)-³H]palmitoyl-2-acyl-GroPCho as substrate; phospholipase A_2 was assayed with 1-palmitoyl-2-[1-¹⁴C]oleoyl-GroPCho as substrate.
(**) 1-acyl-*sn*-GroPCho acyltransferase activity was assayed with [1-¹⁴C]linoleoyl-CoA as substrate; 2-acyl-*sn*-GroPCho acyltransferase activity was assayed with [1-¹⁴C]palmitoyl-CoA as substrate; both activities are expressed as nmol/min/mg protein. — not determined.

distribution quite closely followed that of AChE, prompting us to conclude that the enzymes were predominantly located in the axonal plasma membranes in the four nervous tissues examined. Besides, a possibility exists that these acylation enzymes may be coupled to phospholipases, allowing rapid and efficient pathway of phospholipid transformation. The results obtained support the hypothesis (first in the scheme of Figure 2) that in the axolemma the presence of deacylation-reacylation enzyme reactions might play a role in the selective change of phospholipid molecules belonging to axonal plasma membrane or crossing over there during their axonal transport and associated transfer into other extra-axonal compartments.

RAPIDLY TRANSPORTED PHOSPHOLIPIDS ASSOCIATED WITH AXONAL GROWTH DURING NERVE REGENERATION

Thus far some aspects of the axonal transport of phospholipids in neuronal pathways during normal conditions have been considered. Concerning the ability of neurons to recover after injury, some aspects of the remyelination process in the nerves during axonal regeneration will be analyzed.

Table 2. *Phospholipase and acyl-CoA:lyso-PtdCho acyltransferase activities measured in tissue fraction prepared from rabbit cranio-spinal nerves*

Tissue	Fraction	Phospholipase*		Acyl-CoA:lyso-PtdCho acyltransferase**	
		A_1	A_2	1-acyl-*sn*-GroPCho	2-acyl-*sn*-GroPCho
Hypoglossal nerve	Homogenate	—	—	1.50	0.56
	Myelin fraction 15-24% sucrose)	—	—	0.87	0.56
	Axolemma (25-30% sucrose)	—	—	2.02	1.38
Sciatic nerve	Homogenate	24.8	0.040	1.85	0.63
	Myelinated axons	28.7	0.024	—	—
	Myelin fraction (15-24% sucrose)	9.9	0.029	1.20	0.22
	Axolemma (25-30% sucrose)	18.2	0.044	3.15	1.34

The values are the mean of 2-3 experiments. S.E.M. was less than 20%. GroPCho is glycero-3-phosphocholine.
(*) The enzyme activities (expressed as nmol/h/mg protein) were determined in the NaCl extracts of acetone-dried powders. Phospholipase A_1 was assayed with 1-[9,10(n)-³H]palmitoyl-2-acyl-GroPCho as substrate; phospholipase A_2 was assayed with 1-palmitoyl-2-[1-¹⁴C]oleoyl-GroPCho as substrate.
(**) 1-acyl-*sn*-GroPCho acyltransferase activity was assayed with [1-¹⁴C]linoleoyl-CoA as substrate; 2-acyl-*sn*-GroPCho acyltransferase activity was assayed with [1-¹⁴C]palmitoyl-CoA as substrate; both activities are expressed as nmol/h/mg protein. — not determined.

Late Events during Axonal Regrowth

Experiments on the process of remyelination of regrown axons in the rat sciatic nerve after crushing have been carried out in our laboratory (Alberghina et al., 1983d). As regeneration went on, a progressive increase of the phospholipid, cerebroside, and sulfatide contents of nerve segments distal to the injury approached normal values. The activity of 2', 3'-cyclic nucleotide 3'-phosphodiesterase (CNPase) of myelin fraction purified from distal segment of regenerating sciatic nerve showed a significant increase in the 30-120 day regenerating period. By sequential double labeling, a marked increase of the incorporation of [2-³H]glycerol, injected intraperitoneally, and [Me-¹⁴C]choline, injected intraspinally, into myelin lipids of distal segment of regenerating nerve, compared to normal one, was found. The content of alkenyl-acyl, alkyl-acyl and diacyl types of myelin PtdCho, PtdEtn and PtdSer was also determined in myelin fractions separated from sciatic nerve segments of rats at 12, 25, and 45 days after birth, and of adult rats (6-month-old) 90 days after crush injury (Alberghina et al., 1984). The biosynthesis and metabolic heterogeneity of lipid classes and types were studied by incubation with [1-¹⁴C]acetate of nerve segments of young rats at different ages as well as crushed and sham-operated control nerve segments of adult rats. The analysis of composition and positional distribution in major individual molecular species extracted from light myelin and myelin-related fraction suggest that the metabolism of alkenyl-acyl-GroPEtn and unsaturated species of PtdCho and PtdSer may not be regulated in the same manner during peripheral nerve myelination of developing rat and remyelination of regenerating nerve in the adult animal. The ^{14}C-radioactivity incorporation into lipid classes and alkyl and acyl moieties of the three major phospholipids of sciatic nerve segments during the developmental period investigated revealed that Schwann cells were capable of synthesizing acyl-linked fatty acids in both myelin fractions at a decreasing rate and with different patterns during development. In regenerating sciatic nerve of adult animals the labeling of myelin lipid classes and types of remyelinating nerve segment distal to the crush site was markedly higher than that of sham-operated normal one; however, the magnitude and the pattern of the specific radioactivity never approached those observed during active myelination of the nerve in young animals.

These findings suggest that, in the distal segment of injured nerve, the lipid content as well as the myelin synthesis increase as the regeneration proceeds. Although there are some points of similarity, myelin formation seems not to recapitulate nerve myelin ensheathment occurring during postnatal development.

Neuronal and Glial Response to Axotomy

Despite the obvious importance of lipogenesis for the axon repair, relatively poor attention has been paid to biochemical events associated with lipid biosynthesis and transport in regenerating neuronal tissue.

During regeneration it can be expected that neurons are engaged in active synthesis of axoplasm and axoplasmic constituents. Variation of the metabolic responses that occur in nerve cell bodies after axonal injury may be considered as a key point for understanding the general process of nerve regeneration because the synthesis of macromolecules and phospholipids is largely restricted to the soma. In the rat superior cervical ganglion 3-7 days after cutting the postganglionic nerves, increases of total lipid content as well as enzyme activities of the pentose phosphate pathway, in vivo incorporation of [U-^{14}C]glycerol into lipids of ganglion explants, and steady-state concentrations of metabolites utilized for the lipid biosynthesis, are among the earliest events to occur during the retrograde reaction to axonal injury (Harkonen and Kauffman, 1974; Jerkins and Kauffman, 1983). Unfortunately, these studies essentially dealt with the early degenerative response of ganglia to injury rather than with a regeneration process. Contrary to crush injury, nerve transection close to the ganglion generally is considered a drastic denervation procedure that is much more unlikely to lead to successful axonal regeneration in short periods. Furthermore metabolic changes due to the massive glial reaction in the ganglion (Matthews and Nelson, 1975) were not at all considered previously as a contributing factor in producing the effect.

Adjustment of Lipid Metabolism during Neuron-glia Reaction

The rabbit ventral horn motoneurons and hypoglossal nucleus provide a very useful model for the study of the perikaryal changes associated with axonal injury and nerve regeneration. These tissues contain a relatively dense population of large nerve cell bodies, are well delimited with respect to surrounding nuclei, do not show signs of retrograde degenerative reaction after nerve crush (Sjöstrand, 1966), and can be easily sampled as a whole tissue; furthermore, they are convenient for axotomy of corresponding cranio-spinal nerves.

Ultrastructural and metabolic responses of motoneurons in the hypoglossal nucleus and spinal cord to neurotomy have already been studied mainly focusing, among other changes, on protein and RNA synthesis, enzyme activities associated with neurotransmitter functions as well as energy metabolism (Miani et al., 1961; Sjöstrand, 1966; Lieberman, 1971; Sunner and Sutherland, 1973; Grafstein and McQuarrie, 1978; Wooten et al., 1978; Aldskogius et al., 1980; Cova et al., 1981; Reisert et al., 1984; Smith et al., 1984).

By using this model (Miani et al., 1961; Miani 1962; see Fig. 3), in recent experiments we started the exploration of the motoneuronal metabolic response to axotomy, with specific reference to the alteration of the phospholipid degradation and biosynthesis, at the various time intervals after unilateral crush of the right hypoglossal nerve and cervical-thoracic (C.6, C.7, C.8, Th.1) nerves, at level of brachial plexus of the rabbit.

The measurement of PLA2 activity may be a useful marker of changes in the metabolic status of neurons following alterations in their functional activity, at any stage of regeneration. The specific activities of AChE, phospholipase A_2, and acyl-CoA: 2-acyl-*sn*-glycero-3-phosphocholine acyltransferase, measured in the hypoglossal nucleus as well as in tissue punched out from the ventral horn of spinal cord segments, at various time intervals after nerve crush, are reported in Table 3. AChE showed an early decrease in the hypoglossal nucleus and a later decrease in the spinal cord tissue

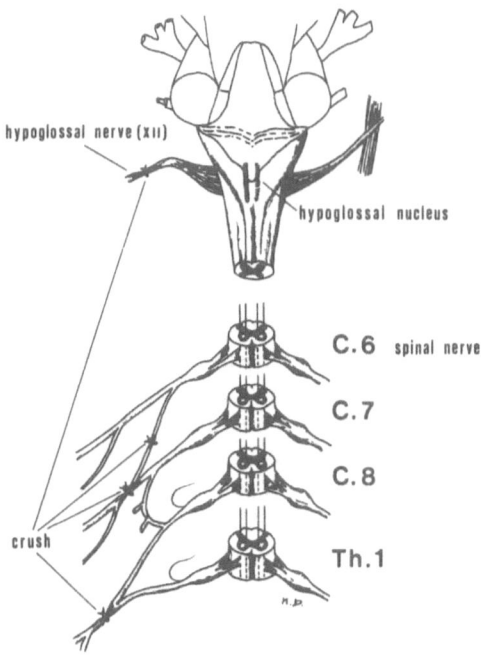

Figure 3. Diagram of medulla oblongata and spinal cord of the rabbit as seen from the dorsal and ventral side, respectively. Illustrated are the locations of monolateral crush site of the hypoglossal (at the level of hyoglossus muscle), C.6, C.7, C.8, Th.1 nerves (at the level of brachial plexus), the hypoglossal nuclei as well as spinal segments that were dissected for biochemical analyses.

Table 3. *Enzyme activities in the homogenate of hypoglossal nucleus and ventral horn spinal motoneurons of sham-operated control and lesioned rabbits in which the right hypoglossal nerve and the right C.6, C.7, C.8, Th.1 nerves were simultaneously crushed at 7,14, or 28 days before dissecting the tissue for enzyme determinations*

Nervous tissue		Days after unilateral nerve crush	AChE* (EC 3.1.1.7)	Phospholipase A_2** (EC 3.1.1.4)	Acyltransferase*** (EC 2.3.1.23)
Hypoglossal nucleus	crushed	7	2.76 ± 0.17	0.105 ± 0.02 ▲	0.40 ± 0.19
		14	3.34 ± 0.38	$0.081 \pm 0,01$ ▲	0.32 ± 0.20
		28	3.00 ± 0.43	0.070 ± 0.02	0.27 ± 0.16§
	sham-operated control	7	3.11 ± 0.29	0.080 ± 0.02	0.32 ± 0.11
		14	3.13 ± 0.19	0.064 ± 0.01	0.54 ± 0.03
		28	3.10 ± 0.42	0.072 ± 0.01	0.56 ± 0.12
Spinal motoneurons	crushed	7	1.94 ± 0.35	0.059 ± 0.02	0.79 ± 0.23
		14	2.42 ± 0.59	0.086 ± 0.01•	0.72 ± 0.31
		28	1.59 ± 0.17•	0.056 ± 0.01	0.27 ± 0.05
	sham-operated control	7	1.84 ± 0.26	0.051 ± 0.01	0.84 ± 0.15
		14	2.14 ± 0.24	0.044 ± 0.01	0.64 ± 0.32
		28	2.40 ± 0.11	0.046 ± 0.01	0.30 ± 0.07

Each value represents the mean ± S.D. of triplicate independent experiments.
(*) μmol/30 min/mg protein, at 25°C.
(**) nmol of substrate (1-palmitoyl-2-[1-^{14}C]oleoyl-GroPCho hydrolyzed/h/mg protein, at 37°C.
(***) nmol of product formed (1-[1-^{14}C]palmitoyl-2-acyl-GroPCho/min/mg protein, at 37°C.
(•) Significantly different from control at $p < 0.001$, (▲) at $p < 0.01$, (§) at $p < 0.05$.

taken from the operated side. This enzyme is representative of a group of transmitter-related enzymes which in general show a drop of activity after axotomy (Lieberman, 1981; Fonnum et al., 1973; Sinicropi et al., 1982) in several neuronal systems. In the hypoglossal nucleus, phospholipase A_2 activity was significantly increased during the first two weeks after injury while the same activity in the operated side of the spinal cord was stimulated at 2 weeks only. Acyltransferase, using [^{14}C]palmitoyl CoA as acyl donor, did not show any change in both tissues after crushing the corresponding nerves, or it was late inhibited in the hypoglossal nucleus only.

It should be noted that the measurements of the enzyme activities have been performed in total homogenates of the hypoglossal nucleus and ventral horn tissue. Which individual cell types show the changes cannot be determined. Morphological studies have shown that the reversible perineuronal glial reaction, appearing within the nuclei after nerve axotomy by crush lesion and lasting for weeks in young and adult rabbit (Sjöstrand, 1966; Torvik and Soreide, 1975), may not be excluded in evaluating metabolic neuronal changes. In this respect, the enhancement of the PLA2 activity observed in the present experiments probably reflects a sign of an abortive phagocytic potential of glial cells rather than an intrinsic neuronal feature alone; this does not exclude that it also can be intrinsic to the neuronal metabolism. In addition, it is possible that contralateral intact motoneurons could be contributing to the metabolic response during ipsilateral axotomy. By using a crush lesion as injury model, contralateral motoneurons seem to fail to give such a response, at least in the hypoglossal nucleus and ventral horn of adult rabbit, especially in the later stage of neuronal regeneration (Sjöstrand, 1966). In adult rats a minimal or transient contralateral reaction, disappearing by day 14, was observed (Kerns and Hinsman, 1973; Reisert et al., 1984). Therefore it seems justified to use a contralateral nucleus as a control in our experiments of regeneration. However, these two considerations taken together indicate that correlated studies on the biochemical and histochemical effects of axotomy must be performed with the identical model.

CONCLUSIONS

The interdisciplinary approaches used over the last six to eight years to study the axonal transport of phospholipids have provided an understanding of neurochemical, physiological, and some pathophysiological aspects of the fast migration of these peculiar compounds within the axons. Much more information is needed about the mechanism that mediates the movement of proteolipid complexes as well as the respective roles of the bidirectional transport components. The expansion of knowledge of the biochemical events occurring in the cell body and associated with the neuron-glia response to axotomy as a whole, should provide further information on the entire neuronal regenerative reaction. On the other hand, work in this area would be greatly advanced not only by detailed studies of the lipid metabolism in motoneurons isolated from adult animals but also by employing autoradiographic and enzymohistochemical techniques; this in an attempt to discriminate the individual contribution of the neurons and satellite cells to neurite outgrowth during peripheral regeneration.

ACKNOWLEDGMENTS

The author wishes to thank all colleagues who contributed with skill and enthusiasm to the experimental work reported here. He would also like to thank Prof. A.M. Giuffrida for her helpful criticism and advice during the preparation of the manuscript and P. Rapisarda, who gave generously of his time in helping to edit the computerized manuscript. This work was supported by grants from the Consiglio Nazionale delle Ricerche (CNR) and Ministero della Pubblica Istruzione (MPI) of Italy.

REFERENCES

Abe T, Haga T, Kurokawa M (1973) Rapid transport of phosphatidylcholine occurring simultaneously with protein transport in the frog sciatic nerve. Biochem J 136: 731-740.

Alberghina M, Viola M, Moro F, Giuffrida AM (1981a) Axonal transport of phospholipids in rabbit optic pathway. Neurochem Res 6: 633-647.

Alberghina M, Viola M, Giuffrida AM (1981b) Rapid axonal transport of phosphatidylinositol in the rabbit optic pathway. J Neurosci Res 6: 723-731.

Alberghina M, Viola M, Giuffrida AM (1982a) Transfer of axonally transported phospholipids into myelin isolated from the rabbit optic pathway. Neurochem Res 7: 139-149.

Alberghina M, Karlsson JO, Giuffrida AM (1982b) Rapid migration of inositol phospholipids with axonally transported substances in the rabbit optic pathway. J Neurochem 39: 223-227.

Alberghina M, Giuffrida AM (1982c) The role of phospholipids in axonal growth and function. In: Giuffrida Stella AM, Gombos G, Benzi G, Bachelard HS (eds): Basic and clinical aspects of molecular neurobiology. Fondaz Intern Menarini Publisher, Milano; pp. 156-172.

Alberghina M, Viola M, Giuffrida AM (1983a) Rapid axonal transport of glycerophospholipids in regenerating hypoglossal nerve of the rabbit. J Neurochem 40: 25-31.

Alberghina M, Moschella F, Viola M, Brancati V, Micali G, Giuffrida AM (1983b) Changes in rapid transport of phospholipids in the rat sciatic nerve during axonal regeneration. J Neurochem 40: 32-38.

Alberghina M, Viola M, Moschella F, Giuffrida AM (1983c) Axonal transport of glycerophospholipids in regenerating sciatic nerve of the rat during aging. J Neurosci Res 9: 393-400.

Alberghina M, Viola M, Moschella F, Giuffrida AM (1983d) Myelination of regenerating sciatic nerve of the rat: lipid components and synthesis of myelin lipids. Neurochem Res 8: 133-149.

Alberghina M, Viola M, Giuffrida AM (1984) Myelination process in the rat sciatic nerve during regeneration and development: molecular species composition and acyl group biosynthesis of choline-, ethanolamine-, and serine-glycerophospholipids of myelin fractions. Neurochem Res 9: 887-901.

Alberghina M, Viola M, Moro F, Giuffrida AM (1985) Remodeling and sorting process of ethanolamine and choline glycerophospholipids during their axonal transport in the rabbit optic pathway. J Neurochem 45: 1333-1340.

Aldskogius H, Barron KD, Regal R (1980) Axon reaction in dorsal motor vagal and hypoglossal neurons of the adult rat. Light microscopy and RNA-cytochemistry. J Comp Neurol 193: 165-177.

Ansell GB, Spanner S (1968) Plasmalogenase activity in normal and demyelinating tissue of the central nervous system. Biochem J 108: 207-209.

Baker RR, Thompson W (1972) Positional distribution and turnover of fatty acids in phosphatidic acid, phosphoinositides, phosphatidylcholine and phosphatidylethanolamine in rat brain in vivo. Biochim Biophys Acta 270: 489-503.

Bell RM, Coleman RA (1980) Enzymes of glycerolipid synthesis in eukaryotes. Ann Rev Biochem 49: 459-487.

Benes F, Higgins JA, Barrnett RJ (1973) Ultrastructural localization of phospholipid synthesis in the rat trigeminal nerve during myelination. J Cell Biol 57: 613-629.

Brunetti M, Droz B, Di Giamberardino L, Koenig HL, Carretero F, Porcellati G (1983) Axonal transport of ethanolamine glycerophospholipids. Preferential accumulation of transported ethanolamine plasmalogen in myelin. Neurochem Pathol 1: 59-80.

Cova JL, Barron KD (1981) Uptake of tritiated leucine by axotomized cervical motoneurons: an autoradiographic study. Exp Mol Pathol 34: 159-169.

Currie JR, Grafstein B, Whitnall MH, Alpert R (1978) Axonal transport of lipid in goldfish optic axons. Neurochem Res 3: 479-492.

DeVries GH, Norton WT (1974) The fatty acid composition of sphingolipids from bovine CNS axons and myelin. J Neurochem 22: 251-257.

DeVries GH, Zmachinski CJ (1980) The lipid composition of rat CNS axolemma-enriched fractions. J Neurochem 34: 424-430.

DeVries GH, Anderson MG, Johnson D (1983) Fractionation of isolated rat CNS myelinated axons by sucrose density gradient centrifugation in a zonal rotor. J Neurochem 40: 1709-1717.

Droz B, Di Giamberardino L, Koenig HL, Boyenval J, Hassig R (1978) Axon-myelin transfer of phospholipid components in the course of their axonal transport as visualized by radioautography. Brain Res 155: 347-353.

Droz B, Di Giamberardino L, Koenig HL (1981) Contribution of axonal transport to the renewal of myelin phospholipids in peripheral nerves. I. Quantitative radioautographic study. Brain Res 219: 57-71.

Dziegielewska KM, Evans CAN, Saunders NR (1980) Rapid effect of nerve injury upon axonal transport of phospholipids. J Physiol 304: 83-98.

Fonnum F, Frizzel M, Sjöstrand J (1973) Transport, turnover and distribution of choline acetyltransferase and acetylcholinesterase in the vagus nerve and hypoglossal nerve of the rabbit. J Neurochem 21: 1109-1120.

Gould RM, Spivack WD, Sinatra RS, Lindquist TD, Ingoglia NA (1982) Axonal transport of choline lipids in normal and regenerating rat sciatic nerve. J Neurochem 39: 1569-1578.

Gould RM, Pant H, Gainer H, Tytell M (1983) Phospholipid synthesis in the squid giant axon: incorporation of lipid precursors. J Neurochem 40: 1293-1299.

Grafstein B, Miller JA, Ledeen RW, Haley J, Specht SC (1975) Axonal transport of phospholipids in goldfish optic system. Exp Neurol 46: 261-281.

Grafstein B, McQuarrie IG (1978) Role of the nerve cell body in axonal regeneration. In: CW Cotman (ed): Neuronal Plasticity. Raven Press, New York; pp. 155-195.

Grafstein B, Forman DS (1980) Intracellular transport in neurons. Physiol Rev 60: 1167-1283.

Guy JR, Bisby MA (1983) Velocity of axonal transport of phospholipid in rat sciatic nerve. Exp Neurol 82: 706-710.

Haley JE, Ledeen RW (1979a) Incorporation of axonally transported substances into myelin lipids. J Neurochem 32: 735-742.

Haley JE, Tirri LJ, Ledeen RW (1979b) Axonal transport of lipids in the rabbit optic system. J Neurochem 32: 727-734.

Harkonen MHA, Kauffman FC (1974) Metabolic alterations in the axotomized superior cervical ganglion of the rat. II. The pentose phosphate pathway. Brain Res 65: 141-157.

Hitzemann R, Loh H (1982) The transport and turnover of phosholipids in the rat nigrostriatal system: Effects of d-amphetamine and haloperidol. Res Comm Chem Pathol 35: 209-227.

Jerkins A, Kauffman FC (1983) Increased lipid content in the rat axotomized superior cervical ganglion. Exp Neurol 79: 347-359.

Kerns JM, Hinsman EJ (1973) Neuroglial response to sciatic neurectomy. I. Light microscopy and autoradiography. J Comp Neurol 151: 237-254.

Kumara-Siri MH, Gould RM (1980) Enzymes of phospholipid synthesis: axonal versus Schwann cell distribution. Brain Res 186: 315-330.

Lands WEM (1960) The enzymatic acylation of lysolecithin. J Biol Chem 235: 2233-2237.

Ledeen RW, Haley JE (1983) Axon-myelin transfer of glycerol-labeled and inorganic phosphate during axonal transport. Brain Res 269: 267-275.

Lee PK, Deskmukh DS, Wisniewski HM, Brockerhoff H (1982) Axonal transport of phosphatidylcholine and two synthetic analogs. Neurochem Int 4: 355-359.

Lieberman AR (1971) The axon reaction: a review of the principal features of perikaryal responses to axon injury. Int Rev Neurobiol 14: 49-124.

Longo FM, Hammerschlag R (1980) Relation of somal lipid synthesis to the fast axonal transport of protein and lipid. Brain Res 193: 471-485.

Matthews M and Nelson V (1975) Detachment of structurally intact nerve ending from chromatolytic neurones of rat superior cervical ganglion during the depression of synaptic transmission induced by post-ganglionic axotomy. J Physiol 245: 91-135.

Miani N, Rizzoli A, Bucciante G (1961) Metabolic and chemical changes in regenerating neurons-II. In vitro rate of incorporation of amino acids into proteins of the nerve cell perikaryon of the C.8 spinal ganglion of rabbit. J Neurochem 7: 161-173.

Miani N (1962) Metabolic and chemical changes in regenerating neurons-III. The rate of incorporation of radioactive phosphate into individual phospholipids of the nerve-cell perikaryon of the C.8 spinal ganglion in vitro. J Neurochem 9: 537-541.

Miani N (1963) Analysis of the somato-axonal movement of phospholipids in the vagus and hypoglossal nerves. J Neurochem 10: 859-874.

Reisert I, Wildemann G, Grab D, Pilgrim CH (1984) The glial reaction in the course of axon regeneration: a stereological study of the rat hypoglossal nucleus. J Comp Neurol 229: 121-128.

Rostas JAP, McGregor A, Jeffrey PL, Austin L (1975) Transport of cholesterol in the chick optic system. J Neurochem 24: 295-302.

Sinicropi DV, Michels K, McIlwain DL (1982) Acetylcholinesterase distribution in axotomized frog motoneurons. J Neurochem 38:1099-1100.

Sjöstrand J (1966) Studies on glial cells in the hypoglossal nucleus of the rabbit during nerve regeneration. Acta Physiol Scand 67: 1-43.

Smith CB, Crane AM, Kadekaro M, Agranoff BW, Sokoloff L (1984) Stimulation of protein synthesis and glucose utilization in the hypoglossal nucleus induced by axotomy. J Neurosci 4: 2489-2496.

Sunner BEH, Sutherland FI (1973) Quantitative electron microscopy on the injured hypoglossal nucleus in the rat. J Neurocytol 2: 315-328.

Sun GY, Sun AY (1972) Phospholipids and acyl groups of synaptosomal and myelin membranes isolated from the cerebral cortex of squirrel monkey (Saimiri sciureus). Biochim Biophys Acta 280: 306-315.

Tang BY, Komiya Y, Austin L (1974) Axoplasmic flow of phospholipids and cholesterol in the sciatic nerve of normal and dystrophic mice. Exp Neurol 43: 13-20.

Toews AD, Goodrum JF, Morell P (1979) Axonal transport of phospholipids in the rat visual system. J Neurochem 32: 1165-1173.

Toews AD, Padilla SS, Roger LJ, Morell P (1980) Axonal transport of glycerophospholipids following intracerebral injection of glycerol into substantia nigra or lateral geniculate body. Neurochem Res 5: 1175-1183.

Toews AD, Morell P (1981) Turnover of axonally transported phospholipids in nerve endings of retinal ganglion cells. J Neurochem 37: 1316-1323.

Toews AD, Saunders BF, Blaker WD, Morell P (1983) Differences in the kinetics of axonal transport for individual lipid classes in rat sciatic nerve. J Neurochem 40: 555-562.

Torvik A, Soreide AJ (1975) The perineuronal glial reaction after axotomy. Brain Res 95: 519-529.

Webster GR, Alpern RJ (1964) Studies on the acylation of lysolecithin by rat brain. Biochem J 90: 35-42.

Woelk H, Porcellati G (1973) Subcellular distribution and kinetic properties of rat brain phospholipases Al and A2. Hoppe-Seyler's Z. Physiol Chem 354: 90-100.

Wooten FG, Park DH, Joh TH, Reis DJ (1978) Immunochemical demonstration of reversible reduction in choline acetyltransferase concentration in rat hypoglossal nucelus after hypoglossal nerve transection. Nature (London) 275: 324-325.

Yoshino JE, Griffin JW, DeVries GH (1983) Identification of an axolemma-enriched fraction from peripheral nerve. J Neurochem 41: 1126-1123.

Phospholipid research and the nervous system
Biochemical and molecular pharmacology
L.A. Horrocks, I. Freysz, G. Toffano (eds)
Fidia Research Series, vol. 4.
Liviana Press, Padova. © 1986

EFFECT OF PHOSPHATIDYLSERINE ON CORTICAL ACETYLCHOLINE RELEASE AND CALCIUM UPTAKE IN ADULT AND AGING RATS

G. Pepeu, L. Giovannelli, M.G. Giovannini and F. Pedata

Department of Pharmacology, University of Florence,
Viale Morgagni 65, 50134 Florence, Italy

The effect of phosphatidylserine (PtdSer) administration (15 mg/kg i.p.) for 30 days was investigated on acetylcholine (ACh) release from electrically stimulated cortical slices and on $^{45}Ca^{2+}$ uptake into K^+-depolarized cortical synaptosomes in 3 and 22-24 month old rats. ACh release in 24 month old rats treated with the solvent was 50% lower than in 3 month old rats at all stimulation frequencies tested. On the contrary, no significant decrease in ACh release was found in the 24 month old rats treated with PtdSer. Similarly, the decrease in $^{45}Ca^{2+}$ uptake found in the 22 month old rats treated with the solvent did not occur in the aging rats chronically treated with PtdSer. The possibility is envisaged that PtdSer may act by influencing age-dependent changes in membrane phospholipid composition.

INTRODUCTION

The administration of phosphatidylserine (PtdSer) liposomes is followed by many neurochemical and pharmacological actions. For reviews see Toffano and Bruni (1980), Bruni and Toffano (1985) and Calderini and Toffano in this volume. Several PtdSer actions involve central neurotransmission mechanisms. PtdSer stimulates brain noradrenaline and dopamine turnover (Toffano et al., 1976), increases both dopamine-sensitive adenylate cyclase activity and cyclic AMP content (Leon et al., 1978), and enhances acetylcholine (ACh) release (Casamenti et al., 1979). The latter authors demonstrated that PtdSer synaptosomes injected i.p. in urethane anaesthetized rats stimulated ACh release from the cerebral cortex through an indirect mechanism presumably involving a dopaminergic link because the effect could be blocked by haloperidol. PtdSer also exerted a weak dopaminergic activity in the striatum as shown by the increase in striatal ACh level following single administrations of 150 and 300 mg/kg i.p. or repeated administration of 50 mg/kg i.p. (Mantovani et al., 1982). This effect was also prevented by haloperidol and is dependent upon a decrease in ACh release. The latter has been observed in striatal slices incubated in Krebs solution containing PtdSer.

Electroencephalographic and behavioral studies also demonstrate an effect of PtdSer on brain cholinergic mechanisms. PtdSer antagonizes the disrupting effect of scopolamine on spontaneous alternation in rats (Pepeu et al., 1980). At the dose of 30 mg/kg i.p. PtdSer reduces or prevents electroencephalographic changes induced by scopolamine in rats and rabbits (Mantovani et al., 1982). These results suggest the possibility that exogenous PtdSer may enhance activity of central cholinergic mechanisms facilitating ACh release.

ACh is strongly impaired in the brain of aging rodents as shown in K^+-depolarized brain slices from 10 and 30 month old mice (Gibson and Peterson, 1981) and in superfused, electrically-stimulated cortical slices from 24 month old rats (Pedata et al., 1983). Peterson and Gibson (1983) suggested that the decrease in ACh release may be related to the marked decrease in Ca^{2+} uptake into brain synaptosomes which they observed in aging rats.

A reduced activity of the cortical and hippocampal cholinergic pathways has been implicated in the memory and cognitive impairment associated in various degrees of severity with physiological aging and Alzheimer's dementia (Bartus et al., 1982).

In this work we investigated if PtdSer liposome administration could correct the decrease in electrically-stimulated ACh release from cortical slices from aging rats. Ca^{2+} uptake into synaptosomes prepared from the cerebral cortex of aging rats was also investigated and compared with the uptake into synaptosomes from young rats and rats treated with PtdSer.

METHODS

Animals

The experiments were carried out with 3 and 22-24 month old male Charles River Wistar rats. The rats were killed by decapitation, the skull was opened, and the right and left parietal cortices were rapidly dissected.

Preparation and Electrical Stimulation of Brain Slices

The cortices were plunged into cold Krebs solution with the following composition (mM): NaCl 118.5, KCl 4.7, $CaCl_2$ 2.5, $MgSO_4$ 1.2, KH_2PO_4 1.2, glucose 10, $NaHCO_3$ 25, and choline 0.02. The slices were prepared and superfused according to the procedure previously described by Pedata et al. (1983) in Perspex superfusion chambers of 0.9 ml volume with Krebs solution gassed with 95% O_2 and 5% CO_2, containing physostigmine sulphate, at the rate of 0.5 ml/min at 37°C.

After 20 min equilibration, the experimental design was the following: 5 min rest; 5 min stimulation at rates of 1, 2 and 5 Hz with rectangular pulses of alternating polarity, using a current strength of 30 mA/cm² and a pulse duration of 4 msec; 10 min rest to allow the washout of all ACh released by electrical stimulation.

The actual net extra release of ACh caused by electrical stimulation was estimated by subtracting the amount of release expected during 15 min rest (calculated by multiplying by 3 the amount released during the 5 min rest preceeding each stimulation period), from the total amount of ACh found in the 15 min of stimulation and washout. In Ca-free Krebs solution no extra release was detectable.

Acetylcholine Assay

ACh content of the superfusate samples was quantified with the isolated guinea-pig ileum perfused with Tyrode solution (Pedata et al., 1983). The identity of ACh as the active substance was routinely checked by adding atropine or by alkaline hydrolysis of the samples. ACh release was expressed as ng/g of wet tissue/min of superfusion $\pm$ S.E.

Calcium Uptake

Ca^{2+} uptake was investigated according to the method of Wu et al. (1982). The cortices were homogenized in 10 volumes (wt/vol) of ice cold 0.32 M sucrose solution and a crude mitochondrial fraction (P_2 fraction) was prepared. Aliquots of the preparation were incubated in Krebs-Ringer phosphate medium in plastic tubes. Ca^{2+} uptake into synaptosomes was initiated by the addition of 55 μl of $^{45}CaCl_2$ (0.7 mCi/μmol) obtained from the Radiochemical Centre (Amersham) in a 1 M solution of KCl. A final concentration of $^{45}Ca^{2+}$ equivalent to 0.6 μCi per sample was obtained. For control 55 μl of $^{45}CaCl_2$ in 1 M solution of NaCl was added in order to maintain the same osmolarity in the two sets of samples.

The reaction was allowed to proceed for 60 sec at 37°C. The uptake was terminated by adding EGTA 5 mM. The mixture was immediately filtered and radioactivity associated with the synaptosomes was measured in a Packard Tricarb (model 577) scintillation spectrometer after extraction with Aquassure (NEN). The rate of Ca^{2+} uptake was expressed as nanomoles of Ca^{2+} per min incubation time per mg of protein.

Ca^{2+} uptake by ruptured synaptosomes was used as a blank value. The results represent the net uptake (subtraction of the blank value from the value of the total Ca^{2+} uptake into intact synaptosomes).

Protein Determination

Protein content of synaptosome preparations was measured by the method of Lowry et al. (1951).

Drugs

Freshly prepared solutions of the following drugs were used: acetylcholine (Sigma), atropine sulphate and physostygmine sulphate (BDH), tetrodotoxin (Biochemia) and morphine sulphate (Carlo Erba). Cyproheptadine was a gift from Merck Sharp and

Dohme. Purified phosphatidylserine and phosphatidylcholine from bovine brain were obtained from Fidia Research Laboratories. The procedure described by Casamenti et al. (1979) was used in order to check the purity of the samples and obtain a liposome suspension (7.5 mg/ml, w/v) in Tris buffer.

RESULTS

The effects of PtdSer on ACh release from electrically stimulated cortical slices are reported in Fig. 1 which illustrates several findings. First, ACh output from the cerebral cortex increased linearly with the increase in stimulation frequency in all groups of rats investigated. Second, at all stimulation frequencies used, ACh output from the

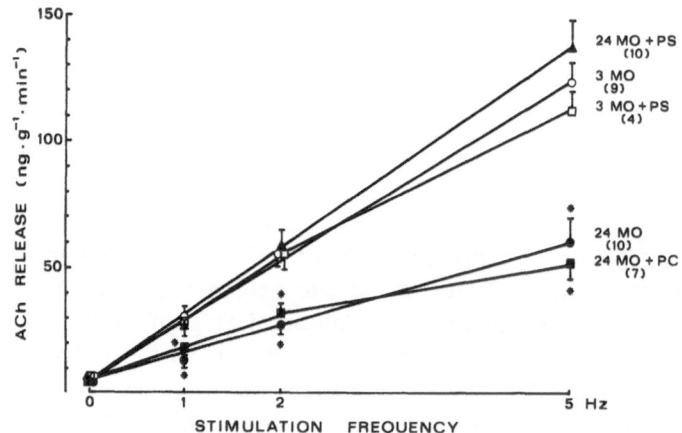

Figure 1. Acetylcholine release from electrically stimulated cortical slices from 3 and 24 month old rats treated with Tris buffer, phosphatidylserine (PS) 15 mg/kg i.p. or phosphatidylcholine (PC) 15 mg/kg i.p. for 30 days.
*Statistically significant difference from 3 months, P <0.01.
Number of rats in parentheses.

slices dissected from 24 month old rats treated with Tris-buffer was significantly lower than the output from slices from 3 month old rats. Third, the difference in ACh output between young and aging rats did not occur when the aging rats were treated with PtdSer 15 mg/kg i.p. for 30 days. In the latter rats ACh output showed a 100% increase at 1, 106% at 2 and 124% at 5 Hz stimulation frequency in comparison with the output from 24 month old rats treated with Tris-buffer. No increase was found in the 24 month old rats treated with a single injection of PtdSer 15 mg/kg i.p. (data not shown) or with PtdCho 15 mg/kg i.p. for 30 days. Finally, the administration of PtdSer 15 mg/kg i.p. for 30 days did not modify ACh release from cortical slices prepared from 3 month old rats.

No difference in ACh release was found during the rest period between young and aging rats and treated and untreated groups.

Table 1 shows that Ca^{2+} uptake into K^+ depolarized cortical synaptosomes was significantly smaller in 22 month than in 3 month old rats. A small difference, not statistically significant, was also found in the synaptosomes incubated in low K^+ medium. In preliminary experiments on rats treated with PtdSer 15 mg/kg i.p. for 30 days, no difference in Ca^{2+} uptake was found between aging and young rats.

Table 1. $^{45}Ca^{2+}$ *uptake into cortical synaptosomes in adult and aging rats*

Incubation time: 1 min.			$^{45}Ca^{2+}$ uptake nmol/mg protein/min	
Age months	no. rats	Treatment	5 mM KCl	60 mM KCl
3	8	Tris buffer	1.22 ± 0.10	3.65 ± 0.33
22	5	Tris buffer	0.96 ± 0.15 (-21%)	$1.91 \pm 0.46^*$ (-48%)
22	5	PtdSer 15 mg/kg × 30 days	1.23 ± 0.23	4.37 ± 0.55

* Statistically significant difference from Tris buffer, $p < 0.01$.

DISCUSSION

The present experiments confirm our previous results (Pedata et al, 1983) that ACh release from electrically-stimulated cortical slices is much smaller in 24 month old than in 3 month old rats. ACh release in the old rats was approximately 50% less than in the young rats. However this difference did not occur if the cortical slices were dissected from 24 month old rats treated for 30 days with PtdSer. Repeated PtdSer administrations appear, therefore, to prevent the decrease in ACh release associated with aging while being inactive in young rats. Neither a single PtdSer injection nor repeated PtdCho administrations were active in the old rats. We also found a significant decrease of Ca^{2+} uptake, over 1 min incubation time, into cortical synaptosomes of old rats. This result confirms the finding of Peterson and Gibson (1983) who found a marked decrease in Ca^{2+} uptake into synaptosomes from the forebrain of Fischer 344 rats incubated for 20 min.

PtdSer enhances histamine release from mast cells evoked by dextran and protein antigen (Goth et al.; 1971; Mongar and Svec, 1972). The increase in histamine release is associated with an increase in intracellular Ca^{2+} concentration (White et al., 1984). It is pertinent to mention that PtdSer is the most effective of several phospholipids examined in increasing Ca^{2+} uptake into a chloroform phase (Feinstein, 1964). PtdSer also increases histamine or norepinephrine-induced tension in aortic smooth muscle and alters the availability and/or exchangeability of a membrane bound Ca^{2+} fraction (Goodman et al., 1976). In the small number of aging rats treated with PtdSer for 30 days in which Ca^{2+} uptake was investigated, no decrease was found in comparison with young rats. PtdSer appears, therefore, to restore membrane permeability to Ca^{2+} fluxes in aging rats.

After i.v. administration of 20 mg/kg of labeled PtdSer, 0.25% of the injected amount was detected in mouse brain (Toffano et al., 1982). It is therefore feasible that

PtdSer may fuse with the membranes and facilitate the Ca^{2+}-mediated coupling mechanism between electrical stimulation and ACh release by influencing the age-dependent changes in membrane phospholipid composition (Hershkowitz, 1983). Further investigations are needed to confirm this hypothesis and ascertain whether this mechanism may involve other neurotransmitters whose release is impaired by aging.

The improvement in acquisition and retention of avoidance tasks observed in aging rats treated with PtdSer (Drago et al., 1981) could depend on the facilitation of cortical ACh release.

ACKNOWLEDGMENTS

This work was supported by CNR grant No. CT 83.02771.56, P.F.M.P.R., within the subproject Aging Mechanisms. We thank the Italian Study Group for Brain Aging for a generous gift of old rats.

REFERENCES

Bartus RT, Dean RL, Beer B, Lippa AS (1982) The cholinergic hypothesis of geriatric memory dysfunction. Science 217: 408-409.

Bruni A, Toffano G (1985) Influence of serine phospholipids on biogenic amine secretion in vivo and in vitro. In: Horrocks LA (ed): Phospholipids in the nervous system. Vol 2: Physiological roles. Raven Press, New York, pp. 21-29.

Casamenti F, Mantovani P, Amaducci L, Pepeu G (1979) Effect of phosphatidylserine on acetylcholine output from the cerebral cortex of the rat. J Neurochem 32: 529-533.

Drago F, Canonico PL, Scapagnini U (1931) Behavioral effects of phospatidylserine in aged rats. Neurobiol Aging 2: 209-213.

Feinstein MB (1964) Reaction of local anesthetics with phospholipids. A possible chemical basis for anesthesia. J Gen Physiol 48: 357-374.

Gibson GE, Peterson C (1981) Aging decreases oxidative metabolism and the release and synthesis of acetylcholine. J Neurochem 39: 978-984.

Goodman FR, Weiss GB, Goth A (1976) Alteration by phosphatidylserine of tension responses and ^{45}Ca distribution in aortic smooth muscle. J Pharmacol Exp Ther 198: 168-175.

Goth A, Adams HR, Knoohuizen M (1971) Effect of calcium on dextran-induced histamine release from isolated mast cells. Br J Pharmac 46: 767-769.

Hershkowitz M (1983) Mechanisms of brain - the role of membrane fluidity. In: Gispen WH, Traber J (eds): Aging of the brain. Elsevier, Amsterdam; pp. 101-114.

Leon A, Benvegnù D, Toffano G, Orlando P, Massari P (1978) Effect of brain cortex phospholipids on adenylate-cyclase acitivity of mouse brain. J Neurochem 30: 23-26.

Lowry OH, Rosebrough NJ, Farr AL, Randall RJ (1951) Protein measurement with the Folin phenol reagent. J Biol Chem 193: 265-285.

Mantovani P, Aporti F, Bonetti AC, Pepeu G (1982) Effects of phosphatidylserine on brain cholinergic mechanisms. In: Horrocks LA, Ansell GB, Porcellati G (eds): Phospholipids in the nervous system. Vol. 1: Metabolism. Raven Press, New York; pp. 165-172.

Mongar JL, Svec P (1972) The effect of phospholipids on anaphylactic histamine release Br J Pharmac 46: 741-752.

Pedata F, Slavikova J, Kotas A, Pepeu G (1983) Acetylcholine release from rat cortical slices during postnatal development and aging. Neurobiol Aging 4: 31-35.

Pepeu G, Gori G, Bartolini L (1980) Pharmacologic and therapeutic perspectives on dementia: an experimental approach. In: Amaducci L, Davison AN, Antuono P (eds): Aging of the brain and dementia. Raven Press, New York; pp. 271-274.

Peterson C, Gibson GE (1983) Aging and 3,4-diaminopyridine alter synaptosomal calcium uptake. J Biol Chem 258: 11482-11486.

Toffano G, Leon A, Benvegnù D, Boarato E, Azzone GF (1976) Effect of brain cortex phospholipids on catecholamine content of mouse brain. Pharmacol Res Comm 8: 581-590.

Toffano G, Bruni A (1980) Pharmacological properties of phospholipid liposomes. Pharmacol Res Comm 12: 829-845.

Toffano G, Battistella A, Mazzari S, Orlando P, Massari P, Giordano C (1982) Fate of phosphatidyl-L-(U-^{14}C)-serine in mice. In: Horrocks LA, Ansell GB, Porcellati G (eds): Metabolism. Raven Press, New York; pp. 173-180.

White JR, Ishizaka T, Ishizaka K, Sha'afi RI (1984) Direct demonstration of increased intracellular concentration of free calcium as measured by quin-2 in stimulated rat peritoneal mast cell. Proc Natl Acad Sci 81: 3978-3982.

Wu PH, Phillis JW, Thierry DL (1982) Adenosine receptor agonists inhibit K^{+}-evoked Ca^{2+} uptake by rat brain cortical synaptosomes. J Neurochem 39: 700-708.

Phospholipid research and the nervous system
Biochemical and molecular pharmacology
L.A. Horrocks, L. Freysz, G. Toffano (eds)
Fidia Research Series, vol. 4.
Liviana Press, Padova. © 1986

SYNTHESIS DE NOVO OF CHOLINE, PRODUCTION OF CHOLINE FROM PHOSPHOLIPIDS, AND EFFECTS OF CDP-CHOLINE ON NERVE CELL SURVIVAL

R. Massarelli, R. Mozzi[1], F. Golly, H. Hattori[2], F. Dainous, J.N. Kanfer[2] and L. Freysz

Centre de Neurochimie du CNRS, U44 de l'INSERM, 5, rue Blaise Pascal, 67084 Strasbourg Cedex, France, [1]Department of Biological Chemistry, The Medical School, University of Perugia, 06100 Perugia, Italy, [2]Department of Biochemistry, Faculty of Medicine, University of Manitoba, Winnipeg R3E OW3, Canada.

The repair of nerve cell membranes may overcome or, at least, alleviate neurological damage and diseases. The mechanisms responsible for such a repair are not known but they appear to be mediated by some components present in the plasma membrane of nerve cells. Much experimental data suggest that glycolipids, especially of the ganglio series present in large amount in neuronal membranes, may mediate one of the steps implicated in the repair and regeneration of nerve cells (De Felice and Ellenberg, 1984). Among other components of plasma membranes the fundamental "bricks" of all animal cell membranes are the phospholipids. Their function has been considered for a long time to be purely structural and this concept has often assumed the idea of passivity. It was customary, until recently, to think of phospholipids simply as a wall of constraint for the cytoplasmic content of the cells and as a useful support for proteins floating in a hydrophobic bilayer.

Such a ptolemaic approach to the role of phospholipids, particularly in nerve cell membranes, has been challenged by several experimental findings. Phospholipids are asymmetrically distributed in the two membrane leaflets (Dominski et al., 1983; Fontaine et al., 1980; Smith and Loh, 1976) producing a distribution of charges (positive outside because of the high proportion of PtdCho) and suggesting a physiological role for such a distribution. Inositol phospholipids, in particular their degradation products, are strictly linked to the mobilization of Ca^{++} (Hawthorne and Azila, 1982; Van Rooijen et al., 1985) hence with one of the most fundamental phenomenon of neurotransmission. Methylation processes: carboxymethylation of proteins (O'Dea et al., 1981) and stepwise methylation of PtdEtn (Mozzi et al., 1982) (even if the latter does not seem to be so universally linked to membrane function as had appeared, Moore et al., 1984), are certainly important phenomena in the dialogue capacities of nerve cells.

To study phospholipid metabolism and their physiological importance, however, presents serious difficulties due to the presence of the post-mortem effects which become dramatic already within a few seconds after death: activation of phospholipases (A_1, A_2, C and D) (Bazan, 1970; Dross and Kewitz, 1972) leading to an overproduction of free choline and free fatty acids thus increasing the synthesis of prostaglandins and leukotrienes, and the degradation of PtdIns affecting the Ca^{++} balance, thus favouring increased membrane fluidity and the production of free radicals.

Such an intricate series of events is further complicated by the choice of an experimental model which must be made when taking into consideration the limitations imposed by animal sacrifice. In this respect, nerve cell cultures have been suggested as a type of ideal model for the study of nerve function. This is partly true because it is possible to study certain biochemical and physiological phenomena in isolated cells, which interact with one another forming functional synapses, synthesizing neurotransmitters and neurohormones much as nervous systems in vivo (Louis et al., 1983).

Cautiously, it should however be considered that a dispersed system of nerve cells, maintained in a minimal essential medium, cannot lead to extrapolations even if it is very tempting. With this in mind we describe in the present review some studies performed in our laboratories on isolated cultures of nerve cells (neurons and glia) from chicken embryo and on rat brain synaptosomes. The findings correspond to observations made in vivo and suggest some hypotheses.

Three major axes of research will be presented: the de novo synthesis of choline containing compounds by the stepwise methylation of ethanolamine containing compounds, the production of free choline by the activation of phospholipase D, and the protective action of CDP-choline against CNS damage.

DE NOVO SYNTHESIS OF CHOLINE

The classical studies of Ansell and Spanner (1975) have shown that intravenous or intraventricular injections in vivo of radioactive ethanolamine or of one of its methylated derivatives do not produce measurable amounts of radioactive choline. Kewitz and his group (Kewitz and Pleul, 1981) have, however, shown that there is more free choline leaving the brain than entering it. These conflicting results have not been resolved, perhaps because of the phenomena which have been summarized above (post-mortem effects).

Nerve cell cultures have then been used to partly overcome the post mortem effects and to observe whether, under "normal" growth conditions of the cultures, cells are able to methylate ethanolamine in its free or phosphorylated form.

Uptake of Ethanolamine

Radioactive ethanolamine may be taken up in chick neuronal and glial cell cultures following saturation kinetics which show only one affinity for the substrate but with a remarkable difference between the two cell types (Massarelli et al., 1986). Neurons have a much higher affinity than glial cells and the latter have a hundred fold higher capacity of transport (Massarelli et al., 1986). The uptake appears to be inhibited by high concentrations of cyanide or ouabain, by low temperature, and by the substitution of Na$^+$ ions with Li$^+$ and Cs$^+$. Among several natural compounds tested, mono- and dimethylethanolamine, as well as choline, appear to be the most active inhibitors of the uptake with little or no difference in specificity between neurons and glial cells (Massarelli et al., 1986).

Methylation of Phosphatidylethanolamine

The pathway for methylation of PtdEtn appears to be similar in neurons and glial cells with few remarkable differences. More PtdCho and lysoPtdCho than PtdMeEtn and PtdMe$_2$Etn were labeled by incubating glial cells with radioactive ethanolamine. In neurons PtdMeEtn contained more label than other methylated phospholipids. Also in glial cells the labeling of ethanolamine plasmalogens relative to PtdEtn was greater than in neurons (Dainous et al., 1982).

The results show that in glial cells, after 5 hrs of incubation with radioactive ethanolamine, 1.5% of the label in phospholipids was present as PtdCho while in neurons after 3 hrs only 0.2% of the radioactivity was present as PtdCho. Such a difference between the two cell types may have a great importance in the economy of choline in the nervous system and may suggest once more in glial cells a storage role of choline for neuronal needs (Dainous et al., 1982).

Analysis of Methylated Acid Soluble Compounds

The analysis of methylated acid soluble compounds in particular, phosphocholine, CDP-choline, free choline, and acetylcholine, has been studied in glial cells and in neurons of chicken cerebral hemispheres after incubation with radioactive ethanolamine or methionine as precursors.

The results showed, after incubation with 10 μM [^{3}H]ethanolamine, that the radioactivity found in the acid soluble fraction was detected as phosphocholine, CDP-choline and some free choline both in neuronal and glial cells respectively (Figs. 1-2). However, only neurons showed the presence of radioactivity which comigrated with acetylcholine on thin layer chromatograms.

This acetylcholine synthesis did not require the reuptake of de novo synthesized choline because the incubation of neurons with [^{3}H]methionine, in the presence of an inhibitor of choline transport, reduced the accumulation of radioactivity in free choline but much less in acetylcholine (Fig. 3). It appears then that acetylcholine may be synthesized from newly synthesized choline labeled directly from ethanolamine or methionine without the need for the newly synthesized choline to get out of the cell and then be taken up.

PRODUCTION OF CHOLINE FROM PHOSPHOLIPIDS VIA PHOSPHOLIPASE D ACTIVATION

Some experimental evidence indicates that free choline may be produced under physiological conditions from the hydrolysis of PtdCho (Blusztajn et al., 1986). Recently, Hattori and Kanfer (1984) have shown that in rat brain synaptosomes a phospholipase

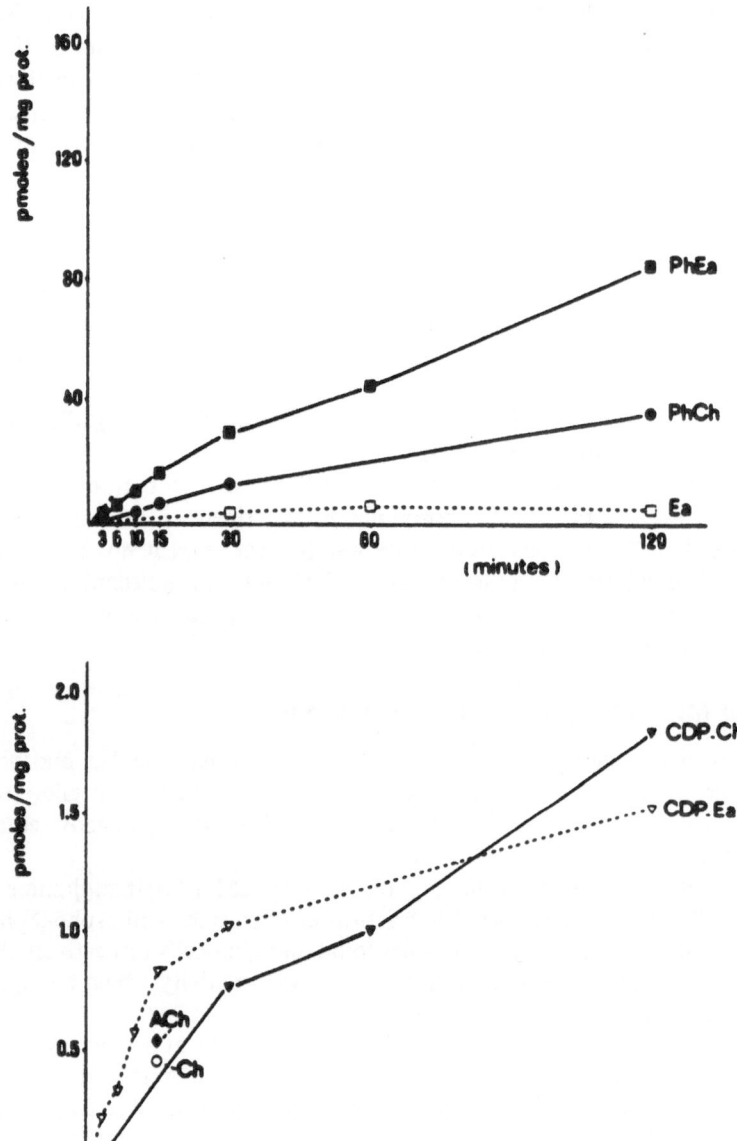

Figure 1. Incorporation of [³H]ethanolamine into acid-soluble compounds of chick neurons in culture. Cells (6 d.i.c.) were incubated with 10 μM [³H]ethanolamine in their growth medium for various periods of time. Cells were washed 4 times with 0.147 M NaCl, the acid soluble compounds extracted with 0.4 M HCl and separated by thin layer chromatography (Marchbanks and Israel, 1971).

PhEa: Phosphoethanolamine; PhCh: Phosphocholine; Ea: Ethanolamine; CDPEa: Cytidine diphosphoethanolamine; CDPCh: Cytidine diphosphocholine; Ch: Choline; ACh: Acetylcholine.

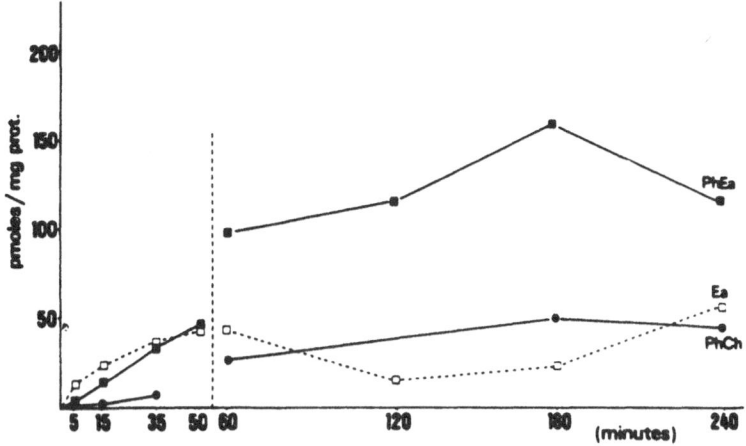

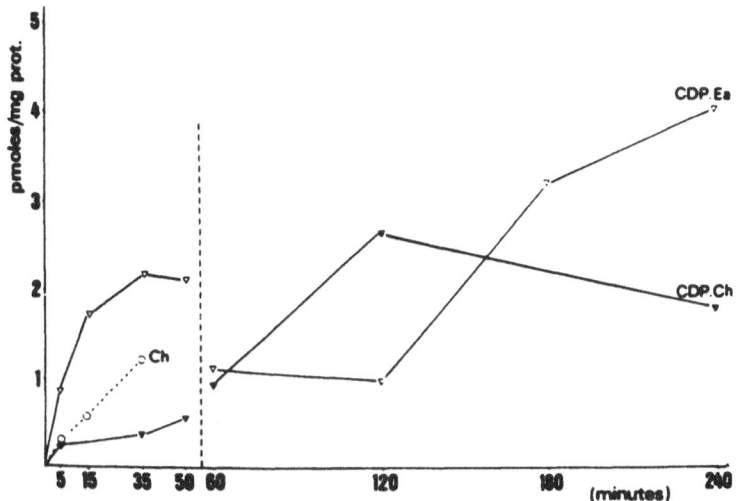

Figure 2. Incorporation of [³H]ethanolamine into acid soluble compounds of chick glial cell cultures. Cells (16 d.i.c.) were incubated with 10 μM [³H]ethanolamine in their growth medium for various periods of time and treated as in Figure 1. Abbreviations as in Figure 1.

D activity is detectable when oleate is added to the organelle suspension and further experiments have shown that the free choline thus produced may be acetylated if acetyl-coenzyme A is added to the incubation mixture.

Preliminary experiments have indicated the presence of a choline acetyltransferase-like activity at the level of plasma membranes, saturated by choline and acetyl-CoA, showing NaCl requirement and an optimum pH of about 7.5. Thus free choline produced by the stimulation of phospholipase D by fatty acids may be transformed, in the presence of acetyl-CoA, into acetylcholine. The nature of this phenomenon and its physiological significance are presently under study.

278

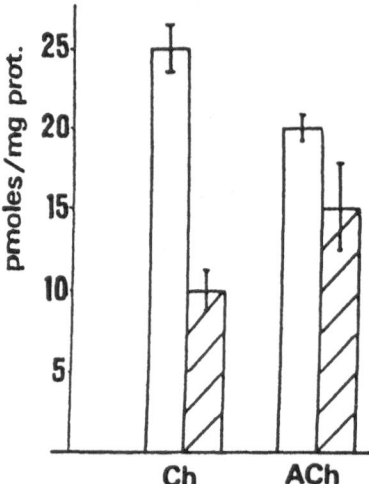

Figure 3. Effect of hemicholinium-3 on incorporation of [³H]methionine into neurons. Neurons (6 d.i.c.) were incubated in Krebs Ringer phosphate solution for 1 hour with 1 mM of [³H]methionine in the presence (hatched bars) or absence (open bars) of 10⁻⁵M-hemicholinium-3. Results are expressed as pmol/mg protein⁻¹/hour⁻¹. Each value represents the average of 3 experiments ± S.D. Abbreviations as in Figure 1.

EFFECT OF CDP-CHOLINE ON NERVE CELL REPAIR

Several clinical studies have indicated that CDP-choline may be used with some success in the treatment of several cerebral injuries ranging from stroke to oedema (Cohadon et al., 1979; 1982). In particular, it appears that ischemic or hypoxic states may be treated with some success by the administration of CDP-choline. This effect may be mediated by a variety of mechanisms which have been studied recently by several laboratories (Boismare et al., 1985; Horrocks and Dorman, 1985). The protective action of the molecule has been implicated in ATPase activities (Cohadon et al., 1979), nucleic acid metabolism (Giuffrida et al., 1985) and especially phospholipid metabolism because of its role as the direct precursor of PtdCho. It is the latter property which has stimulated the study of CDP-choline as a useful substance in the treatment of brain injuries.

If, by an yet unknown series of mechanisms, CDP-choline may be of help in the treatment of brain damage produced by a low pO_2 it should per force be of use in case of low pCO_2. Such cases are less well known and concern mainly the shifts towards basic pH values.

It is rather difficult to produce hypoxia or even more anoxia in cell cultures. The relatively small amount of tissue compared to the large volume of growth medium makes it difficult to eliminate oxygen. Conversely it is relatively easy to reduce CO_2 levels because nerve cell cultures are grown in an atmosphere containing 5% CO_2 to maintain the pH of the medium close to neutrality. A decrease of the CO_2 content to atmospheric values (around 0.03%) leads inevitably tó the degeneration of neurons followed by cell death (Freysz et al., 1985). When neurons however are grown in the presence of 10^{-6} M CDP-choline they resist the damage and survive (Freysz et al., 1985). Such a protective effect upon the cell morphology was quantified with a morphometric study and was shown to be therapeutic as well by the addition of CDP-choline after the hypocapnic shock. It was also shown that the intact molecules exerted the effects and not parts of it such as choline, phosphocholine and cytosine or CDP. Moreover, other nucleotides (AMP, GMP, UMP) did not have any effect even at high concentrations (up to 10^{-4}M).

When the incorporation of [^{3}H]choline was investigated at 7 days in culture (with or without exposure to hypocapnia at 3 days in culture), the neuronal cultures which were pretreated with CDP-choline showed an incorporation of choline into phospholipids similar to that in control cultures. No effect was observed on the methylation of PtdEtn which, anyway, was not affected by the hypocapnic treatment (Mykita et al., 1986).

Further studies have indicated that the effect of CDP-choline is mainly directed upon the activity of phospholipase A_1. Such an effect was confirmed after incubation of neurons with radioactive palmitate or arachidonate in the presence or absence of CDP- choline after hypocapnic treatment. The release of free palmitate is greatly influenced by CDP-choline in hypocapnic cells contrary to that of arachidonate which is very little affected (Table 1), suggesting again (because of the preferential esterification of arachidonate in position 2 and of palmitate in position 1 of the glycerol moiety of phospholipids) that the effect of CDP-choline seems to act mainly upon the activity of phospholipase A_1.

At present it is difficult to give an interpretation for the seemingly specific action of CDP-choline upon phopholipase A_1. Preliminary experiments, however, have suggested the possibility of an indirect action of the compound upon the enzymatic activity mediated by (a) secondary messenger(s).

Table 1. *Effect of CDP-choline on the release of fatty acids in chick neuronal cultures after hypocapnia*

Fatty acid	Palmitic acid			Arachidonic acid		
min of treatment	0	15	180	0	5	180
Control cells	12.2	13.4	16.2	6.7	5.7	5.4
Control cells + CDP-choline 10^{-6}M	10.5	15.8*	19.1*	6.3	5.2	4.9
Hypocapnic cells	12.2	11.7*	10.6*	6.7	6.5	6.7
Hypocapnic cells + CDP-choline 10^{-6}M	10.5	10.5*	7.5*	4.9	6.5	5.8

Neurons were seeded in normal growth medium containing 20% fetal calf serum in presence or absence of 10^{-6}M CDP-choline. After one day the medium was replaced by the same medium containing 0.2 μCi of [^{14}C]palmitic acid or [^{14}C]arachidonic acid. At day 3 the radioactive medium was removed, the cells washed 3 times with unlabeled medium, incubated with the same medium, and submitted to hypocapnia for various times. Control cells were incubated in an atmosphere enriched in CO_2 (5%) and 95% air. Hypocapnic cells were kept in an atmosphere containing 0.03% CO_2 for 15 or 180 min. Immediately after, the medium was removed and the cells extracted with chloroform-methanol 2:1 and lipids separated by thin layer chromatography.

Results are expressed as % of total lipid radioactivity in free fatty acids in cells and represent 3 to 4 separate determinations. Standard variation was less than 10%. *Significantly different from controls, $p < 0.05$.

CONCLUSIONS

The results presented in this review lead to several tentative conclusions:
a) Choline may be synthesized de novo in nerve cell cultures from cerebral hemispheres of chick embryos. The synthetic pathways may follow the well known methylation pathway, i.e., the stepwise methylation of PtdEtn, and it appears that the free choline, which may thus be eventually produced, is transformed into acetylcholine in the neuronal cytoplasm.
b) Free choline may also be produced by the hydrolysis of choline containing phospholipids by the activation of phospholipase D. Such a production of choline leads as well to the synthesis of acetylcholine. Three pathways seem then to be utilized by the nerve cells for acetylcholine biosynthesis: 1) the transport of choline which accounts for about 60% of the free choline content into the cells; 2) the de novo synthesis of choline whose influence in the acetylcholine metabolism must still be quantitatively evaluated, and 3) the production of choline from phospholipids which accounts for about 30% of the cellular free choline content.
c) CDP-choline, the direct precursor of PtdCho in the Kennedy pathway, may be valuable in the clinical treatment of hypocapnia. Such an effect however is not mediated by the addition of excess substrate for the synthesis of PtdCho but may be rather an action of CDP-choline exerted upon phospholipase A_1.

ACKNOWLEDGMENTS

This investigation was supported by grants from the NATO (grant 07582) and the Italian CNR (grant CT 810012504). The skillful secretarial assistance of Ms. C. Thomassin-Orphanides was greatly appreciated.

REFERENCES

Ansell GB and Spanner S (1975) The origin and metabolism of brain choline. In: Waser PG, (ed): Cholinergic mechanisms, Raven Press, New York, pp. 117-129.
Bazan NG (1970) Effects of ischemia and electroconvulsive shock on free fatty acid pool in the brain. Biochim Biophys Acta 218: 1-10.
Blusztajn JK, Holbrook PG, Lakher M, Liscovitch M, Marie J-C, Mauron C, Richardson UI, Tacconi M, Wurtman RJ (1986) Relationships between acetylcholine release and membrane phosphatidylcholine turnover in brain and in cultured cholinergic neurons, this volume.
Boismare F, Saligaut C, Moore N, Chretien P and Daoust M (1985) Effect of CDP choline on brain hypoxia: are dopaminergic receptors involved? In: Zappia V, Kennedy EP, Nilsson BI, Galletti P (eds): Novel Biochemical, Pharmacological and Clinical Aspects of Cytidinediphosphocholine, Elsevier Science Publishing Co Inc, New York, pp. 195-201.
Cohadon F, Rigoulet M, Guerin B and Vandendriessche M (1979) Oedème cérébral vasogénique. Alterations des ATPases membranaires. Restauration par un précurseur des phospholipides. Nouv Presse Med 8: 1589-1591.
Cohadon F, Richer E and Poletto B (1982) Etude d'un précurseur des phospholipides dans le traitement des comas traumatiques. Neurochirurgie 28: 287-291.
Dainous F, Freysz L, Mozzi R, Dreyfus H, Louis JC, Porcellati G and Massarelli R (1982) Synthesis of choline phospholipids in neuronal and glial cell cultures by the methylation pathway. FEBS Letters 146: 221-223.

DeFelice SL and Ellenberg M (1984) Gangliosides-Clinical overview. In: Ledeen RW, Yu RK, Rapport MM, Suzuki K (eds): Gangliosides structure, function, and clinical potential. Plenum Press, New York, pp. 625-628.

Dominski J, Binaglia L, Dreyfus H, Massarelli R, Mersel M and Freysz L (1983) A study on the topological distribution of phospholipids in microsomal membranes of chick brain using phospholipase C and trinitrobenzenesulfonic acid. Biochim Biophys Acta 734: 257-266.

Dross K and Kewitz H (1972) Concentration and origin of choline in the rat brain. Naunyn-Schmiedeberg's Arch Pharmacol 274, 91-106.

Fontaine RN, Harris RA and Schroeder F (1980) Aminophospholipid asymmetry in murine synaptosomal plasma membrane. J Neurochem 34: 269-277.

Freysz L, Golly F, Mykita S, Avola R, Dreyfus H and Massarelli R (1985) Metabolism of neuronal cell cultures: modifications induced by CDP-choline. In: Zappia V, Kennedy EP, Nilsson BI, Galletti P (eds): Novel Biochemical, Pharmacological and Clinical Aspects of Cytidinediphosphocholine, Elsevier Science Publishing Co Inc, New York, pp. 117-129.

Giuffrida AM, Alberghina M, Serra I and Viola M (1985) Biochemical changes of lipid, nucleic acid and protein metabolism in brain regions during hypoxia: effect of CDP-choline. In: Zappia V, Kennedy EP, Nilsson BI. Galletti P (eds) Novel Biochemical, Pharmacological and Clinical Aspects of Cytidinediphosphocoline, Elsevier Science Publishing Co Inc, New York, pp. 217-228.

Hattori H and Kanfer JN (1984) Synaptosomal phospholipase D: potential role in providing choline for acetylcholine synthesis. Biochem Biophys Res Commun 124: 945-949.

Hawthorne JN and Azila N (1982) Phosphatidylinositol and calcium gating: some difficulties. In: Horrocks LA, Ansell GB and Porcellati G (eds): Phospholipids in the Nervous System, vol. 1, Raven Press, New York, pp. 265-270.

Horrocks LA and Dorman RV (1985) Prevention by CDP-choline and CDP-ethanolamine of lipid changes during brain ischemia. In: Zappia V, Kennedy EP, Nilsson BI, Galletti P (eds): Novel Biochemical, Pharmacological and Clinical Aspects of Cytidinediphosphocholine, Elsevier Science Publishing Co Inc, New York, pp. 205-215.

Kewitz H and Pleul O (1981) Synthesis of choline in the brain. In: Pepeu G, Ladinsky H (eds): Cholinergic Mechanisms, Plenum Press, New York, pp. 405-413

Louis JC, Dreyfus H, Wong TY, Vincendon G and Massarelli R (1983) Uptake, transport and metabolism of neurotransmitters in pure neuronal cultures. In: Caputto R, Marsan CA (eds): Neuronal Transmission, Learning and Memory, Raven Press, New York, pp. 49-64.

Marchbanks RM and Israel M (1971) Aspects of acetylcholine metabolism in the electric organ of *Torpedo marmorata*. J Neurochem 18: 439-448.

Massarelli AC, Dainous F, Hoffmann D, Mykita S, Freysz L, Dreyfus H and Massarelli R (1986) Uptake of ethanolamine in neuronal and glial cell cultures. Neurochem Res 11: 29-36.

Moore JP, Johannsson A, Hesketh TR, Smith GA and Metcalfe JC (1984) Calcium signals and phospholipid methylation in eukaryotic cells. Biochem J 221: 675-684.

Mozzi R, Goracci G, Siepi D, Francescangeli E, Andreoli V, Horrocks LA and Porcellati G (1982) Phospholipid synthesis by interconversion reactions in brain tissue. In: Horrocks LA, Ansell GB and Porcellati G (eds): Phospholipids in the Nervous System, vol. 1, Raven Press, New York, pp. 1-12.

Mykita S, Golly F, Dreyfus H, Freysz L and Massarelli R (1986) Effect of CDP Choline on hypocapnic Neurons in culture. J Neurochem, in press

O'Dea RF, Viveros OM and Diliberto EJ (1981) Protein carboxymethylation: role in the regulation of cell functions. Biochem Pharmacol 30: 1163-1168.

Smith AP and Loh HM (1976) The topological distribution of phosphatidylethanolamine and phosphatidylserine in synaptosomal plasma membrane. Proc West Pharmacol Soc 19: 147-151.

Van Rooijen LAA, Fisher SK and Agranoff BW (1985) Biochemical aspects of stimulated turnover of inositol lipids in the nervous system. In: Horrocks LA, Kanfer JN, Porcellati G (eds): Phospholipids in the Nervous System, vol. 2, Raven Press, New York, pp. 31-38.

Phospholipid research and the nervous system
Biochemical and molecular pharmacology
L.A. Horrocks, L. Freysz, G. Toffano (eds)
Fidia Research Series, vol. 4.
Liviana Press, Padova. © 1986

RELATIONSHIPS BETWEEN ACETYLCHOLINE RELEASE AND MEMBRANE PHOSPHATIDYLCHOLINE TURNOVER IN BRAIN AND IN CULTURED CHOLINERGIC NEURONS

Jan Krzysztof Blusztajn, Pamela G. Holbrook, Michael Lakher, Mordechai Liscovitch, Jean-Claude Maire[1], Charlotte Mauron, U. Ingrid Richardson, Mariateresa Tacconi[2], Richard J. Wurtman

Laboratory of Neural and Endocrine Regulation, Department of Applied Biological Sciences, Massachusetts Institute of Technology, Cambridge, Massachusetts, USA; [1]Department de Pharmacologie, Centre Medical Universitaire, Geneva, Switzerland; [2]Istituto di Ricerche Farmacologiche Mario Negri, Milano, Italy

Cholinergic neurons derive their choline from the circulation (choline crosses the blood-brain-barrier via a facilitated diffusion mechanism, Pardridge et al., 1979) or from its synthesis in situ (Blusztajn and Wurtman, 1981). Within cholinergic neurons choline can undergo two transformations: it may be acetylated by choline acetyltransferase (CAT) to form the neurotransmitter acetylcholine (ACh), or it may be incorporated into choline phospholipids (Ch-PL), PtdCho, sphingomyelin, or plasmalogens via the CDP-choline (Kennedy and Weiss, 1956) or base-exchange (Porcellati et al., 1971; Abdel-Latif and Smith, 1972) pathways. We hypothesize that a dynamic equilibrium exists between choline's reversible fluxes into and out of ACh and Ch-PL, and that these fluxes are well regulated in order to maintain both adequate ACh synthesis and the functional integrity of membranes (Ch-PL are the major lipid components of all biological membranes). This chapter describes the relationships between choline, ACh, and Ch-PL in the nervous system, and discusses evidence derived from experiments on brain slices, synaptosomes, and cultured cells.

POOLS OF PHOSPHATIDYLCHOLINE IN BRAIN

PtdCho molecules in brain apparently exist within several, not-mutually-exclusive, populations: particular molecules may be localized within several different membranes at different distances from synapses (e.g. that of the synaptic vesicles vs. mitochondria or plasma membranes); they may be synthesized by three different pathways; or they may be distinguished based on their fatty acyl compositions. We have examined the hypothesis that particular biosynthetic pathways generate PtdCho molecules of characteristic (and differing) fatty acyl compositions. The enzymes mediating these three biosynthetic pathways are (1) *phosphatidylethanolamine N-methyltransferase (PeMT)*, which catalyzes the methylation of PtdEtn using S-adenosylmethionine as a methyl donor; (2) *choline kinase*, that catalyzes the formation of phosphocholine utilizing ATP, *phosphocholine cytidyltransferase*, that catalyzes the formation of cytidine di-phosphocholine (CDP-Ch) from phosphocholine and CTP, and *cholinephospho-transferase*, that catalyzes the transfer of phosphocholine from CDP-Ch to diacylglycerol forming PtdCho; and (3) the *base exchange* enzymes, which reversibly catalyze the exchange of a choline molecule for a molecule of ethanolamine or serine esterified to phosphatidic acid (i.e. PtdEtn, or PtdSer). The fatty acyl compositions of radiolabeled PtdCho molecules newly-synthesized in synaptosomes by these three pathways are relatively similar (all containing large proportion of polyunsaturated fatty acids, PUFA), and yet strikingly different from the bulk of unlabeled synaptosomal PtdCho (which contains little PUFA) (Table 1) (Tacconi and Wurtman, in press; Holbrook and Wurtman unpublished). It is possible that the syntheses occur (regardless of the pathway), in specific membrane domains which contain highly unsaturated phospholipids. Such regions would be expected to have low microviscosity presenting a low energy barrier for the conformational changes in the membrane-bound enzymes, thus facilitating the catalysis. Once synthesized in such domains, a PtdCho molecule would then diffuse to structural regions of the membrane, where it could be deacylated by a phospholipase A_2 and than reacylated with a more saturated fatty acid to become a member of the "bulk" PtdCho pool. Indeed, inhibition of phospholipase A_2 by quinacrine resulted in a 40% stimulation in synaptosomal PtdCho synthesis by base exchange (Holbrook and Wurtman, un-

Table 1. *Molecular species of phosphatidylcholine synthesized by synaptosomal PeMT, base exchange, or cholinephosphotransferase: comparison with the "bulk" pool*

Molecular species	Radioactivity distribution			
	PtdCho phosphate	PeMT	Base exchange	Choline-phospho-transferase
	% of total			
Saturates, mono-and dienes	77.1	3.2	34.4	38.0
Tetraenes	15.8	34.6	28.7	36.9
Penta-and hexaenes	7.1	62.2	37.0	25.0

Rat brain synaptosomes were incubated at 37° C in the presence of 0.01 mM Ado[³H-methyl]Met for 30 minutes (PeMT assay) (Tacconi and Wurtman, in press): or 0.05mM [¹⁴C]choline for 20 minutes (base exchange assay); or 0.5 mM CDP[¹⁴C]choline for 20 minutes (cholinephosphotransferase assay) (Holbrook and Wurtman, unpublished). The reactions were terminated by the addition of chloroform/methanol 2:1 w/v, and the extracts were washed with 0.75% KCl in 50% aqueous methanol. The organic phases were concentrated and chromatographed on thin layer silica gel plates. The radiolabeled PtdCho was scraped off, redissolved in methanol, and rechromatographed on argentation thin layer silica gel plates. The molecular species were identified and the radioactivity in each band was determined. PtdCho phosphate was determined by the method of Svanborg and Svennerholm (1971). The identity of molecular species was verified by determining the fatty acid composition in each band using gas chromatography.
Coefficients of variation were less than 15% for each determination reported.

published). These observations could be interpreted as meaning that the inhibition of phospholipase A_2 arrested the processing of newly-formed PtdCho molecules needed to generate the characteristic highly-saturated "bulk" pool of synaptosomal PtdCho.

Similar conclusions could be reached from our experiments showing that [³H]Ptd Cho synthesized by synaptosomal PeMT was rapidly degraded to free [³H]choline, such that up to 30% of radioactivity transferred from Ado[³H-methyl] Met during a 30 minutes incubation period could be recovered as free choline (Blusztajn and Wurtman, 1981); perhaps the PtdCho was deacylated by phospholipase A_2, generating lysoPtdCho which could not be reacylated in this in vitro preparation, and was instead degraded to free choline.

EFFECTS OF NEURONAL ACTIVITY ON ACETYLCHOLINE RELEASE AND PHOSPHOLIPID TURNOVER

Since CAT has a low affinity for choline and is highly unsaturated with this substrate in vivo (White and Wu, 1973; Blusztajn and Wurtman, 1983), ACh synthesis and release can be increased by treatments that elevate extracellular choline levels. The effect of the choline is enhanced in cholinergic neurons that are firing frequently (Blusztajn and Wurtman, 1983). For example, we measured ACh release from rat striatal slices superfused with or without choline (Maire et al., 1983). In the absence of free choline, ACh was released spontaneously at a rate of 7.5 ± 1.3 pmol/mg protein/min (mean ± S.D.). Electrical field stimulation (15 Hz for 30 min) accelerated this release (25.6 ± 5.9 pmol/mg protein/min), and addition of choline (20 μM) to the superfusate significantly enhanced both the spontaneous (22.7 ± 5.7 pmol/mg protein/min) and the electrically evoked (37.4 ± 6.7 pmol/mg protein/min) release of the transmitter. Although the amount of ACh in the tissue did not depend on extracellular choline concentration (in this range of choline concentrations), tissue choline contents did increase when choline levels in the superfusate were raised.

The breakdown of membrane PtdCho in neurons can also be accelerated by frequent firing, especially when extracellular choline is unavailable. This was first shown by Parducz et al. (1976) who electrically stimulated the preganglionic trunk of the cat's superior cervical ganglion (in the presence of hemicholinium-3) and observed that the number of synaptic vesicles decreased to 18% that of controls, while ganglionic PtdCho levels fell to 69% of controls. (Other phospholipids were not affected). These observations indicate that when adequate extracellular choline is not available (e.g. after inhibition of its uptake by hemicholinium-3), vesicular PtdCho is used to supply choline for ACh synthesis. We have observed similar phenomena in our striatal slice preparations: in the absence of exogenous choline, the combined efflux of choline + ACh into the superfusate was 75 pmol/mg protein/min; however the decrease in choline + ACh within the tissue was only 16 pmol/mg protein/min. Thus, an endogenous pool of choline, present within a larger molecule, must have provided the free choline to sustain ACh synthesis and tissue choline and ACh levels. The only known compounds whose pool sizes would be sufficient for this purpose are the Ch-PL. Evidence that these compounds are indeed the source of the free choline is provided by the finding that the amount of phosphate in the phospholipids of the slices also decreased (by 23%) when the slices were stimulated (Mauron and Wurtman, in preparation), i.e., suggesting that neuronal activity accelerated the metabolic degradation of phospholipids. Addition of exogenous choline (20 μM) to the superfusate totally prevented the loss of phospholipid phosphate from stimulated slices. This suggests either that the apparent accelerations in phospholipid degradation reflected increased need for choline in order to maintain ACh release from cholinergic neurons (i.e. shifting the equilibrium between ACh, choline, and Ch-PL, towards ACh); or a general increase in phospholipid turnover, related to neuronal firing. In the former case the exogenous choline presumably was used to maintain high ACh synthesis, and thus prevented the shift in the equilibrium; in the latter, the choline presumably provided sufficient substrate to allow the rate of Ch-PL synthesis to catch up with its accelerated degradation. Preliminary data suggest that when slices are stimulated in the absence of choline, the rate of PtdCho synthesis by the de novo methylation pathway is increased (Lakher and Wurtman, unpublished). Perhaps this reflects a mechanism for sustaining membrane PtdCho levels (i.e. when PtdCho degradation is accelerated due to firing), or one to maintain the PtdEtn/PtdCho ratio within the membrane (PtdCho degradation would tend to increase this ratio; PeMT activation would decrease it by converting PtdEtn to PtdCho).

Apparently, the choline liberated from striatal phospholipids must first enter the extracellular space, and be transported into cholinergic neurons by the high-affinity choline uptake system, before it can be converted to ACh: addition of hemicholinium-3 (which blocks this uptake process) to the superfusate suppresses the release of ACh by electrical stimulation and decreases striatal ACh levels, even when these are compared with ACh's release from, and its levels in, tissues incubated without free coline.

RELATIONSHIPS BETWEEN ACETYLCHOLINE AND CHOLINE PHOSPHOLIPID SYNTHESIS IN A CHOLINERGIC CELL LINE, NG108-15.

We have studied the relationships between Ch-PL and ACh levels in a purely cholinergic cell line (Hamprecht, 1977), the neuroblastoma × glioma hybrid, NG108-15. The synthesis of ACh in these cells was found to vary with extracellular choline concentrations. When the cells were incubated for one hour in the presence of various [^{3}H-methyl]choline concentrations, the accumulation of [^{3}H-methyl]ACh exhibited saturable kinetics, with an apparent K_m of 193 ± 34 μM, and Vmax of 268 ± 23 pmol/mg protein/hr (Fig. 1). At the same time, incorporation of the labeled choline into phosphocholine (an intermediate in PtdCho synthesis) proceeded at a much higher rate (apparent Vmax = 16.8 ± 1.2 nmol/mg protein/hr) and had a lower apparent K_m (14.9 ± 5.4 μM) (Fig. 2). Thus choline was more likely to be used in these cells for PtdCho than for ACh synthesis (because choline kinase was more likely than CAT to attack the choline). We hypothesized that a treatment that inhibited choline kinase activity in the cells might enhance their formation of ACh, and observed that addition of ethanolamine to the incubation medium (0.5 mM), which suppressed formation of phosphocholine (Fig. 3) and PtdCho (Fig. 4) (in a competitive fashion), did indeed enhance the labeling of [^{3}H-methyl]ACh (Fig. 3). These data demonstrate that the dynamic equilibrium between ACh, choline, and phospholipids can be shifted towards ACh when choline's utilization for phospholipid synthesis is slowed.

CONCLUSIONS

The turnover of Ch-PL is dependent upon neuronal firing and, in cholinergic neurons, on ACh output. When choline is in short supply and the neuron is firing rapidly,

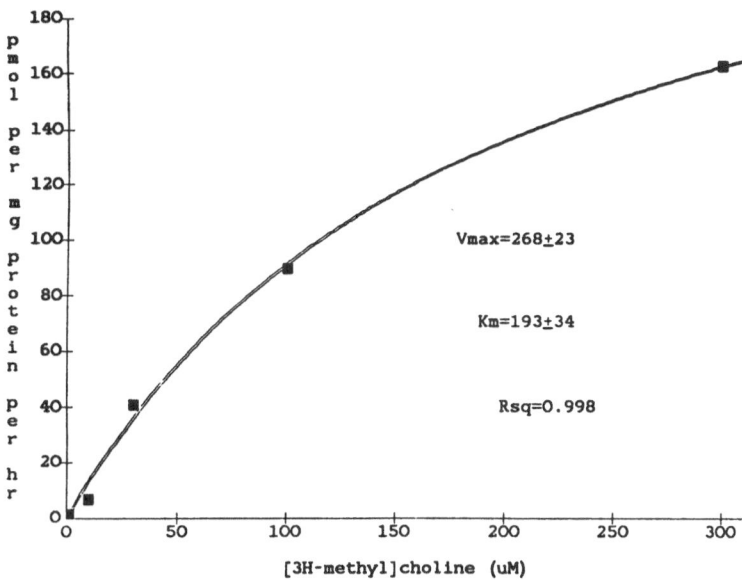

Effect of [3H-methyl]choline on the accumulation of [3H-methyl]acetylcholine in NG108-15

Pmol per mg protein per hr (y-axis)

Vmax=268±23

Km=193±34

Rsq=0.998

[3H-methyl]choline (uM)

Figure 1. ACh synthesis in NG108-15 cells: Dependence on the extracellular choline concentrations. NG108-15 cells (passage 21) were subcultured on 35 mm diameter dishes and allowed to grow for one day in the Dulbecco's Modified Eagle Medium containing 5% newborn bovine serum. For the next two days, cells were grown in a serum-free medium (N2) (Bottenstein and Sato, 1979) containing 1 mM dibutyryl-cAMP; most cells developed neurites due to the presence of the cAMP. The cells were then incubated for one hour in N2 medium containing 7 μCi of choline at various concentrations. Incubations were terminated, after the removal of the media, by the addition of 0.7 ml of methanol, and the cells were then scraped off the dishes. 1.4 ml of chloroform was then added, and the extracts were washed with 0.7 ml of water. Both the organic and aqueous layers were collected and dried under a vacuum. The dry residues of the aqueous extracts were reconstituted in 0.06 ml of water and filtered. 0.02 ml of the filtrates were subjected to high performance liquid chromatography (Liscovitch et al., 1985), the radioactivities in the fractions eluting at the same retention time as authentic ACh were determined, the amounts of [3H-methyl] ACh calculated assuming that the specific radioactivity of labeled choline was that of the medium. A rectangular hyperbola was fitted to the data. The inset shows the apparent Vmax, K_m and the standard errors of their determinations, and a regression coefficient.

degradation of Ch-PL apparently is accelerated. This allows the neuron to maintain ACh synthesis, but also results in the loss of phospholipids, a process which could, if maintained for a prolonged period, cause abnormalities in membrane functions. It is possible to prevent the phospholipid loss while maintaining high ACh output by supplementing the neurons with choline. Moreover, a homeostatic mechanism, operating within the neuron, might act to accelerate de novo PtdCho synthesis (by PeMT) during periods of accelerated PtdCho turnover.

Newly-synthesized PtdCho in nerve terminals is highly enriched with PUFA. This pool of PtdCho is then deacylated by phospholipase A_2 and further reacylated with *a more saturated fatty acid* before entering the "bulk" PtdCho pool. Studies now under-

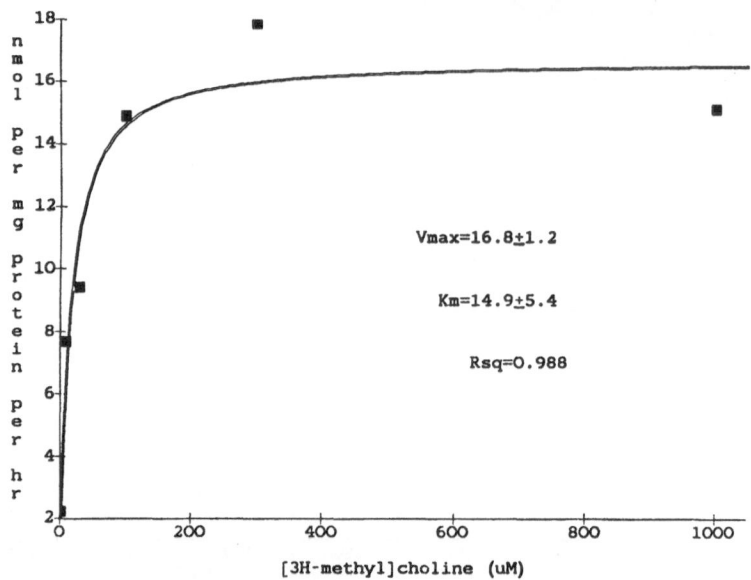

Figure 2. Phosphocholine synthesis in the NG108-15 cells: Dependence on the choline concentrations. The experiments were carried out as described in the legend to Figure 1. Phosphocholine peaks were collected and their radioactivity determined.

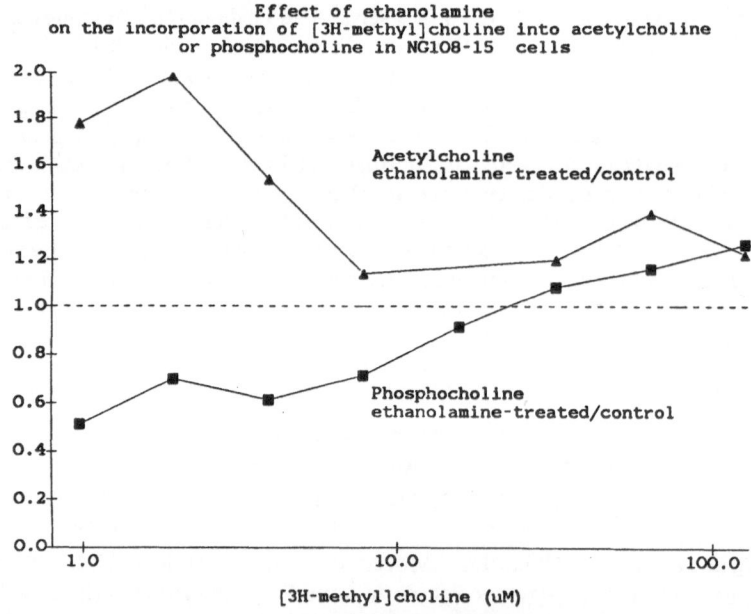

Figure 3. Effect of ethanolamine on the ACh and phosphocholine synthesis in NG108-15 cells. The experiments were carried out as described in the legends to Figures 1 and 2. Ethanolamine (0.5 mM) was added during the labeling period only. The results are the ratios of the amounts of labeled ACh or phosphocholine synthesized in the presence or absence of ethanolamine.

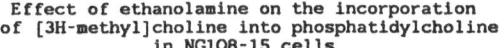

Effect of ethanolamine on the incorporation
of [3H-methyl]choline into phosphatidylcholine
in NG108-15 cells

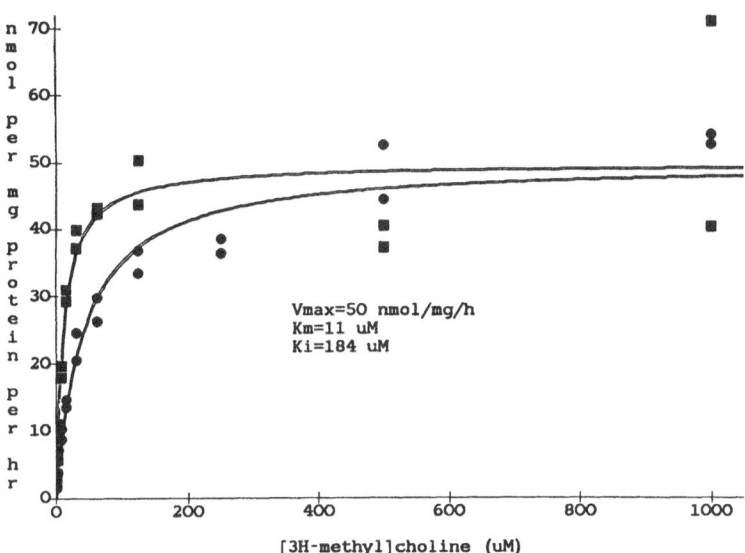

[3H-methyl]choline (uM)

Figure 4. Competitive inhibition by ethanolamine of the synthesis of PtdCho in NG108-15 cells. The experiments were carried out as described in the legends to Figures 1,2 and 3. The dry residues of organic phases were redissolved in chloroform/methanol 1:1. PtdCho was purified by thin layer chromatography, its radioactivity determined, and the amounts formed calculated. The graph shows the experimental points (squares for controls; circles for ethanolamine-treated) and the rectangular hyperbolae that are the results of best fits to the experimental points. The inset shows the apparent Vmax and K_m for choline, and a Ki for ethanolamine.

way are designed to determine the fatty acid composition of the PtdCho pool that is degraded during accelerated neuronal firing, and to determine which phospholipases mediate this process.

REFERENCES

Abdel-Latif AA, Smith JP (1972) Incorporation in vitro of [^{14}C] choline into phosphatidylcholine of rat synaptosomes and the effect of calcium ions. Biochem Pharmacol 21:436-439.

Blusztajn JK, Wurtman RJ (1981) Biosynthesis of choline by a preparation enriched in synaptosomes from rat brain. Nature 290:417-418.

Blusztajn JK, Wurtman RJ (1983) Choline and cholinergic neurons. Science 221:614-620.

Bottenstein JE, Sato GH (1979) Growth of a rat neuroblastoma cell line in serum-free supplemented medium. Proc Natl Acad Sci USA 76:514-517.

Hamprecht B (1977) Structural, physiological, biochemical, and pharmacoloical properties of neuroblastoma-glioma cell hybrids in cell culture. In: Bourne GH, Danielli JF (eds): International Review of Cytology. vol 49. Academic Press, New York, pp. 99-170.

Kennedy EP, Weiss SB (1956) The function of cytidine coenzymes in the biosynthesis of phospholipids. J Biol Chem 222:193-214.

Liscovitch M, Blusztajn JK, Wurtman RJ (1985) High performance liquid chromatography of water-soluble choline metabolites. [Abstract 62] Trans Am Soc Neurochem 16:123.

Maire J-C, Tacconi MT, Wurtman RJ (1983) Source of choline for the release of choline and acetylcholine from brain slices. [Abstract] Soc Neurosci 9:283.8.

Pardridge WM, Cornford EM, Braun LD, Oldendorf WH (1979) Transport of choline and choline analogues through the blood-brain-barrier. In: Barbeau A, Growdon JH, Wurtman RJ (eds): Nutrition and the brain. vol 5. Raven Press, New York, pp. 25-34.

Parducz A, Kiss Z, Joo F (1976) Changes of the phosphatidylcholine content and the number of synaptic vesicles in relation to neurohumoral transmission in sympathetic ganglia. Experientia 32: 1520-1521.

Porcellati G, Arienti G, Pirotta M, Giorgini D (1971) Base-exchange reaction for the synthesis of phospholipids in nervous tissue: the incorporation of serine and ethanolamine into phospholipids of isolated brain microsomes. J Neurochem 18: 1395-1417.

Svanborg A, Svennerholm L (1961) Plasma total cholesterol, triglycerides, phospholipids, and free fatty acids in a healthy Scandinavian population. Acta Med Scand 169:43-49.

Tacconi M-T, Wurtman RJ (1985) Phosphatidylcholine produced in rat synaptosomes by N-methylation is enriched in polyunsaturated fatty acids. Proc Natl Acad Sci USA; 82: 4828-4831.

White HL, Wu JC (1973) Kinetics of choline acetyltransferase (EC 2.3.1.6) from human and other mammalian central and peripheral nervous tissue. J Neurochem 20: 297-307.

SUBJECT INDEX

293